SELECTED PAPERS

THIRD EDITION

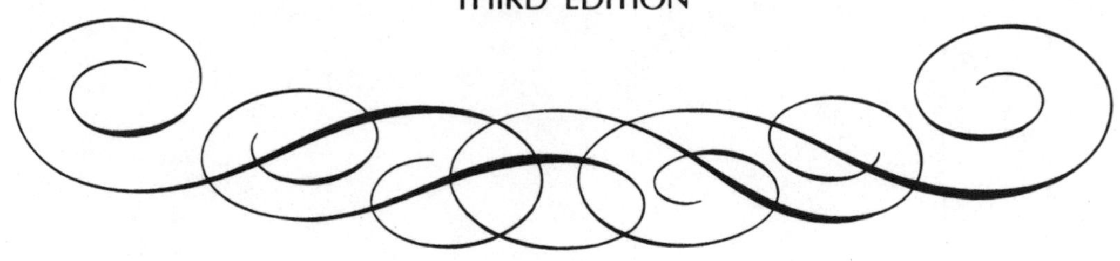

THE JOURNAL *of* IRREPRODUCIBLE RESULTS

Edited by
George H. Scherr, Ph.D.

SECOND PRINTING	1981
THIRD PRINTING	1983
THIRD EDITION	1986

ISBN: 0-930376-42-0
Lib. of Cong. Cat. No.: 70-5814

This edition published by Dorset Press, a division of Marboro Books Corp., by arrangement with The Journal of Irreproducible Results

Selected Papers

* * *

THE JOURNAL *of* IRREPRODUCIBLE RESULTS

THIRD EDITION

A selection of superb and irreproducible research from the illustrious and irreproducible archives of the Society for Basic Irreproducible Research

PREFACE
TO THE THIRD EDITION

THE JOURNAL OF IRREPRODUCIBLE RESULTS has become a forum for the humorous, satirical, and critique writings concerning those in whom the hallmark of achievement has resulted in jargon, pomposity, verbosity and obfuscation to befuddle the unitiated and preserve the mystique that often clothes the culprit or the ignorant. That these attributes are widely acknowledged and broad in scope is recognized by the fact that the JOURNAL is currently circulated to 55 countries.

The JOURNAL boasts an editorial board of highly competent representatives of their own discipline, all of whom serve without compensation to pass on the acceptability of contributed articles.

The efforts of people in all walks of life are oftentimes characterized as 'irreproducible'. Either the proposal is so unique that most people refuse to accept its validity (the Science Adviser to President Franklin D. Roosevelt insisted that an atomic bomb was totally unfeasible) or, the work is so contrived that no one else can reproduce the results (or may not want to).

We have selected for this edition those musings that are representative of a wide breadth of fields of interest. Many of the articles and presentations contained herein which are contrived, will tax the credibility of those articles which are taken from the literature or are representative of communications from some of the ostensibly solemn writings in the realm of law, medicine, engineering, accounting, teaching, and innumerable others.

Based on the wide use of mice as a model for the study of human diseases, can we accept the report from a University in Turkey that high-level noise such as found in discos, causes homosexuality in mice? If so, how do we explain the fact that the same experiments found that the same music merely caused deafness in pigs? Would the same experiments turn out differentlyif they were performed in Hoboken, New Jersey? Obviously, there is still a lot of good irreproducible work to be done.

GHS

CONTENTS

1. SOCIAL CONCERNS

Luckily

irreproducible

research

about

vast

SOCIAL CONCERNS

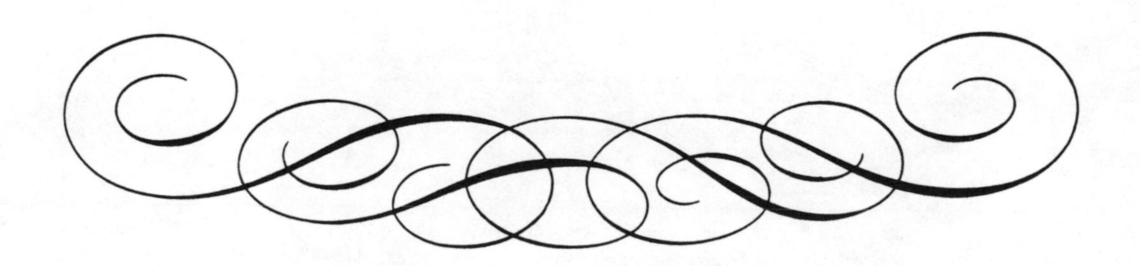

SOCIAL LAWS:
From The Complex To The Simple

Don L.F. Nilsen
Alleen Pace Nilsen
Tempe, AZ

We don't know who Murphy was, but we do know that he was a genius because what he said makes sense. It was Murphy who postulated the sociological principle, "If anything can go wrong, it will," and who then expanded this principle with, "When left to themselves, things always go from bad to worse," and "Nature always sides with the hidden flaw." (Martin 1973, p. 4).

A.J. Barton of the National Science foundation extended Murphy's law to read, "If anything can go wrong, it will, and even if it can't, it might." O'Toole—we don't know his first name either—offered the commentary, "Murphy was an optimist." Paul Dickson gives other names for Murphy's Law: *Thermodamnics,* and *Klugemanship,* and suggests that "Err is basically a synonym for Murphy, but those who quote him over the better-known prophet insist he is as real as Murphy. The basis for their argument: (1) His spirit, like Murphy's, is everywhere, and (2) Err is human."

According to Dickson, Murphy's Law has undergone empirical testing on two different occasions, first in an academic and second in a religious environment.

> One day a teacher named Murphy wanted to demonstrate the laws of probability to his math class. He had thirty of his students spread peanut butter on slices of bread, then toss the bread into the air to see if half would fall on the dry side and half on the buttered side. As it turned out, twenty-nine of the slices landed peanut-butter side on the floor, while the thirtieth stuck to the ceiling.

In the second test, a man conducted the Murphy experiment and all of his bread landed buttered side up.

> He ran straight away to his rabbi to report this deviance from one of the basic rules of the universe. At first the rabbi would not believe him but finally became convinced that it had happened. However, he didn't feel qualified to deal with the question and passed it along to one of the world's leading Talmudic scholars. After months of waiting, the scholar finally came up with an answer: "The bread must have been buttered on the wrong side."

Roy W. Walters' *Law of Management* is that, "If you're already in a hole, there's no use to continue digging." Damon Runyon's law is, "In all human affairs, the odds are always six to five against."

Why have such observations or laws become popular?

First, people like them because they are witty, compact, and unequivocal. They're easy to understand and their creators speak with such confidence that they inspire confidence. Many laws are short enough to go on a bumper sticker, which seems to fit our current yen for efficiency. They're sort of a *Readers' Digest* approach to books of quotations and famous sayings. But some of the things found in the old quotation books could easily qualify as today's laws, for example, Henry David Thoreau's advice, "If you see a man approaching you with an obvious intent of doing you good, you should run for your life."

The founding father of the modern laws appears to be C. Northcote Parkinson with his 1957 *Parkinson's Law* and his 1960 *The Law and the Profits. The Peter Principle: Why Things always Go Wrong,* was published in the late 1960's, but most of the other books and articles listed in the bibliography have copyrights in the 1970s and 1980s.

Their popularity correlates with the increasing complexities of modern living. When people hear or read social laws, they get a glimpse into the macrocosmic patterning of society. It may be just an illusion, but the laws nevertheless give people the feeling that they can deal with insurmountable obstacles if they just start with these manageable pieces. It's very satisfying to reduce complexity to simple and understandable terms. To illustrate this, Paul Dixon cited an anecdote about the late comedian Robert Burns.

> Burns got a perfectly symmetrical plank and balanced it across a sawhorse. He would then put a hog on one end of the plank and begin piling rocks on the other end until the plank was again perfectly balanced across the sawhorse. At this point he would carefully guess the weight of the rocks.

The social laws have a similar effect. They don't change things; they simply make us feel better by giving us the illusion that we understand what's happening. Once something has a name, or a law about it, then in our minds it can be moved from the category of the unknown to that of

THE PALACE AT 4 A.M., Alberto Giacometti, The Museum of Modern Art, NY

known and we aren't so afraid of it.

This is especially true of those aspects of our lives that are the newest and hardest to understand, such as statistics and research. We are relieved to learn that there's reason for our confusion and that others are as confused as we are as shown by Marshall's Generalized Iceberg Theorem, "Seven-eighths of *everything* can't be seen," and Paul Herbig's Principle of Breaucratic Tinkertoys, "If it can be understood, it's not finished yet."

On an almost daily basis we read or hear statements preceded by such phrases as "Research proves...," "Conclusive evidence shows...," and "Investigators discover..." In nearly all such instances we have neither the time nor the background knowledge to judge the validity of the research and so we are vaguely distrustful and pleased to have our doubts corroborated by such laws as the following:

> *The Finagle Factor* is charactrized by changing the universe to fit the equation.
>
> *The bouguerre Factor* changes the equation to fit the universe.
>
> *The Diddle Factor* changes things so that the equation and the universe appear to fit, without requiring any real change in either.

Paul Dixon explains that the latter is also known as the "smoothing" or "soothing" factor, "mathematically somewhat similar to a damping factor; it has the charactristic of eliminating the differences by dropping the subject under discussion to zero importance."

Skinner's Constant is "that quantity which, when multiplied by, divided by, added to, or subtracted from the answer you get, gives you the answer you should have gotten." This is sometimes referred to as "Flannegan's Finagling Factor, or DeBunk's Universal Variable Constant." In *Malice in Blunderland,* Thomas. L. Martin also presents *Finagle's Laws:* "The likelihood of something happening is inversely proportional to its desirability," (Martin 1973, p. 10) and "Once a job is fouled up, anything done to improve it only makes it worse."

It makes people feel smart to read a scholarly sounding rule and to realize that they knew it all the time. For example, Frederick E. Terman's advice that "If you want your track team to win the high jump you should find *one person* who can jump *seven feet,* and not *seven people* who can jump *one foot,*" Martin Friedman's advice to a non-swimmer, "Never walk across a river just because it has an *average* depth of four feet," and Paul A. Samuelson's observation that "The stock market has called nine of the last five recessions."

It's also satisfying for people to have one of their personal experiences documented in print. Arthur Block listed several such laws in his *Murphy's Law, Book II:*

> *The Law of the Letter:* The best way to inspire fresh thoughts is to seal the letter.
>
> *Stewart's Law:* It is easier to get forgiveness than permission.

Over the past three decades, the employment picture has probably changed more than any other aspect of American life. Today few of us are self-employed. We are no longer our own bosses nor do we work alone because what needs doing is too complicated to be handled by an individual. We have to work in partnerships and on teams and committees. We don't get to see a task through from beginning to end, and there's always the chance that our good work will be ruined by someone further down the line or won't be recognized by our bosses. This understandably makes people apprehensive, hence the popularity of career-related observations. Especially those that speak to the problems of groups dynamics.

"If you can't convince them confuse them," (Martin 1973, p. 74) and Dean Acheson observed "A memorandum is written not to inform the reader but to protect the writer." Television Producer Gene Roddenberry was being interviewed for *TV Guide* and was confronted with the statement that, "Ninety percent of TV is junk." He responded with what became an instant law, "Ninety percent of *everything* is junk." His answer was appealing not only because of its truthfulness, but also because readers could empathize with the way he was put on the spot to defend a profession very little of which was under his direct control.

Probably the most famous work-related law is from *The Peter-Principle* which states that in every hierarchy—government, business or whatever—each employee tends to rise to a level of incompetence, that is, as long as people are doing well, they will continue to be promoted. This means that every post is likely to be filled by an employee not quite competent to execute the duties.

It would be interesting to find out whether people think of themselves or of their fellow workers when they first hear the Peter Principle. Judging from the existence and popularity of other anti-boss laws, most people think of their supervisors:

> Being boss doesn't make you right; it only makes you boss.
>
> —Milton Metz
>
> Everything must be done immediately even if it doesn't have to be.
>
> —Larry Kane
>
> Officials advance subordinates, not rivals.
>
> —Cyril Northcote Parkinson
>
> (Martin 1973, p. 51)

Historians appreciate these laws all quoted by Paul Dixon:

> *Fiedler's Forecasting Rule:* When presenting a forecast, give them a number or give them a date, but never both.
>
> *Ryan's Law:* Anyone making three correct guesses consecutively will be established as an expert.
>
> *The Historian's Rule:* Any event, once it has occurred, can be made to appear inevitable by a competent historian. ■

SELECTED BIBLIOGRAPHY

Block, Arthur. 1980. *Murphy's Law book II: More Reasons Why Things Go Wrong.* New York: Price, Stern, Sloan Publishers.

Caldwell, George S. 1966. *Good Old Harry—The Wit and Wisdom of Harry S. Truman.* New York: Hawthorne Books.

Dickson, Paul. 1978. *The Officail Rules.* New York: Delta.

Donelson, Ken. 1981. "Laws for Life, Laws for Teaching," *English Journal,* Vol. 70, No. 6, pp. 30-31.

Friedman, Milton. 1972. "Milton Friedman on Averages Without Their components", *Newsweek,* January 10.

Kane, Larry. 1975. *New York Times,* October 19.

Kohn, Alexander. 1976. *The Journal of Irreproducible Results, Selected Papers,* JIR Publishers.

Martin, Thomas L. Jr. 1973. *Malice in Blunderland,* New York: McGraw-Hill.

Parkinson, C. Northcote. 1957. *Parkinson's Law,* Boston: Houghton-Mifflin.

Parkinson, C. Northcote, 1960. *The Law and the Profits,* Boston: Houghton-Mifflin.

Peter, Lawrence J., and Raymond Hall. 1969. *The Peter Principle: Why Things Always Go Wrong,* New York: Morrow.

Pocheptsov, G.G. 1974. *Language and Humour: A Collection of Linguistically Based Jokes, Anecdotes, Etc.* Kiev, Russia: Vysca Skola Publishers.

Roddenberry, Gene. 1974. *TV Guide,* April 27.

Runyon, Damon. 1964. *New York Times,* December 20.

JOURNAL COVERAGE CHANGES

Current Contents (Oct. 31, 1973)

Bulletin of Suicidology (ceased publication)

ON INTERNAL STRESSES DUE TO A RANDOM DISTRIBUTION OF DISLOCATIONS

Scripta Metallurgica. 2(1):9 (1968)

I wish to note the following: A physically meaningful random distribution of dislocations can be defined only by assuming the *positions of the dislocations are completely random.*

AN ANIMAL MODEL FOR INFIDELITY IN MAN

A study was performed at Ohio State University designed to Study social behavior in rats (J-Psychonomic Sci 26:171). "Rats were housed alone, in paris, or with a tennis ball for 10 days before the start of 10 days of testing for attraction to tennis balls or to other rats. Rats were more attracted to other rats than to tennis balls, attraction to both tennis balls increased over days, and, and housing with either a ball or a rat reduced attraction to that object but not to the other".

This means that to a rat, who has had to live with another rat, even a tennis ball looks good after awhile.

E PLURIBUS URANIUM

CHARLES T. STEWART, JR.
Dept. of Economics
George Washington University
Washington, D. C. 20006

Today Berengaria officially went on the uranium standard, setting the value of its currency, the benarus, at 12 units per microcurie. South Africa is the lone holdout for a better conversion rate between gold and uranium. Since it has both metals, its indecision seems to be the result of uncertainty as to where its interests lie, rather than dissatisfaction with the International Monetary Fund conversion ratio.

Thus comes to a close the most momentous chapter in world monetary history since the invention of check-kiting. The revolution was initiated by the United States in 1984 as a master stroke in its psychological offensive against the Soviet Bloc. It was the first nation to go on the uranium standard (although Russia now claims otherwise). By this move it created a vast new market for its uranium stockpile, estimated to constitute 70% of the total world supply of uranium metal, eliminated its balance of payments difficulties for many years to come, and greatly increased international liquidity.

The last two results might have been achieved by raising the dollar price of gold, but the United States was unwilling to do this because it feared loss of face and was unwilling to hand the Russians a windfall profit on their unknown but enormous pot of gold.

The main objective of the conversion, however, was to further disarmament aims and to eliminate the risks of nuclear war. This objective was fully achieved. The Russians were put under heavy pressure to convert their nuclear warheads into currency, and did so in large number. Although they are believed to retain a few warheads, their first harvest failure is expected to lead to total nuclear disarmament. In the meantime, the Russians, with their well-known peasant attachment to cold cash, would never think of blowing up their hard-earned hoard in a mushroom cloud. Nuclear testing has ceased. India has gone conventional.

The new standard incorporates a foolproof inspection, warning, and control system. The conversion of monetary uranium into nuclear warheads by any country is promptly reflected in the foreign exchange markets if not in domestic price indices. The deflationary aspects of wars and preparations for war make them unatractive to all business, and practically impossible to finance. Conversely, the uranium standard has banished all fear of an economic collapse as a result of disarmament. The automatic increase in money supply as warheads are converted to currency sets off an investment boom and an inflationary fever.

In retrospect, it is hard to understand mankind's prolonged love affair with the yellow metal of such limited uses other than personal adornment. Perhaps it illustrates what Whitehead called the "fallacy of misplaced concreteness." Certainly it is clear today that people valued gold because others valued and were willing to pay for it, because still others valued it, ad infinitum, and for no other reason. This unwitting conspiracy, this keeping up with the Joneses, gave rise to a durable fashion, lasting thousands of years, but now comes to an end.

The advantages of uranium are obvious enough. Its value is not based on a fickle fashion, on the persistence of an illusion, or on human propensity for invidious comparisons. In an insecure world its virtues are unique. Not only is it a store of value but also a store of power. It can earn interest without moral qualms by generating electricity while it is serving as a monetary reserve. It lacks the stigma of petty bourgeois conservatism, of the banking profession and the capitalist persuasion. It is ideologically neutral.

The uranium standard now provides an effective technique for changing the velocity of circulation at will. Threatening recessions can be prevented by the simple expedient of increasing the proportion of salaries paid in radioactive uranium coinage. This "hot money" is immediately spent and just as promptly respent. Thus we have, in the old game of musical chairs, and the new monetary metal, a final cure for economic instability.

Not the least of its virtues from the American viewpoint is that it is found in abundance in the Western United States, and that its mining and processing is now a major support of a number of states with sparse populations but dense political representation. ■

ON THE RELATIVE MERITS OF SURGICAL VERSUS MEDICAL MANAGEMENT OF CORONARY ARTERY DISEASE

Lloyd Rucker
Orange, CA

Savo Radulovic's SURGERY

Cardiologists and cardio-thoracic surgeons have virtually acknowledged that there never will be a concurrent double blind matched controlled study of the medical versus surgical management of coronary artery disease.[1] Therefore it remains for each primary care physician to decide for himself as to the relative efficacy of the alternatives. We present here a simple yet effective rationale for dealing with these troublesome patients.

One very effective way of approaching sticky problems like these is the cost: benefit ratio, that is, costs to patient vs. benefits to doctor.[2] Consider this, the physician who opts to manage his patient medically accrues certain benefits. The maximum maturation of patient investment will occur over a five year period (since most patients will be dead after five years). Considering yearly complete physicals at $100.00 a shot with every two month reviews for renewal of medication at $30.00 per visit, office accounts total a measley $1250.00. If you own your own pharmacy-lab you might toss in an extra grand for medicine, x-rays and bloodwork.

Alternatively, consider this. If, as some claim, operation really extends life for the majority of patients, for instance to ten years, then assuming the same quality of follow-up, a $4000.00 base is assumed. However, even without any significant extension of life, the following benefits appear: 1) cost of treadmill $200.00, (with thallium $500.00), 2) cost of cardiac catheterization $2000.00, 3) surgeons fees $5000.00, 4) pump time for cardiologist $300.00-$500.00, 5) post-op care visits variable depending on course, but assume minimum of $250.00, 6) operating room time and pump time if privately owned hospital $5000.00, 7) anesthesiologist fees $1000.00.

This all appears fairly straightforward. Any first year medical student could add up the numbers and decide at first glance; even the layman ought to be able to decide. However, thing are not so simple as they seem. Two separate factors require consideration. First, the cost: benefits ratio. The ideal is unity.[3] Unity indicates that the MD or his group is exacting the highest possible percentage of the total take. That is, whatever the patient pays, they get ahold of. The simpler the set-up, the easier unity is to approach.

On the other hand, you need to consider the absolute level of benefit. Unity is fine, but 100% of $100.00 just doesn't approach 10% of ten or twenty thousand dollars. So you have to individualize. A GP needs to consider that if he refers his patient, he may never see him again. So a cost:benefit ration of one yields an absolute take of $2250.00. If however, the GP happens to own the hospital, then a cost:benefit of even 6:1 wouldn't be too shabby, after the accrued from operating time and the ideal unity may be sacrificed for maximal absolute benefit. The surgeon operates at the other end of the spectrum. If he opts for medical management and refers the patient back to the cardiologist, his cost:benefit ratio is 2250/150 (whatever the hell that is) and the absolute benefit is maybe 150 bucks for consultation. The surgeon as you can see, doesn't have much trouble deciding on the operative approach.

In the past, cardiologists have had some healthy skepticism about the benefits of operation. However, as cardiology has expanded into the fields of radiology, nuclear medicine and anesthiology, both the cost: benefit ratio and absolute benefits have risen dramatically. Very few cardiologists now quibble with the operative approach. As new procedures are devised, these figures will rise. Now that it takes five or six expensive procedures to diagnose conditions formerly evaluated with one or two, the direct cardiologist benefit has risen dramatically.

Conclusion: It is apparent that any smart operator, from GP to CT surgeon can accrue significant benefit from the coronary artery bypass operation. We feel that this procedure deserves the esteemed position it has acquired in the armamentarium of medical care, and that it will become more entrenched as general practitioners and cardiologists become facile with the cost: benefit approach. ■

[1]Hurst, J.W., personal communication
[2]Karlinsky, A., my Estate Planner, personal communication
[3](God), personal communication

Anaylsis of Modest Brand Pantyhose Forthcoming Advertising Campaign

A Deitrich Motivational survey, and a 984 base statistical study for demographical breakout have been run through the computer, with the following results.

(1) Men like to look at female legs. (2) Many men (68.5%) prefer to view legs covered with a transparent fiber. (3) Women are aware of this. (4) Women (86.5%) wear pantyhose or stockings in order to appear attractive to men. (5) Women will sacrifice durability to appearance, but are interested in both. (6) Women are much impressed with testimonial advertising.

Therefore we suggest a testimonal advertising campaign. Illustrations will be in good taste, showing only legs to upper thigh.

Beauty will be suggested by illustrations, which should be actual phtographs of legs in various positions. Face may be inset, to show who is giving the testimonial. The face need not necessarily belong to the legs, though both should be attractive.

We give two suggested testimonials, both of which are in our files. We have releases for pictures and testimonial copy.

Copy 1

"In my business I take off my pantyhose from six to twelve times a night, yet Modest Brand Pantyhose gives me a full three months service."

Copy 2

"I go out with younger men, attracted to my pantyhose, yet the triple reinforcement at top of Modest Brand Pantyhose even allows me to go to a drive-in movie and sit on a back seat without fear of tear or penetration. My life would be different if I didn't wear Modest Brand."

This is a very brief summary of our projected campaign, which is certain to be a winner and a breakthrough in creative advertising.

We are now preparing media lists, budget, comps, TV storyboards, and analysis of statistical data, and our 120 page total analysis will be in your hands within the week.

We will supply sixty copies for distribution. ■

The gunk in Lake Erie's reducible.
'Twould be good if the lake were more usable.
To stop the pollution,
Would be the solution,
And again make Lake Irreproducible.

Robert J. Bates
Greensboro, NC

RIVER SCENE, John Frederick Kensett, Metropolitan Museum of Art

WANTED:
A Pharmacologist For M.I.5

Augeas Incredulus*
Middlesex, England

"Fifteen grains...are the maximum dose. Fifteen grains is fine. It stimulates the heart, very good. More than fifteen grains is very bad. The heart goes pouf! It stops to beat. The guy falls dead. It looks just like heart failure. No one is suspecting filthy play".[1]

You wouldn't think they were talking about fifteen grains of *strophanthin*—and by injection! Martindale[2] gives the maximum dose as 1/60 grain, which shows how much these thriller writers really need help. In fact until he started his research for this paper, Augeas hadn't realised there were so many dangerous authors about! He didn't have to look far; every one of the following examples was found in his local public library.

There are several very common misconceptions. The first is that a drug can pass from the tissues to the brain in a matter of seconds. Since the *intravenous* arm-brain time is about half a minute, this is pushing things a bit, even if, as it seems reasonable to expect, secret agents do have a permanently hyperdynamic circulation. There is, for example[3], the K.G.B. Colonel who, having unmasked the British spy, offers a deal if he will murder the Colonel's wife; the weapon is to be a fountain pen that fires an eight-centimeter steel dart *"tipped with one of our synthetic poisons. The victim is totally paralysed thirty seconds after the dart has punctured the skin. Death follows in about two minutes maximum...it doesn't matter where it hits her"*. In real life, of the known relaxant drugs, an intramuscular injection of suxamethonium (Scoline), for example, will produce paralysis in young babies in about two minutes, but the dose/weight ratio is four times that used intravenously in the adult, and the duration of action is also about four times as long. The quantity required for an adult on this basis would be at least 200 mg., which is more than could be accommodated on the tip of a dart, even if one could guarantee instant solubility and absorption. In fact, recent investigations, following the historic descriptions of Waterton[4] in the last century, have shown that one of the limitations of the classic drug, curare, as used by the South American Indian tribes on arrows, spears, and blow-gun darts as a weapon for hunting and warfare, is its relative ineffectiveness except for small animals[5]. Five to ten darts are needed to paralyse a wild pig, and many tribes confine the use of the blow-gun, the *sarabatana,* to hunting small, arboreal animals; it is an ideal weapon for killing monkeys, which, when shot by non-curarized darts, stiffen their tail muscles and thus remain suspended from high trees, from which they can be removed only with difficulty. Curare, by paralysing the muscles,

causes the animal to fall. In the field of human conflict, Koch-Grünberg[6] mentions a rather unusual weapon used by the Yahuna. *"They fix on their elbows and wrists long thorns of curarized paxiuba palm with which they fatally wound their enemy in hand-to-hand fighting. Especially the women defend themselves in this way from attacks by strangers".* but we will return to relaxant drugs later, merely remarking that most Indian tribes have abondoned their traditional weapons in favour of those of the civilized world[12].

Sometimes our writers try to explain—which only makes things worse[7].

'Vilsky clamped a black cheroot between his thick lips and turned his head sideways to say, 'Excuse me, would you be able to oblige me with a light, please?'

The man with silver hair turned to glance at him. 'Delighted,' he said easily, accompanying the word with a generous smile. He produced a lighter. 'Thank you,' Vilksy said grudgingly.

'Not at all,' the tall man said, flashing a final smile. As he was transferring his homburg back to the hand in which it had originally been he let the hat fall. 'How clumsy of me,' he drawled, and bent down to pick it up.

Vilsky winced as something jabbed into his calf, and for one moment he thought what a clumsy fool the other man was, then the shock hit him that he had been tricked. Something was happening. Vilsky tried to shout and raise his arms, but nothing happened. The last thing he saw before the blackness came down was the out-of-focus smiling face of the man with silver hair".

In a footnote we are told that *"The drug used was Paraldehyde—15 ml. This is the drug in general use to restrain violent patients in lunatic asylums, and its effect is instantaneous (intramuscular injection)".* but it takes time to inject 15 ml. of anything into anyone, even into the hospitable calf of an unsuspecting agent, with a very sluggish withdrawal reflex, whom we are told is *the top* man in the Russian secret service; and the effect of intramuscular paraldehyde is hardly instantaneous.

This particular author, however also knows about intravenous injections[8].

" 'I took the opportunity of pulling the shroud right back when nobody was looking. I found no signs of cuts or bruises but there was a pinprick in the crook of his left arm, such as would be left by an intravenous injection. The hole, when I did a bit of Sherlock Holmes stuff on it with a magnifying-glass, hadn't sealed, so it had been done just before Penn died'.

'Interesting', Simon commented. 'It could have been an injection to silence him, or the truth drug—which might have killed him. Know much about the truth drug?'

'Not a thing. When I was in the Metropolitan Police we used to use our truncheons.'.

'Well, it's dangerous to use. It consists of an intravenous injection of 150 mg. Sodium pentothal, and usually profuse sweating, trembling, laboured breathing and vomiting are normal reactions. Death can easily occur if administered by novices. A medical practitioner should always be present, with an oxygen cylinder and mask at hand, together with equipment to perform an emergency tracheotomy if necessary. What I am getting at is that a person or persons unknown knew that he had some secret knowledge and gave him an S.P. injection, during which he died. They then threw him in the river' ".

Relaxant drugs are used to produce paralysis, but conveniently spare the muscles of respiration. Ex-Superintendent Snow, having been rendered unconscious, finds himself on the Zurich express to Milan[10].

"In turn, he tried to move his left hand and both feet. Sweat began to moisten his hands as he realised the truth. He was not tied or bound in any way, but he could not move a muscle. He was paralysed...

He was lying full length on the compartment seat under a blanket. He was not tied up but he couldn't move even a finger, so they must have drugged him. But wait a minute, if he was drugged, surely he must be unconscious? It became a nightmare. Then he heard the voice, a man's voice speaking in Italian...

He was conscious but unable to move. So he had been injected with a certain type of drug. It clicked in his mind like the turn of a key. They had used a paralytic drug. He would not even be able to speak...

He opened his eyes slightly and the ceiling light shimmered between his lashes.

He could open his eyes so perhaps they hadn't given him enough of the drug. What had they injected into him? A terrifying thought slipped into his mind, that it would be a corpse they intended carrying off the train. Grimly he fought for self-control'.

Snow is taken from the train by stretcher. After some hours...

"it struck him with shock force. He had been able to move. He flexed his right hand. The nail of his index finger scraped against something rough. The hand was still powerless but the finger had moved. He lay still, experiencing a wave of relief...

He tried moving his left hand. He found he could move the whole of his index finger. His feet were still quite lifeless. Then he closed his eyes quickly as he heard a door open.

The door closed again and two pairs of feet walked across a stone floor. When they stopped he felt sure they were beside the stretcher, looking down at him. He heard the nun's voice first.

'When will he need the next dose?'

'In about four hours'. It was an unknown man's voice, light coloured in tone.

'Will that be soon enough?'

'It doesn't matter if you take a chance'.

'You're not the one who's administering it'.

'I did it on the train'.

Snow listened to the feet walk away, the door open and close...

Then he looked at the clock again. They would be back before 7:30 a.m. It would be a race against time. There might be only minutes in it...

At 5:00 a.m. he turned his ankle to one side. The rest of his leg was still lifeless. At 5:30 a.m. he found he could twist both ankles.

The clock hands had just registered 6:00 a.m. when he managed to move the whole of his right hand. Half an hour later achieved the same feat with his left hand...

By 7:00 a.m. feeling was flowing back into his arms and legs but he could only wriggle slightly. He knew now that the straps were still in position. Making a tremendous effort, he tried to force his knees upwards but the straps across his ankles defeated him. He collapsed backwards. Then he heard the door open. They had come back''.

Fortunately, they hadn't. But the muscles of the eyelids are among the first to be affected by relaxants and would not be spared by an underdose. A fully paralysing dose of a long-acting relaxant, tubarine or pancuronium, would cause death from respiratory paralysis, but in the classic experiment of Prescott,[10] who allowed himself to be curarized in the conscious state to the level of diaphragmatic paralysis with 30 mg. of tubarine, and was then ventilated artificially, muscular power had returned completely in thirty minutes.

Since, in the case of ex-Superintendent Snow, the effects of the drug lasted, as far as one can tell, for anything up to eight hours, we have made a happy discovery that takes us momentarily into the higher realms of literary criticism. In his essay on Philoctetes, Edmund Wilson[12] deals with the myth of the wounded hero, and *"the conception of superior strength as inseparable from disability"*. Descending precipitately to our previous level, we observe that it is fashionable for detectives to suffer from a disability, such as being blind; excessively obese; a woman; a member of the aristocracy; French; Belgian, with little grey cells; Chinese; an Australian aboriginal; a bachelor violinist and cocaine addict; a doctor; a Catholic Priest; or a Reform Rabbi; and there is even one story[13] where the Inspector of Nurse Training Schools to the General Nursing Council turns out *not* to be the detective; but in ex-Supt. Snow we now have our first detective who suffers from myasthenia gravis—and do not laugh, for there is confirmatory evidence. Snow's wife left him in deeply mysterious circumstances that are barely alluded to from time to time but weigh on him heavily; and now we have unravelled that mystery, too. After this, the detective who suffers from simple indigestion is old hat, and a new era opens before us—detectives with thalassaemia, porphyria, nasal polyps, urinary retention with overflow, or on home dialysis, where the possibilities for technical descriptions, and for sabotage, are endless; and whoever put the primaquine into the canteen soup must have known that the assistant commissioner suffered from glucose—6—phosphate dehydrogenase deficiency. ■

REFERENCES

[1]**Ames, D.** (1955). "Landscape with Corpse". London, Hodder & Stoughton.

[2]**Martindale** (1941). The Extra Pharamacopoeia, 22nd Edn., Vol. 1. p. 927. The Pharmaceutical Press.

[3]**Porter, J.** (1966). "Sour Cream with Everything". London, Cape.

[4]**Waterton, c.** (1825). Wanderings in South America. London, Everyman Edn.

[5]**Ribiero, B.G., and De Melo Carvalho, J.C.** (1959). "Curare; a weapon for hunting and warfare", in "Curare and Curare-like Agents", ed. Bovet, D., *et al.* Amsterdam, Elsevier Publishing company.

[6]**Koch-Grünberg**, T. (1908). Jagd und Waffen bei den Indianern Nordwestbrasilians, Blobus (Braunschweig), 93. 197 and 215.

[7]**Smith, L.** (1968). "Fear and the Dead Man". London, Collins.

[8]**Smith, L.** (1968). *op. cit.*

[9]**Sawkins, R.** (1967). "Snow in Paradise". London, Heinemann.

[10]**Prescott, F., Organe, G., and Rowbotham, S.** (1946). Lancet, ii, 80.

[11]**Capone** *et al.* (1929). Galileo revised: studies on projectiles and falling bodies. Chicago, St. Valentine Press.

[12]**Wilson, E.** (1974). The Wound and the Bow. London, W.H. Allen.

[13]**James, P.D.** (1971). "Shroud for a Nightingale". London, Faber & Faber.

*Dr. D. Zuck

Reprinted with permission from the Enfield Medical Gazette

PROBLEMS IN THE URBAN ENVIRONMENT

Harvard-MIT Joint Center for Urban Studies (1968).

"The trouble with the poor is that they don't have enough money."

Breeding the Bald, Plastic-working Ape at the Tau Ceti IV Planetary Zoo

Translated from an article appearing in Interstellar Zoo Yearbook

Elaine Radford
New Orleans, LA

DETAIL OF THE CENTER PANEL OF THE GARDEN OF DELIGHTS, by Hieronymus Bosch, The Prado, Madrid

Xylene (524) Xylxyl & Elwhy ($3) Ex Curator of Apes and Ape Keeper, Tau Ceti IV Planetary Zoo, Zthn, Certx, Tau Ceti IV, Sirius Sector

Because of the observation that most apes require fair-sized colonies in order to thrive (Zellon (1)), few zoos in the past have attempted to breed the Bald, Plastic-working ape *Homo sapiens narcissus*. (And none has reported success.) Although *H. s. narcissus* is unique among the Soll III subhumanoids in its possession of a largely naked body and some assorted primitive technical skills, its nasty temper and enormous space requirements have made establishment of large breeding colonies an impractical proposition for even the largest zoos. We would like to suggest an alternative—a "model environment" technique that may result in breeding even with a party as small as a single couple.

In Standard Year 57,000, three Bald, Plastic-working apes were taken from a densely populated nest located on a northern continent of Sol III, the sole planet of origin. Two of the specimens were slightly post-pubescent adults of opposite sex. (As is typical of multi-celled inhabitants of Type 8 biospheres, only two sexes are present.) The third specimen, an older male, died in transit as a result of unknown causes. Immediately upon acclimating the remaining two, we transferred them to a permanent exhibit some 200 zibdits high, 100 zibdits wide, and 150 zibdits deep.

The model environment technique involved mimicking as closely as possible the natural surroundings of the specimen upon display. To this end, we transported from Sol III a large earthern dwelling, along with its encircling garden of various non-flowering grasses. A small adjunct contained a non-functioning vehicle of the type often used by the apes for overland transport, a device for trimming the grasses, and an assortment of simple tools and chemicals. The interior of the main living quarters was equipped with a number of fascinating primitive machines—among them a refrigeration unit, a food preparation device, an audio-visual display panel, and an electronic gaming opponent.* Simple furnishings such as bed, couch, table, and chairs were also provided. A fairly complex plumbing network, to be described in detail in an upcoming paper, proved essential to the psychological if not physiological well-being of the Bald Plastic-worker. Diet, however, was not at all problematical. Like the proverbial subhumanoid, the bald ape will eat just about anything.

The entire enclosure was secured within a sealing field. We might also note that, due to the well-documented vicious nature of *H. s. narcissus* (Xy1(2)), we did not attempt to place any other animals on display in this cage.

We should like to digress for a moment to stress the importance of monitoring devices installed secretly within the apes' living environment. We suggest that the Bald Plastic-worker requires the illusion of privacy in order to mate. Failure of previous breeding attempts is partially due to the fact that even the dullest of apes cannot help but be aware that he is being watched by vastly superior beings. We

---

*Although the "idiot savant" mechanical abilities of *H. s. narcissus* are not rooted in true intelligence, the species' facility with technical production has progressed to the point of threatening all life forms on Sol III with extinction. Our interest in breeding this difficult ape stems from the realization that, not many years hence, captive specimens may be the only ones available.

avoided this problem by surrounding the dwelling and yard in an invisible visual-aural detector net. Hence we were in an excellent position to study unobserved the unself-conscious activities of the ape in a near-natural situation.

Aggressive behavior possibly related to courtship commenced as early as one-tenth of a standard year after placing the pair in their final accomodations. Loud vocalizations, punching, and slapping constituted most of the initial exchange, although the tiff concluded with the male and female embracing one another, sniffling, and excreting prodigious quantities of moisture from a duct located within the eyes. (This peculiar use of the eye-washing mechanism, dubbed "crying", has been treated in depth by Darlb (3).) On this occasion mating was not attempted, but subsequent domestic quarrels were almost invariably terminated by copulation.

Like all Sol III subhumaniforms, the Bald Plastic-worker carries the developing fetus in the lower abdomen of the female for an extended period of time prior to bearing a single live young. We were elated when, not four-tenths of a standard year from acquisition, the swelling belly of the female gave us our first clue that insemination had been successful. However, the prospective mother soon developed a complex of worrisome symptoms, including vomiting, disinterest in her mate, and overall weakness. We treated the problem with dosages of full-spectrum light and increased nutritional supplementation. Anecdotal evidence from an abortive breeding at Sirian II planetary Zoo had suggested that such unusual food combinations as a semi-frozen milk derivative garnished with preserved cucumber are desirable additions to the gravid ape's diet; we found that such treats were accepted without much enthusiasm by our specimen.

At any rate, although the female remained somewhat debilitated, she was able to carry the young to term. We confirmed the gestation period noted by Zellon (1), i.e. six-tenths of a standard year.

The mother gave birth on standard date 121/57,001. Due to the anomalously large head of a normal Bald Plastic-worker infant, labor tends to be a protracted process that consumes most of a standard day. We manned the audio-visual room throughout the birth, prepared to deliver assistance should it seem necessary. Fortunately the lusty cries of the mother assured us that our intervention would not be required.

Throughout labor, the male paced house and yard restlessly, often shouting and shaking a fist at the sky. Zellon, in a private communication, hazarded that such rituals are evidence of primitive culture; we ourselves have yet to reach a conclusion as to the purpose of this seemingly meaningless activity.

A scan of external genitalia revealed that the newborn was female.

For several standard days the mother suckled her young with fluid from two engorged glands located on her upper chest. (The adult female *H. s. narcissus* is usually characterized by both large upper body protrusions and upright stance, despite the obvious physical drawbacks to such an arrangement. See Darlb (4) for a possible evolutionary explanation.) As both mother and baby seemed to be thriving, monitoring was returned to automatic mode. Not two days after implementing this policy, the mother inexplicably smothered the juvenile by placing a pillow over its face. A replay of the sacrifice (?) recording shows that the killing was immediately followed by loud vocalizations on the part of both parents. The father once again repeatedly shook his fist skyward as he shouted and engaged in "crying" activity.

Despite this disappointing setback, we are satisfied that we have hit upon a promising method for breeding the Bald Plastic-working ape. The achievement of a first captive birth in a pair held for under a year indicates that the "model environment" technique is far more viable than the normal colony approach. We are confident that we shall eventually be able to report a captive rearing of this heretofore problematical species. ■

REFERENCES

1. **Zellon, (XC) A. (56,983):** Apes and ape-forms in the Sirius Sector, Int. Zoo. Yb. 1,917:1223-1456.
2. **Xyl, (524) G. (56,992):** An attempt at placing *Homo sapiens narcissus* in a mixed exhibit, Cetan Zoo Report, 25B:12-19.
3. **Darlb, (½¼$) U. (56,991):** The function of the tear gland in *Homo sapiens narcissus*, Review of Sol System Biology, 139:200B-200X.
4. **Darlb, ½¼$) U. (56,992):** The function and structure of the mammary gland in *Homo sapiens narcissus*, R. of S.S. Bio., 142:21B-XX.

The two reviews which follow were of the same article and appeared on the same page of Computing Reviews, November 1982.

All in all, the collection of ideas presented in the paper is a major contribution towards solving the problem of protecting shared information storage in a decentralized system, in a well thought-out, coherent manner using encryption. This reviewer feels that the paper should be read by all individuals who are interested in using encryption as a mechanism for providing controlled access to information stored on computers.

This poorly organized, notationally obscure, imcomplete, and inaccurate paper does little to enhance the cryptographic state of the art since it fails pedagogically. As a result, its valid technical content is totally lost. Unless you enjoy the masochistic exercise required to decipher this text and uncover the underlying cryptographic organizational concepts, don't bother reading it.

Submitted by:
Bernard A. Sobel
Baton Rouge, LA

Man Makes Himself?

N. A. DREKOPF*
New York City

Most anthropologists will agree that one of the more exciting developments in our discipline in recent years has been the discovery of the science of primatology. We concede that work on animal behavior has been done in the past by zoologists, comparative psychologists, ethnologists, and an extinct breed of scholar known as the "naturalist" (e.g., Darwin, Wallace, Agassiz, etc.), but their efforts have been insignificant when compared to the edge-cutting research now being done by anthropologists. Only members of our discipline have had the courage and imagination to extrapolate from the behavior of baboons, macaques, gorillas, and chimpanzees to a line of inquiry that has shed light upon the social evolution of early man. They have thus developed a methodology which allows us to determine the parameters of primitive human existence through analysis of the possibilities and limitations inherent in the organic equipment and behavioral inventory of our precursors. It is this Inferential Method which I intend to faithfully apply in the present paper.

It has been convincingly argued that hands are the father of man. Now, all primates have considerable manual dexterity, but it is agreed that the evolution of the full potential of the forelimbs is dependent upon the abandonment of quadrupedal motion and the lifting of the hands from the ground. Quite clearly, this is not a characteristic of the terrestrial monkeys and apes, all of which (or *whom*, depending upon one's opinion as to their capacity for culture) use their forelimbs in locomotion, and most scholars see arboreal existence as a precondition of the evolution of man. Attention has properly been turned to the arboreal primates in our search for insight into human evolution, but the observation of these creatures in the wild is made difficult by the fact that the human observer cannot follow them and usually cannot even see them.

The research upon which this paper is based suffered from this limitation, and, regretfully, it was necessary to observe arboreal monkeys in captivity. Except for brief trips to the Bronx Zoo, for comparative purposes, all the data were collected at Kornbluth's Katskill Kongo, a game farm in Grossinger, New York. The primate population at this research station included three spider monkeys, one capuchin, and two squirrel monkeys. (Mr. Kornbluth also had in his collection an aged hyena, a descented skunk, and two stuffed owls.) The monkeys lived largely on knishes thrown to them by tourists; since this is probably not characteristic in their natural habitat, I will not dwell heavily on feeding. The focus of this paper, however, is upon the use of the hands, and it is worthwhile to note at this juncture that I observed one squirrel monkey catch with his right hand a piece of halvah thrown from a distance of fifty feet. All the animals observed exhibited considerable manual dexterity, an ability made possible by the fact that they were usually in a sitting posture. Thus, though they do not have true bipedal gait, they very rarely used their forelimbs in locomotion. In fact, they moved around very little at all due to a limitation in space that was made necessary by the recent expansion of Kornbluth's Kottage Kolony, where I resided while in the field.

The monkeys observed by me at KKK only employed their hands in eating during 5% of the time. This again is an artificial limitation which must be corrected if we are to properly interpret the wild state. That the animals spent so little time in feeding was largely a function of meteorological conditions. Rainy and cool weather during the summer in which the field work was conducted drastically lowered the number of tourists, and therefore the knishes, and the A.S.P.C.A. ultimately closed Kornbluth's Katskill Kongo after half the animals had died. Mr. Kornbluth has since declared bankruptcy, a great loss to primatological research.

Even with the above slight deviation from natural conditions, a startling fact was noted. Approximately 40% of the manual movements of the monkeys were oriented to scratching and delousing (perhaps a higher figure than in the natural state due to the conditions of the cage), but, and this should be carefully noted, *55% of hand use was in masturbation*. It has long been known that this practice is common among monkeys, but I believe that this is the first time in which hard figures have been compiled. Frequency of masturbation varied from one squirrel monkey that masturbated on the average of 130 times daily to a spider monkey that communed with himself 723 times during a 24-hour period.[1] It was noted that towards the end of each day fatigue impelled the latter animal to use his prehensile tail for the purpose. This behavior, which I term *caudurbation*, has not previously been reported in the literature. These inordinately high rates of self-congress do not necessarily imply that most of the monkey day was taken up in such activity, for each episode lasted only three and one-half seconds.

It is possible now to consider the implications of these finds for evolution using the Inferential Method outlined in the introduction of this paper. (A tabular presentation of the full data will appear in a book to be published shortly by Pincus-Hall, Inc.) Man, it is agreed, developed culture through the use of his hands in the making of tools. There is also little doubt that the monkey hand, as we know it, is just about as evolved as was man's at the time when he made his breakthrough to humanity. The difference between the proto-human and the monkey lay exactly in the differential *uses* of the forelimbs by each primate. Our thesis that there is not all that much difference between the monkeys and man leads to a query of the usual assumption that the ancestry of monkeys and of man became differentiated early in the Tertiary Era. I would suggest instead that the two lines parted company in the Pliocene. The inferential basis for this statement is contained in the data presented above. I submit that man and the monkey had reached approximately the same stage of evolution during the Pliocene period (there is very strong support among certain eminent physical anthropologists for such parallelism), but man made tools with his newly evolved manual equipment whereas the monkey masturbated. The result was that this almost human creature rapidly degenerated, becoming the fuzzy and unintelligent animal that we now see in the zoo. While I will grant that occasional, even daily, masturbation has not produced marked deterioration among *Homo sapiens,* one can only wonder at the evolutionary consequences if men were to do so hundreds of times a day as reported in this paper for monkeys. Given these considerations it would perhaps be more profitable to look upon the monkey not as a prehuman, but as insane. It thus becomes necessary to reclassify the monkey as being a member of the genus *Homo*. Sapient he is not, however, so I will suggest the term *Homo onanismus drekopfii*, a name that at once combines his close relationship to man with his principal activity and at the same time incorporates the name of the writer.[2]

It may now be asked why man took the direction of tool making and *Homo onanismus* directed his interests inwards. The answer is really very simple: female monkeys remained victims of the estrus cycle while the human woman gained control over her generative abilities. During most of the year, the male *Homo onanismus* had no forms of gratification other than those provided from his own resources, a routine which was only occasionally broken by a female coming into heat.[3] Infrequent though these occasions may have been, biological compulsion required the female to present herself in a subordinate manner, and penetrability of the identity was maximized. Lacking choice alternatives, she never advanced to the position of a social person, unlike her human counterpart. The ultimate key to understanding humanity, then, is not that the *Homo sapiens* females are in heat all the time: they are *not*. Rather, they are able to choose exactly when to go into heat and are thereby able to control the males. The female stages this with sufficient frequency that man chooses to use his hands for externally oriented work, usually instigated by women. The female is therefore ultimately responsible for the evolution of culture. In conclusion, we may correct V. Gordon Childe's famous title. Man did *not* make himself—women made men—only monkeys make themselves. ■

---

* Mr. Drekopf is the alter ego of Dr. Robert F. Murphy, Department of Anthropology, Columbia University, who it should be noted, has confined his primate researches to occasional trips to the zoo with his children.

1 The only female in the troop was the capuchin monkey. This, however, seems to have little bearing upon the data or the conclusions given below.

2 This will strike some readers as immodest, but I should stress that the theory outlined in this paper has never been presented before, and the wording of my reclassification indicates only that I bear sole responsibility for it. I wish to restate my obligation to others, however, for the basic methodology that has produced these conclusions. Pioneering though my theory may be, I am optimistic that even more startling results will follow the further application of this method.

3 The inference could be challenged by citing the availability of homosexual outlets to the monkeys in my sample population. Patterned and regular homosexuality is, however, confined to *Homo sapiens,* and there would seem to be excellent ecological reasons for this. These derive from population considerations, but not from the point of view of the birth rate, as would be the inclination of most ecologists. Rather, we should consider the rate of morbidity and the accompanying fact that most forms of homosexuality are highly unsanitary. Primatologists have observed that of all the nonhuman primates only the gorilla fouls himself, but not even a gorilla would foul himself by another gorilla.

EYES ONLY[1]

RUSSELL BAKER

WASHINGTON, MARCH 11 — Datelines:

March 8, 1972 — The Pentagon said today it is looking for a new type of paper that cannot be Xeroxed or otherwise duplicated. The point is to stop Government secrets from leaking to newspapers. Present paper stocks can all be easily Xeroxed by Government people with a stake in publicizing them and printed before you can say Jack Anderson.

May 5, 1972 — The Pentagon asked Congress today for $983,000 to start a feasibility study of Leakpruf, a new paper made of plutonium-reactor shavings and horsehair, Spokesmen said early tests indicated that Leakpruf was Xerox-proof, but that much research was needed.

Oct. 16, 1972 — The Pentagon asked Congress today for $18.3 million to develop a method for shaving a plutonium reactor. Plutonium-reactor shavings and horsehair are the ingredients of Leakpruf, an experimental paper required for national security.

Jan. 28, 1973 — Controversy has arisen between the Pentagon and the Atomic Energy Commission over who has the right to shave a plutonium reactor. The Botchko Corporation, which holds the contract to develop Leakpruf, wants to try a new shaving technique on an A.E.C. reactor. The A.E.C. contends that Botchko's security clearance does not entitle it to engage in reactor shaving.

March 17, 1973 — Embattled Botchko Corporation executives want an additional $42 million to get enough plutonium-reactor shavings to produce a prototype of Leakpruf, the controversial new paper. The A.E.C. is suing Botchko for $172 million for pain and suffering sustained by one of its plutonium reactors which was severely nicked during an experimental shaving by Botchko last month.

April 11, 1974 — Senate investigators of the Pentagon's Leakpruf contract with Botcko were told today that $900 million was wasted in futile attempts to shave a plutonium reactor with a straight razor. Pentagon officials revealed, however, that the problem has since been solved — an electric razor did the job for $39.95 — and that Botchko can start work on a prototype of Leakpruf if Congress votes $1.5 billion needed to rescue the company from bankruptcy.

Nov. 3, 1974 — The Pentagon annouced today that it was launching a crash program to produce horsehair, an element needed in the manufacture of the controversial new Leakpruf paper. Spokesmen insisted that Botchko, Inc., Leakpruf's developer, had not forgotten about the horsehair when the project began, but had counted on finding a ready supply in the U.S. Cavalry. The Pentagon said that rather than re-create the U.S. Cavalry for this one project, it had contracted, for $795 million, with Gasso da Morte, the International chemical cartel, to develop a new synthetic horsehair.

July 24, 1975 — The Botchko Corporation told House investigators today its first prototype ream of Leakpruf had failed to resist Xeroxing because synthetic horsehair would not bond properly with plutonium-reactor shavings. To make the product work, the company said, would require genuine organic horsehair.

Aug. 14, 1975 — The Army will ask for a $6-billion budget increase next year to reactivate the U.S. Cavalry.

Oct. 9, 1977 — The Secretary of Defense, it was learned today, was among 42 high Government officials who suffered radiation sickness after participating in the recent press-conference demonstration of the Pentagon's new miracle paper, Leakpruf. A Xerox machine on which the demonstration was conducted has melted. The historic paper sample that could not be Xeroxed has turned into a glowing mass of horsehair with a half life of 8,000 years.

Feb. 11, 1978 — The President denied today that B-52's are dropping Leakpruf on Vietnam, but said he could not discuss what was being dropped on Laos.

June 9, 1978 — The Pentagon unveiled today a new device for stopping Xerox machines from copying secret documents. It is milk. All secret documents from now on will be written in milk, which does not show on a Xerox coy, but can be read by holding a match under the original paper. The discovery, made by Costplusco, a subsidiary of General Messes, cost $2.5-billion.

July 6, 1978 — Xerox reported today development of a milk-copier unit which, fitted in a copying machine at an added cost of $19.87, makes it possible to reproduce words written invisibly in milk. A Xerox executive said his son had invented the device while playing with his toy chemistry set and an old light switch.

Nov. 18, 1978 — The Pentagon asked Congress today for $983,000 to develop a new milk product that could not be Xeroxed or otherwise duplicated. ■

---

[1] From *The New York Times*, 3/12/72.

A GUIDE TO CORRECT BARKING ABROAD: A Review

by MARY WARE

Shi Pu's study *A Guide to Correct Barking Abroad* is an arresting work which demands the careful attention of any canine planning travel abroad. A dictionary of useful terms in several hundred canine languages, this work fills a painful hiatus in the linguistic canon heretofore available.

All too few animals heed the fact that the rewards of travel increase when one knows something of the language of the host country. Pets in particular should familiarize themselves with the sounds and structures of the languages of the countries in which they are to stay. How many every year spend up to six months in quarantine without understanding a bark barked or a yap yipped? If they were to study before leaving their homelands, they would save themselves much boredom in the kennel, where a little pleasant conversation is the only recreation outside meals. They could also prevent the inconvenience of having to rely on gestures. So often the wave of a paw or flap of a wing can be misconstrued. Knowledge of the language of a country also promotes better international relations through the grass-roots relationships with the host-country nationals. Animals, being generally closer to the grass than anyone else, are an untapped source for improved international understanding in this troubled time of ours.

A review of this work must mention that the study goes beyond a mere listing of terms. Special cultural notes are given to aid in sensitizing the visiting dog to the *faux pas* he might make, so unnecessary in our tense era. Various quotations from the dictionary listings can only indicate in part the scope covered.

ARGENTINA
Gua-gua /Gwa-gwa/ (The dog visiting Spanish-speaking countries is cautioned against assuming that the Spanish language is the same the world over. A cursory examination of the first half dozen items will reveal the fallacy of this assumption.)

BRAZIL
Au-au /Ow-ow/

CHINA
汪汪 n. b. Hong Kong; at the time of preparation of this manuscript, dogs from Red China were unavailable as informants.
/Won-won/ (Do not confuse with 咪咪 /mai-mai/ said by Chinese cats and roughly equivalent to the U. S. Standard *meow*.)

COLOMBIA
Guau-guau /Huow-huow/ (This sound is rather difficult for non-native speakers, but by diphthongizing the *uo* and drawing the sound out slowly, a comprehensible approximation can be made.)

COSTA RICA
Guau guau (*sic.*) /Gwow-gwow/

CUBA
Jau-jau /How-how/ (Although the revolution has produced some slang variants, the standard has remained constant.)

EL SALVADOR
Guau-guau /Woaw-woaw/

ESTONIA
Auh-auh /Aw-aw/

FINLAND
Hau-hau /How-how/

GUATEMALA
Bow-wow /Bow-wow/ n. b. The full import of the similarity between the Guatemalan and U. S. barks is still being researched.

ISRAEL
הַו הַו n. b. Before June, 1966, Hebrew-speaking sector.
/How-how/

LEBANON
عوْ عوْ /Haw-haw/

NIGERIA: Calabar area
Wai-wai /Waing-waing/

NIGERIA: Western area, Yoruba language
Gbogbo /Gbo-gbo/ (The multiplicity of languages within a single country complicates communication immeasurably; however, it is hoped that the increasing use of English as a *lingua franca* will lessen this difficulty. It would be unfortunate, nevertheless, if the traditional forms were lost in the process. A society is currently being formed to minimize this danger.)

PERU
Gua-gua-gua /Wow-wow-wow/ (Peruvian dogs are known for their over reaction to even the simplist situations.)

PHILIPPINES
Bow-wow /Bow-wow/ (American influence; regretably the beautiful, n a t i v e sounds have been forgotten by even the oldest inhabitants.)

SYRIA
عَوْ عَوْ /Haw-haw/ (A Dog speaking Arabic has a better chance of making a socially acceptable comment to Arabic speakers from various countries than a dog speaking Spanish in different Spanish-speaking countries. See Lebanon; Argentina, note; however, the visitor should listen carefully to the inflection of the host country national.)

THAILAND
/Hong-hong/

UKRAINA
Браше /Breshe/

VIETNAM n. b. The southern section.
Gâū gâū /Go-go/ (French and American contacts with the canine population have apparently been minimal.)

To help the animal bent on greater rewards in foreign travel, Shi Pu has urged publication of this compendium in paperback rather than in the better-grossing hard-cover edition, explaining that the "small size makes it [the book] easy to carry in the mouth or beak."[1] The pet is advised to try to learn the languages of as many countries as possible, as the quarantine location, a miniature United Nations, will contain animals from a variety of linguistic backgrounds other than that of the host country.[2]

The approach is sensibly aimed at the general canine rather than at those already initiated into the mysteries of linguistic scholarship, as it is the general reader that has special need for this study. This method of presentation is explained in the introduction:

> The concentration of this phrase book is on greetings, which are the first words needed. The transliterations are into the sound system of American animals as they in particular expect everyone else to speak their language and constitute, therefore, the group that must be reached with greatest urgency. A more comprehensive listing of American dog terms is included so that American animals need no longer be embarrassed by the criticism that they do not know even their own language properly. Although pronunciation is considered the most important linguistic aspect for conversational barking, the really conscientious animal will want to study the written forms presented as well.[3]

Yet the study is backed by erudite research for which Shi Pu is eminently qualified, as can be readily seen by the description of the method used:

> The dictionary has been prepared after several years of intensive research through English as a Second Language classes in various cities, including dozens of sessions in which the students brought their dogs to class to record their voices. The dogs had to be brought singly to avoid their accents' becoming impaired through contact with dogs of other linguistic backgrounds. To keep foreign influences at a minimum, foreign students' dogs were preferred to dogs currently in quarantine.[4]

It is indeed touching that the book is dedicated to the dogs "who so freely donated their time to this study."[5]

While it is not Shi Pu's way to use current work to advertize former works, a study of *A Guide to Correct Barking Abroad* cannot end without mentioning related studies by this scholar; for the animal planning travel abroad would do well to study the languages of species other than his own so that he can recognize them without having to crane his neck into many awkward positions, thus risking an attack of lumbago so uncomfortable in cramped quarters. Available by the same author are *The Vocabulary of Cats, Bird Sounds: An Elementary Phrase Book,* and *The Languages of Larger Animals.* A catalogue of tapes recorded by native speaker canines of superior educational background is also procurable upon request.

---

[1] Shi Pu, *A Guide to Correct Barking Abroad,* New York: Animus Animalorum Scribendi, 1972, p. vi.
[2] *Ibid.,* p. vii.
[3] *Ibid.,* p. xi-xii.
[4] *Ibid.,* p. xxix.
[5] *Ibid.,* p. xxiii.

UNSAFE AT ZERO SPEED

Charles Osterberg
9312 Gue Road
Damascus, MD 20750

It had to happen. In response to a suit brought by an irate motorist caught in a radar trap, the court ruled that President Carter did not have the authority to restrict the highways speed to 55 mph. The president then asked Congress to set such a limit. Congress, as is its want, decided to hold hearings before acting. The response from interested parties was large.

Testimony indicated that reduction of the speed limit to 55 mph had indeed saved lives and gasoline, and should be re-imposed by Congress.

This was immediately challenged by I.Q. Nadir, speaking on behalf of 225 million Americans, who pointed out that the loss of lives from highway accidents at 55 mph was enormous, morally wrong, and unacceptable to him. To make his point, he showed gory movies of head-on collisions at 55, 45, 35, and 25 mph. "Twenty-five mph." he said, "while not perfect, would be an acceptable and reasonable compromise, since deaths would be minimized and battery driven cars would be able to compete with gasoline cars." The 55 mph limit "traded human lives for earnings for thebig business interests in Detroit, with their huge investment in highspeed gasoline engines. They were using this subsidy, given them by the 55 mph speed limit, to freeze out competition from battery driven cars, which performed best up to 25 mph."

"Automobile Parts and Other Metal" by John Chamberlain, Museum of Modern Art, New York

A spokesman for the American Bicycle Association felt the occasion called for a drastic rethinking of our way of life. Reducing the highway speed from 55 mph to 25 mph was at best a halfway measure—one of degree, not substance. The speed limit should be reduced to 15 mph. While this appeared to be only more of the same, it was not. Most persons killed while riding bicycles were in urban areas with 25 mph speed limits, so the proposed reduction to 25 mph did absolutely nothing for the growing millions of bicyclists—really, it was only a cruel hoax! But at 15 mph, no one would be passing anyone. This major cause of accidents would be eliminated and the death rate would drop sharply—far out of proportion to the small reduction in speed. After all, if we're really serious about the sanctity of human life, we should strive for a zero death rate on the highways.

To which Prof. Thirdclass of the U. of Pillsbury, stated that to reach zero death rate on the highways, which was certainly a legitimate goal, we need only set a speed limit of zero mph. His data showed that death rates increased linearly with highway speed limits, and the line passing through the data points, if extended backwards, passed through zero at zero mph. In fact, if he extrapolated even further to negative auto speeds, he got negative auto deaths, and could only conclude, from his data, that if automobiles went backwards rather than forwards, lives would be created, not lost.

At this a representative of the Malthusian Society protested about Prof. Thirdclass's alarming data. Apparently, cars backing up on the highways could lead to a "wanton creation of life," something the overcrowded Earth could best do without. His society preferred a 100 mph minimum forward limit, which, his calculations showed, would cause the auto death rate to equal the birth rate, leading to a natural balance which, everyone knew, was preferred by Mother Nature. He admitted that there would be a temporary increase in gasoline consumption, but, in the long run, this limit would actually save gas as drivers were eliminated by attrition. Besides, if Congress gave them the mandate, Detroit could build cars that would give as good mileage at 100 mph as at 55, or even 15.

Dr. Hoffman, of the Union of Uninformed Scientists, felt that Prof. Thirdclass, whose work he normally admired, had in this case erred in stating that auto deaths would drop to zero if auto speeds were reduced to zero. Certainly this was an understandable mistake. There just weren't good data on auto deaths at 2 or 3 or even 5 mph, so the "tail of the curve" was subject to errors. But intuitively he felt that, while autos traveling at 55 mph might be randomly distributed throughout the nation's highways, at zero mph they would be closely clustered together in parking lots, with a much greater probability of interaction. His calculations showed that the relationship was not a straight line, as Prof. Thirdclass had mistakenly assumed, but rather, because of the "cluster effect" there was a sharp secondary increase in deaths at speeds approaching zero mph. Probably a lazy J-shaped curve would describe it best, with the low point—the nadir, to make a small joke—at about 4 to 6 mph. Unfortunately, he concluded, automobiles were unsafe at any speed and should be banned.

Dr. Hoffman then went on to publicly apologize for his early role in the development of the automobile, which he said was, pound for pound, the most deadly machine in the world. (Readers may not be aware that Dr. Hoffman is father of the seatbelt-ignition interlock system, hailed in the 70's as the foremost safety development of the 20th century, until it was banned by pressure from the anti-safety lobby in Congress. In its brief existance, the interlock is credited with saving millions of gallons of gasoline and hundreds of lives, as thousands of cars sat idle in driveways because puzzled drivers couldn't figure out how to start them.)

Dr. Bernie Commonenuff of Lincoln University said he was delighted with the previous testimony because it was certainly within our capabilities to build a solar car that could approach and perhaps achieve 6 mph on a sunny day in Arizona. these could be assembled from kits in the backyard by the average handyman thus eliminating Detroit and big business. While the solar collectors needed were too large to allow two cars to pass, or to fit under an underpass, these problems were considered minor as the technology was developing rapidly. All that was really needed was a subsidy from the government and the 6 mph speed limit to make solar autos a viable option. Trees along the highway that shaded the sun would have to be cut down, he admitted, but, as he pointed out, there is no free lunch. At any rate, Dr. Hoffman's data showing minimum deaths at 4 to 6 mph was welcome news because, with a bit of development and government support, the solar industry could achieve that goal, which was a reasonable one.

Congress acted decisively, calling for a 5 year moratorium. It was thus left to the Supreme Court to decide whether the speed limit stayed at 55, dropped to 25, or zero, or infinity. ■

TRANSLATION

The article by Dr. M.H. Levin in the March 14 Almanac contains some rather technical language, e.g.

"A feature of the urban ecosystem is the conspicuous transportation network . . . portions of this vital circulatory system may vunction sluggishly at times. Fossil fuel burners bearing the human component to their places of employment within the city may reduce the speed at which fuels, manufactured products and services are imported into the urban ecosystem." In the interest of improved interdisciplinary communication which the University so fervently desires, I submit a translation of the above:

"Automobiles carrying people to work may get in the way of trucks."

Further insights into the functioning of the urban ecosystem are eagerly awaited.

—*Michael Cohen,* Associate Professor of Physics

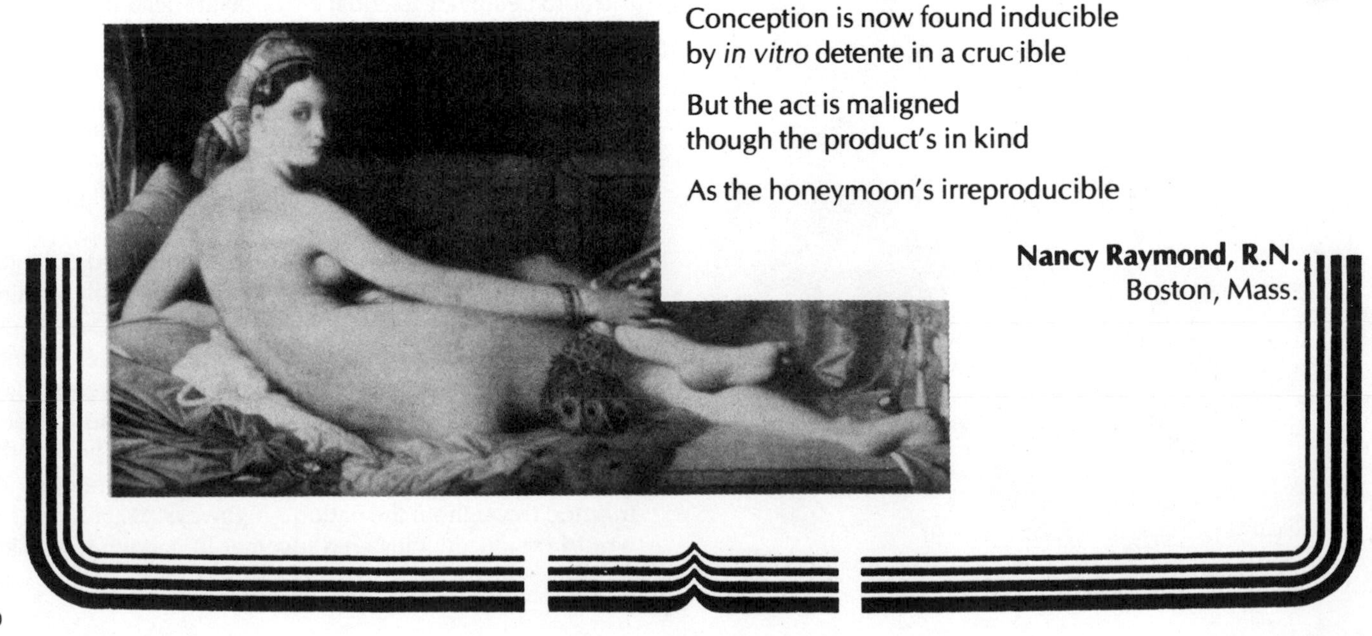

Conception is now found inducible
by *in vitro* detente in a cruc ible

But the act is maligned
though the product's in kind

As the honeymoon's irreproducible

Nancy Raymond, R.N.
Boston, Mass.

Reducing Automobile Accidents

By JOHN L. S. HICKEY*

Recently reported research[1] by the National Safety Program has provided a significant clue which can, if properly exploited, reduce and perhaps completely eliminate automobile accidents. The germ of the breakthrough lies in the NSP finding that "75% of automobile accidents occur within 40 miles of home".[2] Now there are five ways in which one may react to this statement. Three of the reactions are elementary:

1. Normal reaction: "I should be just as vigilant driving near home as when on the highway".
2. Statistician's reaction: "Since about 90% of the driving occurs within 40 miles of home, and only 75% of the accidents, this a "safe" area and I can relax while driving.
3. Reverse reaction: "I'd better drive as fast as possible to get out of the 40-mile "danger" zone into the surrounding "safe" zone".

The other two reactions will be explored in some detail, as reaction #4 provides the means to reduce accidents, and reaction #5 can, at some small risk, eliminate them completely. Both employ the method of Data Enrichment[3] previously reported in the Journal of Irreproducible Results.

The fourth reaction is: "If 75% of automobile accidents occur within 40 miles of home, it can be seen through use of the data enrichment method[4] that the farther one is from home, the smaller the chance of an accident. Therefore, *I will register my car at a "home" 500 miles away and never go near there*". This process could easily be developed into a national program for providing simulated or substitute home for all drivers, perhaps with a title like Car Registration at Substitute Homes (CRASH). Obviously, if no one ever drives within 40 miles of their "home" accidents will inevitably be reduced 75%.

The fifth reaction requires a little background discussion. As we all know, accident rates vary from locality to locality; one can therefore expand on the fourth reaction and surmise that, instead of registering their car at a randomly chosen "home" 500 miles away, the safer thing to do would be to register it in a place much farther away *and which has a low automobile accident rate.* A place immediately comes to mind — the South Pole. It is very far away, and there is only one automobile there[5]. Thus the automobile collision rate is necessarily zero, and automobiles registered in Antarctica will sustain this rate. Again, this concept could be extended into a nationwide program under which every automobile would be registered at a single center located at the South Pole, which could be named Central Location for Accident Prevention Through Registering Automobiles Polarilly (CLAPTRAP). Through this method, the automobile accident would become a thing of the past. The small risk? If another automobile were taken to Antarctica and collided with the first one, every car in the country would suddenly be registered in an area with *a 100% automobile accident rate.* ■

[1] Reported via public service time on TV and radio.
[2] It may be 80% within 50 miles of home. The exact figures do not change the concept.
[3] Lewis, H. R., J.I.R. 15:1 (1966).
[4] Calculations not reproduced here.
[5] A Volkswagen: see full page ads in Life, almost any 1967 issue.
* Nom de plume for Mike Robrain.

SYNTHETIC HAPPINESS

G. VAN DEN BERGH
Haarlem*

After many years of research, a team of scientific workers under Professor Sadler of the Amsterdam University has for the first time in history succeeded in producing synthetic happiness. Although the quantity produced is still negligible, there have been already some successful experiments on men, and the mass production of synthetic happiness is only a question of time[1].

The research team took as a starting point the common experience that happiness is not where you look for it. Yet, instead of giving up that seemingly unsurmountable problem, they redoubled their efforts and that with success.

The first thought of Professor Sadler proved to be unfortunately useless. If simply all feelings, thoughts, experiences and situations were to be studied, the problem could be solved. But it soon appeared that even the best electronic computer would need 3.1×10^{11} years to complete the task, provided that the number of possible combinations remained constant during that period of time. As the number of possibilities increases every year by about 10 million, the advocated method did not seem to be very promising for the nearest future.

Eventually the scientists succeeded in constructing a rather simple electronic brain, which seeks where it does not seek. Thus happiness could easily be located and isolated. It proved to be very volatile and extremely difficult to analyze chemically.

Another problem which was investigated was how to produce happiness in usable form. Experience has shown that an uncontrolled explosion of happiness has effects similar to those known from other explosive substances including the atom bomb. Blindness and deafness are the most common known effects, in more serious cases the brain is blown out and a serious disorder of the heart ensues.

Some pressure groups in the Western world suggest the immediate production of a HAP-bomb, before the Russians get the priority. The production of happiness for peaceful purposes has not yet been seriously delayed by this threat. The scientists have happiness under control and may produce any amount of it in reasonable time. Before, however, the economic exploitation of happiness can begin, there is still another problem to solve, because happiness cannot be bought. It has not yet been possible to sell any amount of happiness, without rapid denaturation of it.

There are different approaches to the problem of selling happiness. Some lawyers assert there is no problem in constructing a form of buying and selling, which is not buying and selling. Professor Sadler is trying to obtain a transaction-resistant happiness by a chemical method. His sponsor, General Happiness Inc., has again voted 1.5 million dollars for this research.

In the meantime, there seem to be also serious political implications in the exploitation of happiness. Political observers state the possibility of a complete social revolution after which everybody would have as much happiness as he liked. Some political parties have already urged their governments to take appropriate measures to prevent this.

So for a while we shall still have to do with ordinary happiness. ■

* It is reported that the author has found non-synthetic happiness.

[1] Huxley: *Brave New World*, 1947, Albatross Ltd., London.

EVOLUTION OF SCIENTIFIC THOUGHT

FRANK ANDERSON
Miami, Florida

Prof. Ludwig Botchall
Dept. Of Geophysics
Tifton University
March 20, 1971

Prof. Karl Von Vettkrotch
Dept. Of Geology
Pfledering Inst.

Dear Karl,

Since my new grant I've been researching a method of controlling seasons via population selection and I'm forwarding this theory for your perusal and in hopes that you may have some suggestions.

It appeared to me as if from a dream that if one could elicit the cooperation of the world's total population for a mere few minutes one could control the earth's rotational velocity. I propose having everyone face in the opposite direction of the earth's rotation and at the same precise moment all run at top speed for five minutes. This would slow the earth's rotational speed by a factor TP^x. X being the frictional force average/individual and TP as total population.

How does this sound to you?

Your friend,
Prof. Ludwig Botchall

Prof. Karl Von Vettkrotch
Dept. Of Geology
Pfledering Inst.
April 14, 1971

Prof. Ludwig Botchall
Dept. Of Geophysics
Tifton University

Dear Lud,

In answer to your letter dated March 20, 1971, I played with that theory several years ago as a means of avoiding daylight-savings time and I met a few stumbling blocks. 1. It seems that the mainland of Communist China would slam into California and the Chinese population would end up in the ocean as their continent slipped out from under them. (Not a bad idea though if you could get them to co-operate blindly). 2. Tidal upheavels would be immense. 3. Linearity would be difficult due to population distribution factors.

Have you considered these factors?

Karl Vettkrotch

Prof. Ludwig Botchall
Dept. Of Geophysics
Tifton University
April 17, 1971

Prof. Karl Von Vettkrotch
Dept. Of Geology
Pfledering Inst.

Dear Karl,

By gosh you're right. You never cease to amaze me with your overall grasp, and you led me to consider some other ramifications. I feel that the proper mathematical computations of continent mass/total population would allow one to control the coefficient of frictional force necessary to effect rotational lag without continental shifting (although your China postulate sounds exciting). It's apparent that this frictional coefficient would vary with variation in land mass and population. Our computer here is on the blink again. Can you help?

Ludwig Botchall

Prof. Karl Von Vettkrotch
Dept. Of Geology
Pfledering Inst.
May 19, 1971

Prof. Ludwig Botchall
Dept. Of Geophysics
Tifton University

Dear Lud,

Sorry to take so long but our computer was tied up with end of the month billings and I just finished my run. I have really lost interest in the climate control theory but I've become overwhelmed with the possibility of rejoining the continents into the original. When deriving your computations I came across some data on earth spin energies and the results are frightening. I came up with the following calculations for defining the differential rotational inertia of a continent (ΔI)

$$\Delta I = \iiint \Delta O\, R^4 (\sin \phi - \sin^3 \phi \cos^2 \Theta)\, dR\, d\phi\, d\Theta (\pm)\, p^f$$

Where ΔO = density contrast between adjacent continental crustal and oceanic layers

R = continental Radius

ϕ = longitude

Θ = latitude

$(\pm)p^f$ = population friction and is negative or positive depending on direction population is running

Integration would lead to two factors

1. ($1.59 \times 10^{39}\ CM^2$) depending on shape of crustal layer.
2. A $\pi/3$ function relative to hemispherical size.

The factor (p^f) could become a constant (k) by controlling population friction. This could be accomplished by having all participants wear golf shoes to minimize frictional differences caused by terrain.

Lud, I'm personally fearful of this experiment. My calculations show that the continents would definitely rejoin with a force equal to 2×10^7 earthquakes and would back us up to the triassic period in geological time. I'm not sure I would care to go back that far even if I did survive the earthquake forces.

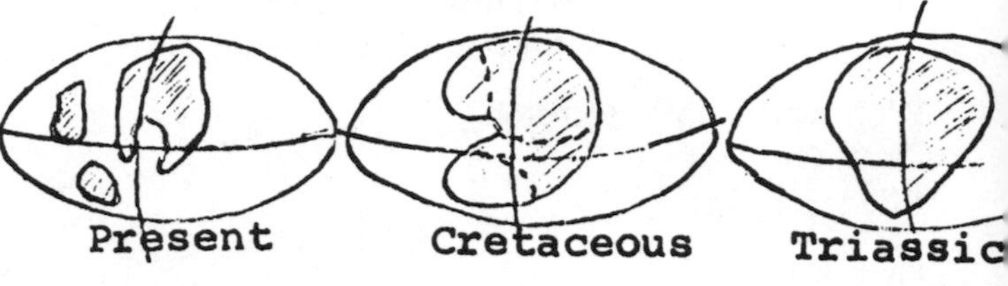

I prefer to abandon the experiment and respectfully suggest that you do the same.

Your friend,
Karl Von Vettkrotch

P.S. My present worry is that all the "junk" we're leaving on the moon is going to alter its polar function. We as scientists should insist that Nasa make some effort to distribute it more evenly. ■

Here's a pip, reported by the Consulting Engineers Council: the latest Occupational Health and Safety Standards has outdone itself in governmentalise, they define the word "exit" as: "Exit is the portion of a means of egress which is separated from all other spaces of the building or structure by construction or equipment as required in this subpart to provide a protected way to travel to the exit discharge." They then had to define "exit discharge" as: "Exit discharge is that portion of a means of egress between the termination of an exit and a public way." Webster does it so much more easily, to him an exit is "A way out of an enclosed place or space."

Engineering Education: April 1972/779

A Survey of Irregular Pedestrian Movement Transverse to Vehicular Flow

R. F. WILDE
Harvard Computer Center

INTRODUCTION

A study of available traffic literature indicates that while considerable attention is devoted to vehicular traffic, comparatively little is devoted to pedestrians[1]. The subject of this paper* is Irregular Pedestrian Movement Transverse to Vehicular Flow (IPMTVF), commonly referred to by the public as "jaywalking."

What is jaywalking?

A clarification of the term IPMTVF will first be given. There is a certain probability, depending upon traffic conditions and his psychological frame of mind, that a pedestrian arriving at a signalized intersection during its red phase, will attempt to cross against the vehicular flow. IPMTVF, or "jaywalking," refers then to this sort of perambulation. (The word "Irregular" in the definition, with its connotation of "random" and "illegal," is felt to be particularly suitable in this context).

Taking the survey:

The simple two-phase intersection used in the survey (44th Str. and 3rd Avenue in Manhattan) was chosen primarily for its convenience. An observer was stationed at the corner with stopwatch and notebook. Since IPMTVF is usually undertaken in defiance of established authority it was felt necessary to disguise the observer as a vagrant. 419.5[2] jaywalkers were studied and timed under varying traffic conditions. Jaywalkers exhibiting bound motion (e.g. children holding their mother's hand) were not included. Unfortunately, owing to the impenetrability of the observer's disguise the final third of the survey had to be carried out one block to the east. Analysis of the results, however, indicated that geographical variation over a distance of this magnitude was not statistically significant.

* Term paper for Mass. Institute of Technocracy, course G1143A (Macrobiotic Traffic).

[1] R. F. Wilde, Evolution of protective clothing in the Metropolitan pedestrian, Grape Press 1968, p. 47.

[2] The fractional individual, when approximately half-way to his destination, became unsuccessful, but was deemed typical and hence included.

RESULTS

1. The IPMTVF-pedestrian:

It soon became apparent that IPMTVF behavior could be broadly divided into three groups:

Type A: the bluffer;
Type B: the gambler;
L.O.L.: the unconscious.

An L.O.L. (Little Old Lady) usually fails to observe the light or notice the traffic and is sometime mistaken for a Type A until panicked by the horns of the oncoming cars. Note that, in spite of the terminology, both sexes may well be included under this classification. Another term in common is A.M.P. (Absent-Minded-Professor), not to mention a number of less flattering appellations.

It was hoped that some means would be found of predicting an individual pedestrian's behavior, as it is of primary importance to the motorist to avoid delay. Unfortunately this could not be done. Stratification by age, sex, and color of hair were sometimes observed, as when a Type A under stress would undergo a rapid metamorphosis into a Type B.

2. The IPMTVF fraction:

Let us now consider a number of pedestrians arriving more or less randomly at an intersection. Although, as previously pointed out, the behavior of an individual cannot be foretold with accuracy, a definite IPMTVF pattern does emerge when the behavior of many persons is averaged. The following diagram illustrates the form of curve obtained when the fraction of accumulated pedestrians attempting to cross illegally (the IPMTVF fraction, denoted here by F (t)) is plotted against time.

The high value and steep slope of F (t) for small values of t is due to the large number of pedestrians who get caught half-way across the street as the light changes and the rapidity with which they are eliminated from the scene. The end of the main vehicular platoon of traffic (slightly anticipated by the horde of impatient pedestrians) results in a new surge of IPMTVF activity. A final upswing occurs when the amber phase for the opposing vehicles is observed.

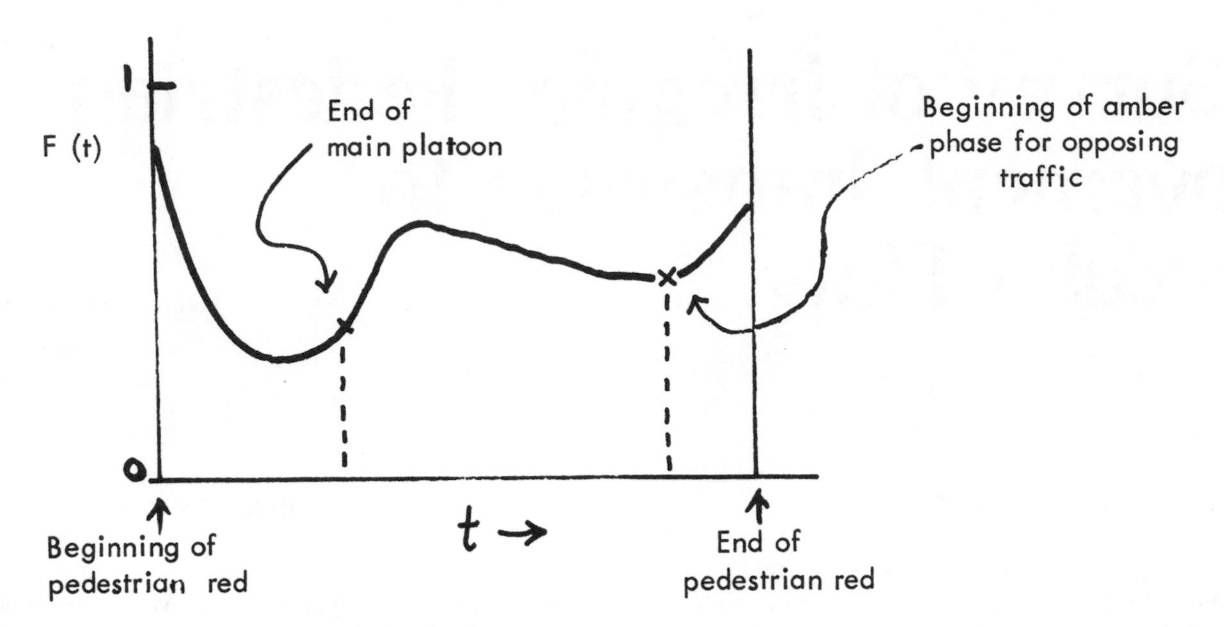

The curve just described was found to vary widely with such factors as time of day, cycle length, prevailing winds, and congestion both pedestrian and vehicular. More consistent results were obtained when a composite factor herein referred to as the *fright factor* was taken into consideration.

By definition, the fright factor

$$FF = k\left(c + \frac{v\,d\,l}{f}\right)$$

where v = average vehicular velocity
d = traffic density (equivalent cars per cubic foot)
l = length of main platoon (miles)
f = footing factor, decreases with insecure footing[3]
k = constant to be computed by regression for each intersection
c = velocity of light (light years/year)

The *standard IPMTVF fraction* is then given by

$$S.F.\ (t)^{+} = \frac{F(t)}{FF}$$

Although S.F. is of no particular value whatsoever, its curve remains remarkably constant, all things considered.

3. The IPMTVF path:

Investigation of the IPMTVF path was first undertaken during attempts to estimate IPMTVF velocity. It was soon noticed, however, that the path followed during IPMTVF was rarely a straight line perpendicular to the curbs. The diagram below shows a typical path.

Intensive observations of IPMTVF behavior indicated that the path *varies according to the obstacles encountered* (dogs, cars, pedestrians, diverse hazards underfoot). Furthermore, it was almost conclusively demonstrated that the path depended to some extent on the subsequent course taken by the pedestrian[4]. It has also been suggested to the author that some of the more exaggerated ambulatory aberrations were influenced by varying degree of alcoholic sedation[5], but this hypothesis could not be verified in detail. ■

[3] A footing factor of zero should not be used in practice.

[4] This remarkable result would indicate that *the future influences the past*—refutation of the principle of causality on the macroscopic scale.

[5] Unidentified Manhattan taxi driver, private communication.

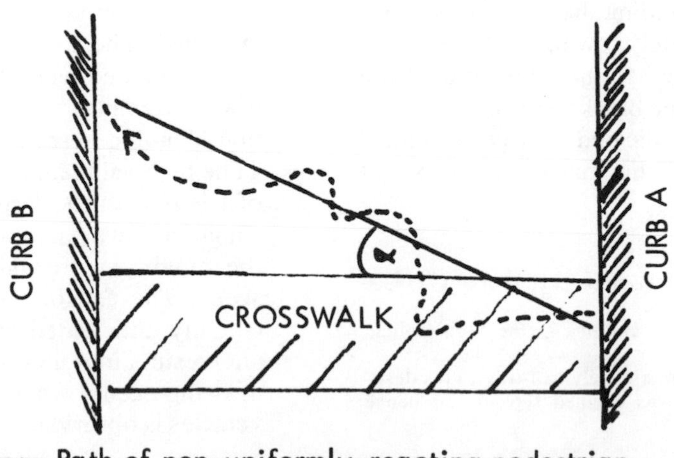

--- Path of non-uniformly-reacting pedestrian

Chemically Chaste Campus

A recent newspaper article told of the growing demands of the student body at a well-known Eastern university for freedom of access to birth control measures, to wit "The Pill." Here at Truly Progressive University, the item was acknowledged with smiles of condescension. In this, a truly progressive university, the arguments pro and con were passe, strictly antebellum.

We had first come to grips with the problem several years ago and certainly a brief outline of the methods we used and the stumbling blocks we met on our way to a solution will serve as an invaluable framework for less advanced colleges to follow. The sagas of brave men challenging the frontiers of social science have ever been a fascination of mine. Thus, I find it doubly satisfying to recount an advance in which I played a personal role.

It all began in the late 1950's when educators across the country displayed increasing alarm over the swelling number of illegitimate pregnancies noted on the best of campuses. The statistics were unquestionable but there was little agreement over the basic cause of the increase.

It was here at T.P.U., I'm proud to say, that the initial advance was made. At one of our faculty meetings, a stalwart professor of Physiology bravely stood up and stated what he felt to be the heart of the matter. His speech was rather liberally spiced with anatomical terms and loose comparisons between some of our students and certain members of the hare family.

The minutes of the meeting were later burned and the professor shortly sought another post. For us, however, the air had been cleared and the enemy, so to speak, discovered.

Open discussions on the subject became the rule of the day. The program was constantly modified and enlarged. After two years of this bombardment of intensive information the results were tabulated. Alas, they showed a larger number of better educated unwed mothers. An advance surely, but not an answer.

We moved on.

The faculty meetings had never been so well attended and became the scenes of titanic struggles between the proponents of various control measures. Those few who persisted in implicating members of the bird or insect family were shouted down. The suggested lines of attack showed unlimited imagination and ran the gamut from compulsory chasity belt, with the master key held by the dean of women, to immediate expulsion for even conversing with the opposite sex.

The struggle between the moderates and extremists waxed and waned. A compromise seemed impossible. Then came the pharmacological breakthrough of the second trimester of the century from a distillate of horse . . . er, Equine urine, came the "Pill."

The administration happily suplied the desired products free of obligation. Had it been up to the men to take the pill it might have worked. The coeds, however, giving lip service to an outdated moral code, refused to forearm themselves with the necessary regularity. The laboratory at the Health Service, where a positive pregnancy test induced massive ovulation in the female of the frog family, was up to its neck in excited frogs.

"Take your pill" buttons and protest marches flourished. Mottoes were slashed on every available facade. "Pills prevent people," "Pill-less passion produces pregnancy," to mention a few.

Male super-egoes were shaken and shattered. The Mental Health Department was a pandemonium of crying men. "If you can't trust your girl to protect you, then what?" they asked. "Look at me," they pleaded, "I'm only a lapse in mommy's memory."

The effects on campus life were dramatic. Twenty per cent of the Junior class volunteered for Viet Nam. The director of the Health Service was repeatedly hung in effigy. Unrest seethed through the classes and gloom enveloped the university. Finally even the "Winter Carnival" was declared a stag party.

It was in this mentally stupifying atmosphere that the star erupted and from that day forward the name of Dean Hubert Hereford Howser has a permanent place among the annals of college men. That memorable day when he rose above the common man to give birth to a plan that spanned both human and animal physiology and set T.P.U. once more on the road of progress.

Later in an interview for Time magazine the details were elucidated: Dean Howser related that while shoveling out the University barn (he doubled for the janitor on weekends) he was trying to avoid the hind foot of a heifer that seemed intent on smashing his thigh bone when suddenly he was struck as if by lightning. His fertile mind conceived the entire stratagem in a nonce.

It was ingenious, workable, and it destined Howser to immortality: feed the hormone to the kines, he reasoned. Give them enough so that a therapeutic dose comes through in the milk. Everyone drinks milk.

The coeds chug-a-lug their milk, chomp up cheese, and are passionate about puddings. It seemed foolproof, perhaps even student proof.

Twenty-one days of each month the cows were slipped a measured dose of medicated hay. The resultant fortified milk was passed on to commissary to be fed to the coeds. The effect was heartwarming to see. Not only was pregnancy no longer a problem but all the females on campus were now set on the same menstrual cycle.

The Infirmary set up a crash program once a month for "Dysmenorrhea Day." From the twenty-fifth to the thirtieth of each month Midol was available on the dinner menu. Classes in Women's Physical Education were simply cancelled if they coincided with "starting day." In recounting the system's benefits before the faculty club, the Director of Health stuttered and finally cried like a baby as he outlined the range of the plan.

We have now been fully activated for a period of three years and I'm happy to report that the tranquility of the campus is only rarely disrupted by unwanted and unwarranted pregnancy. These occur in girls to whom milk products in any form are anathema. We've agreed among us that any little runt as unAmerican as that can just take her chances.

Honor forces me to admit there were some minor shortcomings. These all arose in the male students who could not be dissuaded from partaking of milk products. The kinks were minor and rather easily disposed of, but for the sake of completeness I will list them and our solution to them.

(1) Only five per cent of our male graduates can pass their draft physicals.

No gripes there, eh?

(2) Once a month some of the men get extremely irritable.

We have encouraged them to take Midol. This seems to control that problem.

(3) The enrollment in the Art College has tripled and the sale of male cosmetics has skyrocketed.

To each his own, I say.

(4) In spite of the epic efforts of our animal psychiatrist, our prize Angus bull killed himself in a frenzy of frustration.

A steak is a steak is a steak.

(5) Surprisingly, the major body of complaints have come from the female students. It seems we are left with a campus of lusty fully protected females and the men aren't really interested.

Hell, you can't have everything. ■

Dwan T. Wick
A STUDY OF LEARNING FROM TELEVISION WHEN PICTURE QUALITY IS DEFICIENT
East Texas State University, 1966.

Found "significant difference in the learning from normal T.V. reception conditions and when there was snow or ignition interference in the T.V. picture".

If ego takes control of "id"
And superego is the lid
Then "id" is pretty far below
So, seems that something's bound to blow
Up and away so "Id" can be
Itself, again, unconsciously.

Joan B. Richmond

The Clinical use of Spondiac Vulgarisms in the Assessment of Hearing Aids in Sound Field Hearing Test

ROBERT L. FRANKENBERGER, Ph.D.

Speech and Hearing Laboratory,
Western Illinois University,
Macomb, Illinois 61455

In order that persons with hearing losses may hear better, the usual routine is the purchase of a hearing aid. However, the purchase of a hearing aid is not a simple matter. One needs the advice and counsel of a clinical audiologist to guard against being bilked by an unethical hearing aid salesman. The selection of a hearing aid is akin to the purchase of a new car. All one really needs is basic transportation, but the salesman mentions color coordinated pistons, white walls, automatic antennas, dual mufflers and speakers and other components not necessary to the task of getting to and from one place to another.

There are as many types of hearing aids as there are types of cars, or, for that matter, hearing losses. One should exercise judicious care in the choice. The best hearing aid can be selected by the re-testing of the person's hearing while wearing the unit. The unit that gives the wearer the best hearing is usually selected; regardless of the color coordinated wires and batteries or price. One reliable way to test hearing is the utterance of spondee words (six lists of thirty-five two-syllable words with equal stress on each syllable, such as "doorknob, campstool, ragmop," etc.) in a sound field environment. Thirty-five words are usually presented with a decrement of 1 dB less than the preceding word so that the audiologist can cover a range of thirty-five dB in intensity.

Sound field testing is the presentation of the spondees in a free field environment without earphones but with the hearing aid worn as one would usually wear it in a non-test situation.

Some of the thirty-five words in the list have better responses than others. This should not be. Upon investigation of responses, when the patient repeats the word he thinks he hears, the audiologist must sometimes resist a powerful urge to giggle. Giggling in a sound treated chamber, with the patient seated facing one, seems to diminish one's dignity as well as threaten the professional aura that most hard-of-hearing patients give the clinical audiologist. The words most commonly "heard," or at least repeated, seem to be two-syllable vulgarisms that apparently sound much like the innocuous test words. Words such as "duckpond, horseshoe, hothouse, and storehouse," for instance, are mistaken for common four letter vulgarisms.

With the relative freedom of language in books, the theater, movies, and campus riots, clients often do not hesitate to use expletives if they think they hear them. Older female patients seem to hear these words much better than their uttered counterparts in the test in spite of not being able to hear preceding or following words at the same, or higher, intensity level. One patient, in fact, heard nothing but vulgarisms, in spite of the fact that the innocuous word list was administered. She reported that she heard these words all the time and that was why she needed her hearing tested. She later told the psychologist that someone was sending these words over the radio to her hearing aid. (He suspects that our clinic is the sending station of these words, since other of his patients report similar aberrations.)

A new spondee word list was created, sprinkled with common vulgarisms and expletives. Words were taken from plays, movies, books, and rest room walls.* A list of words accompanies the manuscript. (Inclusion or exclusion of the words from this article is editorial responsibility; however, if he declines, a word list will be sent in a plain brown wrapper if the reader would contact the author.) All words were presented and checked for equal syllabic stress and accurate attenuation and placed on magnetic recording tape to assure standardization. Two methods of response were offered: (1) "repeat the words you hear," (2) "write the words you hear."

One finding was clear. Hearing for spondaic vulgarisms is better than for standard or innocuous spondees. A number of no responses were found however. These instances were most commonly found in prim, geriatric females, men of the cloth, clinic secretarial staff, and a couple of female grad assistants who unwittingly volunteered for the final test. Psychogalvanic skin response (PGSR) would be one way of confirming if the "no response" was or was not a response.

This new clinical test is offered as an addition to already established tests of hearing. It is our opinion that more accurate results are obtained and proper selection, purchase and use of a hearing aid are assured. ■

Traffic Jams for Fun and Profit

*Robert Twintone tests the revolutionary Jamwagen 2000**

I have just returned from testing a car as with-it as Dusty Springfield, as slim as a virgin okapi, as comfortable as an old man's trousers, as appealing as graft, and yet as stationary as the Eddystone Light. The first car to be produced with immobility in mind, the Jamwagen 2000 manages to give an impression of relaxed power when standing firmly still on its revolutionary square wheels. Finished in drip-dry polystyrene, in a wide variety of tartans, and with siliconised paper bumpers, the Jamwagen isn't just a good-looker. Lift the bonnet, and prepare to gasp with the sort of surprise that would have confronted Menelaus had he looked on the Barbican development, or Ptolemy, faced with a sliced loaf! Beneath the racy mocklizard lining, in the space previously filled with a useless engine, is a glistening 4-tap, overhead-valve coffee-urn and food-mixer combined, complete with near-china mugs and a triple-jointed spoon-fork appliance. The wickerwork laundry baskets are linked by Fonk-Worthington lightweight con-rods to an alloy-head spin-dryer, close-coupled to independent front TV and an inflatable kiddies' paddling pool. There are compartments for books, records, dogs, a roulette wheel, an infra-red spit, and a complete 8-mm. home-movie outfit. A single touch on the tasselled dash-button, and the stainless breadbin opens automatically, firing slices into the toaster at a rate of 0–50 in 3 minutes. I found the twin egg-boilers sluggish in second gear, and it is irritating to find that the dishwasher tends to overheat in warm weather, but otherwise the car is a mechanical masterpiece.

Inside, the senses squeak and gibber in a mixture of awe and love, at the care lavished on the plastic ocelot upholstery, the deeppile, scurf-free yak-hair carpets, the succulent peach mirrors, and the glittering minichandeliers. No expense has been spared to make this the most expensive car of its kind. For once, I found myself unable to cavil at the size of the ashtrays, which have a dynaflow flush-system that instantly converts them into sparkling baby-baths, each with its individual rubber mini-duck on an edible chain. (The Speedwell conversion with high-pressure taps and twin-cam quick-lift soap-dishes is recommended for experienced bathers only). At the touch of a gilt handle, the 8 seats turn into 4 beds, to the tune of "Scotland the Brave", and the sink-unit swivels outwards to become a washable plastic flowergarden, I found the lavatory lacking in legroom, but the manufacturers assure me that they will be able to get the bugs out before the first model is carried triumphantly onto our roads. ■

*Reprinted by permission of PUNCH (Sept. 23, 1964, p. 454).

The History of Thought as a Closed Linear System**
More Empty Economic Boxes

YEHUDA SPEWER*

Adam Smith was the first Keynesian, and not Thomas Robert Malthus, despite all prostitutuins to the contrary. It was Smith in his *Stability of the Wealthy Nations* who showed, when emasculating the Paradox of the Feeble Bees, that public vice was virtuous to private parts, thus avoiding, for the the first time, the Fallacy of Competition[1]. Malthus merely introduced a simple logarithmic transformation to show that reproduction could not be ejaculated indefinitely into the future without running into the positive checks. Multiplying through by -1, of course, converts the positive checks to negative balances, and reversing the axes is all that is required to get the liquidity trap — slim evidence that Malthus preceeded Smith. Indeed, assuming a hypothetical but reasonable value for P when $t=0$, and extrapolating Malthus' first law back suggests that Smith preceeded Malthus by 50×10^{-2} economists[2]. That Malthus' own First Law disproves his main claim to fame was shown by Ricardo in his superfluous chapter "The Wages of Irony", and it is very surprising that this confusion should not have died with the Sages Fund Doctrine.

Of course, the strongest evidence that Smith initiated the Keynesian Revulsion lies in Book V, "Death Duties of a Sovereign." It will be remembered that in that book, Smith shows that gold transmits general gluttons of goods from one country to the other via "wagon ruts in the sky". Here we have the beginnings of the International Monetary Feud as well as Boeing's first law of General Dynamics. (Smith was well aware that his roots were complex. They remained so until Walras cut them off, but, of course static existence had to wait for twentieth century lemmings[3]). J.B. Say, of course, frustrated the application of Smith's idea, thus founding a French tradition.

There is an old Chinese Proverb which reads: "Wrong ideas are forever in doctoral designations, but Wong must try".[4] ■

---

* Eugene Smolensky mapped the printout to bond. He, the Computation Center, and its Chief Programmer, Binary Fudgit, share the responsibility for all errors. Responsibility for the rest is, of course, mine.

** EEH/Second Series, Vol. 6. No. 1 ©Graduate Program in Economic History, University of Wisconsin, 1968.

1 Geoffrey Stigler, *In the Land of Canaan* (Irving and Co., Homoken, New Jersey, 1903), 6371.

2 The technique was first used by S. Kuzns. See his *"Numbers Without End", Surcease of Sorrow and Cultural Veneration,* Part II, Vols. II, III, IV, V, VI, VII, iii-iv.

3 A. Lance and J.P. Satyr, *"Existentionalism of a Competitive Equilibrium", Limits With Random Error,* 11110101101, 1ff. Of course, closed convex subsets were not normal in a pin factory in Smith's time. The pointless pin had not yet been invented. Smith's avoidance of Euclidean space was a polar position to which all the classicists were cosiners.

4 I am indebted to my old draftsman for correcting my initial rendering.

On Responsibility

The trouble of the present time
slips and shifts upon my sloping shoulders,
and I think I need
some new venue:
perhaps to troll for hours
to catch pelagic fish,
not for its own significance,
but just for the halibut.

Anthony Thompson
Livermore, CA 94550

A PROPOSAL FOR A NEW IMPROVED STATISTICALLY EFFICIENT CALENDAR

David T. Mage
North Carolina

INTRODUCTION

The statistical analysis of air pollution as a time series is complicated unnecessarily by the interaction of the daily and annual cyclic processes of nature which are out of phase with the 7-day cycle of human activity. If the natural and human cycles could be brought into phase, then simplified analyses of time series may be possible which could lead to significant cost savings by a reduction in the number of statisticians needed to analyze these data, and a savings of the computer costs of making corrections for aperiodicities. Additional cost savings in energy can be made by judicious choices of calendar adjustments such as moving the leap year correction of an extra day from February when heating is required to the spring when neither heating nor cooling would be required.

APPLICATION

Emissions of air pollutants by human activities have fundamental cycles related to needs for energy. The energy needs for heating in the winter and cooling in the summer represent annual cycles. The daily needs for energy have a cyclic component based upon low demand at night when most people are asleep and high demand during the day when most people are awake. The 7-day work week cycle (5 days on, 2 days off) is in phase with the daily cycles that humans developed to match the diurnal cycles of sunrise and sunset, since 7 is an integer multiple of one. Unfortunately, the annual cycles are out of phase with the weekly cycles because neither 365 nor 366 are integer multiples of 7. Consequently, for air pollution analyses, March 23, 1981 which occurred on a Monday, and March 23, 1980 which occurred on a Sunday, are statistically different since a workday Monday has distinctly different traffic patterns and energy utilization patterns than a nonwork Sunday. This is the phase difference that leads to complications in air pollution analyses.

In any consecutive four year period there will be 3 nonleap years and 1 leap year resulting in 1461 days which is not an integer multiple of 7. The smallest common multiple of four and seven is 28, and (if we momentarily neglect the centennial adjustment of no leap year in 00 years, 1900, 2000...) therefore, March 23rd in year N will fall on the same day of the week as March 23rd in year N+28. If we could modify the calendar to make human activities in phase with the annual cycles then all March 23rds could fall on the same day of the week which would simplify predictions of pollution for that day.

The suggested new improved calendar shown in Table 1 has several advantages which should be evident.

TABLE 1. A New Improved Calendar Days of the Months

	JAN	FEB	MAR	APRIL	MAY	JUNE	JULY	AUG	SEP	OCT	NOV	DEC
Present Calendar	31	28$^{(0,1)}$	31	30	31	30	31	31	30	31	30	31
New Improved Calendar	28	28	28	35	35$^{(1,2)}$	28	28	28	35	35	28	28

$^{(0,1)}$In ordinary year add 0, in leap year add 1 except in years ending in 00.
$^{(1,2)}$In ordinary year add 1, in leap year add 2 except in years ending in 00.

Statistical Advantages of the New Improved Calendar

All months in the new calendar have either 4 or 5 7-day weeks which will make the day of the week the same from year to year, providing the new leap days which are added in May are national holidays, perhaps called Fundays, which are neither a normal work day nor a weekend day. If January 1 is a Sunday, then May 35th will always be a Saturday and June 1st will always be a Sunday, and the intervening period of one or two days will constitute a "lost weekend" which keeps the cycles in phase. Since each month will have the identical number of weekday and weekend days from year to year, the analysis for air pollution trends can be made without corrections for varying numbers of workdays. For example, March 1980 had 21 work days, March 1981 has 22 work days, and March 1982 will have 23 work days. If we would look at the average concentrations of carbon monoxide for March from 1980 through 1982, we might see an upward trend simply due to the artifact of an increased vehicle utilization. Since CO is primarily emitted from motor vehicles, work day CO levels are higher than weekend CO levels. The new calendar would eliminate this artifact since all March months would have 20 work days.

Cost Savings and Other Advantages

The new improved calendar is energy efficient since it reduces the numbers of days in the winter heating season and the summer air conditioning season. The shorter winter will also reduce the chance of frost damage to the Florida citrus crops, and the shorter summer will alleviate drought conditions by reducing the rate of moisture transpiration from plants. There will be other advantages, such as the extended fall season which will provide an opportunity for colleges to increase the football schedules and earn more income. Not to be overlooked is the elimination of the need for a new calendar each year which will save trees and annual replacement costs.

In summary, this new improved calendar should make the analysis of environmental data much simpler, and it can provide a cost saving to the public by producing a reduction in energy and resource demands. ■

Das Gesellschaftwitzbuch [*The Corporate Joke Book*]

M.B. ETTINGER

This report trifles with Das Geselleschaftwitzbuch. I shall also refer to this nonthing as the Corporate Joke Book or the CJB when I lapse into acronymese. Inherently, the CJB is such a fearfully endangered species it has never existed long enough to have a physiology or establish a history. The failure of this apparently pedestrian event to occur gives many trivial insights into Gesellschaftgestalt.

A corporation exists to make money for stockholders, and it can persist only as long as it concurrently serves a socially acceptable function. To accomplish these dissonant missions the corporation requires capital, employees management and usually consumes nonrenewable resources. Employees demand money and increasingly, a sporting chance to survive the hazards of doing a day's work. Managers require money and ego gratification except that it is called job satisfaction when sought by employees without exceptional skills.

Corporate management has to get its jollies out of pacifying the public with pleasingly priced goods and services, pleasing the stockholders with profitability, and appeasing and nurturing the employees with money and gratification. All these greedy parties insist that their piece of the action grows increasingly pleasing, a requirement which may be satisfied only while the corporate universe imitates the total universe and expands continuously.

The nonexistence of the CJB is best understood by examining the incompatibility of such a corporate member with the physiology of the corporation. In the real world multiple case histories demonstrate that a CJB will induce an immediate anaphylactic reaction of the host corporation if a CJB gains parasitic attachment thereto. For example, consider Gesellschaftgestalt problems associated with my Leonardo joke.

Leonardo trained chimpanzees for a city zoo famous for enhancing the interest of its collection with shows featuring young animals still learning their acts. Leonardo's chimps tap danced, did needlepoint, rode bicycles and unicycles, presented a great atonal rock band and were prevented from seeking public office only by age rules and diffeculties in establishing citizenship. These feats came easily in spite of the fact that Professor Leonardo was demonstrably the ultimate clod, qualified for instructional duties only in a teacher's college.

Clearly Leonardo attained monumental instructional success with a total absence of personal talent. Jealous animal trainers, corporate managers and guys in the Dean business anxious to build parallel success stories with untalented staff finally called on Leonardo's boss, a distinguished savant named Earl, and demanded the logistic secrets and training doctrine responsible for Leonardo's success. Earl very graciously shared his supervisory insight. He said "If you can explain it to Leonardo, he can explain it to the monkeys."

While the Leonardo witz afforded opportunity for animal trainers and educators to take offense, neither group exerted much clout in the protest arena. These guys could, therefore, be disregarded with impunity pending legislation which confers status on them as disadvantaged minorities. This story was generally well received and for years no one overtly objected to the bit.

Boss Leonardo suggested that if I planned to continue to tell the story I had better change Leonardo's name to Jake. To avoid further confrontation, I pragmatically changed Leonardo's name to Rumplestelskin until Boss Leonardo was out of my picture. With any luck at all, I'll never have a boss named Leonardo again.

If you're editing a Gesellschaftwitzbuch, you are systematically alert to the sensitive Leonardo who resent a corporate Leonardo joke. Some of these guys sign orders; others will be talented brats the Gesellschaft wants to hire; still others will feel that the Leonardo witz is a sad attempt to belittle a great Italian genius. For Corporation purposes the joke must concern a guy named Rumplestilskin, or a guy named Bob. (Roberto if you operate in Spain.)

The Corporate anecdote cannot put down anyone because of nationality, race, color, creed, or sex. Such material is both illegal and counterproductive, and probing for loopholes may perturb both customer and employee relations. For instance, employees of scheduled airlines protest cutesy slogans like "Fly Me", "We want to do more for you", Let us wiggle our tail for you", etc. Furthermore, many pflegmatic customers of the airlines feign distaste for a ticket to Cleveland which includes erotic options.

Even normal and legal sex is not a suitable topic for the Gesellschaftwitzbuch of Fortune's Five Hundred. Husband and wife jokes irk the endemic disgruntled spouse present in customer and employee populations at about the 90% level. Anatomical jokes are particularly bad and no concern wishes to jeopardize the collective good will of people with big noses, noisey dentures, bow legs or jaundiced chitin by being pert on the subject of megabeaks, decibel dentures, etc. Labor relations jokes are sure to launch a flood of grievance hearings which cost about 3×10^{-3} megabucks per hearing. Political quips are toxic to the Gesellschaft hopeful of getting a sympathetic hearing at the local, state or national level regardless of who won, stole, or purchased a plurality in the last election.

People with money really don't "laugh all the way to the bank". They're too worried about being robbed, kidnapped, or taxed. Nobody equates his poverty to comedy even though he may be prepared to snicker at the next rerun showing Mr. Charles Chaplin dining on unseasoned boiled brogans. Naturally, financially oriented journals, magazines and newspapers are not bullish on humor.

Fortune, Forbes, Barrons, the Wall Street Journal, etc., are most concerned with facts and the opinions of philosophers with a good track record for fiscal ESP. The Dow Joneses (Barrons,WSJ) love puns and the worse the pun, the more the Dow-Jones editors adore it. In addition, the Wall Sreet Journal has a marginal tolerance of wit provided it is quarantined on the editorial page and the source of the wit is extramural. Gesellschaftgestalt can tolerate an occasional bad pun and a little carefully screened extramural wit, but the Gesellschaft has established empirically that disseminating such material is ungood for the balance sheet.

One of the many good reasons Corporate officers avoid humor relates to gesellschaft toutsmanship. Whenever possible, every large outfit sends its ranking huckster to address "Financial Analysts" to portray a glowing picture of the future profits its dedicated people are about to produce. Financial Analysts prefer Corporations which take their own efforts very seriously. The FA's advise and inform money managers who have no sense of the absurd and are antagonistic to suggestions that such capacity would be helpful. FA's are dedicated to searching for outfits staffed by sober, responsible, money geniuses in depth. (A Gesellschaft doing well is deemed risky if it is only one heart beat away from having boobs take over management.)

Cutesy kids are another publicly tolerated topic for corporate humor. The "funny paper" has its collection of cutesy kids and there was once a TV comedian type who made a good living out of cutesy kids. This cutesy kid specialist is now retired or unemployed - I don't know what happened to him. Most adults get all the opportunity they want to marvel at the kids syndicated in the funny paper or thrust upon them by fatuous parents bursting to talk about little Jack and/or Jill. The babies may be cute, but people equipped with teenagers are usually a puddle of parental concern with little ability to find anything funny. Obviously, the juvenile joke may be acceptable, but it just isn't good Gesellschaftgestalt.

One might think that old jokes from a bastion of regressive stuffiness like the Reader's Digest would be safe. Let's analyze one of them. If memory serves me properly, the joke went like this. In the days when the triumphant American army was occupying Berlin in strength, a disturbed young wife wrote her military husband and demanded to know what the German girls had that American girls didn't have. Sergeant Husband wrote back: "Nothing, but they got it here". What does this joke do for a corporate sponsor?

1. A few German customers will be reminded of days they would like to forget.

2. The wives of business travelers will be stimulated to speculate on the behavior of husbands on the road.

3. A few corporate business travelers will feel encouraged by their management to seek ways to reduce the tedium and loneliness which is the lot of the chronic business traveler.

4. A collection of vigorous but bored business travelers will be persuaded that now is a great time to see if they can get an assignment requiring less travel so they can stay home where "it" is.

5. Etc. Etc. Etc. The joke isn't good Gesellschaftgestalt.

Sometimes a joke may be adapted for a corporate purpose. Years ago when I was a federal employee I received a handmade trophy noting my membership in a group called "The Mushroom Club". The membership of this group consisted of middle managers who were kept in the dark and fed well rotted horse manure. I told this joke to the local Rotarians. A few weeks later I encountered this same basic story, adapted this time for use by a top corporate manager. The story now had a different point. If you kept your middle managers in the dark and fed them with stale horse shit, you could expect them to proliferate all over the place - like mushrooms.

So, the point of it all is that humor can serve Gesellschaftgestalt but only in a very parochial way. However one must be incredibly imaginative to anticipate the neurotic hordes seeking a Gesellschaftgestalt source of grievance. ■

NOT-SAFE applauds the NHTSA decision to require an extra rear brake light on automobiles by 1985. However, a more comprehensive program is needed to reduce injuries even further.

In addition to a "third" tail light by 1985, there should be a "fifth" light required by 1989, a "twelfth" light by 1993 and by the year 1999 a full set of brightly colored blinking strobes would protect all potential rear-end victims. To assure compliance with this program, and all future safety regulations, monthly vehicle inspections will be required.

Other cautionary measures which should be implemented are; 1. Welded shut doors (many fatalities occur when the door pops open). 2. Remove all radio and stereo equipment which might distract driver and cause a mishap. 3. Oatmeal filled front bumpers capable of withstanding a 70 MPH head-on collision and providing an emergency food supply for survivors. 4. Radar guided lasers attached to rear bumper which would incinerate the tires of any auto following too closely. 5. A loudspeaker (aimed directly at tail-gaters) which amplifies screeching brakes and a simulated tire smoke for effect. 6. Maximum capacity for all gasoline tanks of ONE gallon (fire safety). This would *force* fuel conservation (people would be too busy searching for the next fuel stop to worry about anything else). 7. National speed limit of 15 MPH. Those who exceeded this limit would be "reminded" to slow down by a special tamper-proof electro-shock device implanted under the driver's seat. (Ten thousand volts can be a great safety "tool"!) 8. A computer-locked ignition switch which requires a coded sequence of letters to be entered CORRECTLY within 8 seconds. We suggest that the following universal combination be adopted by manufacturers...LIVELLAFOTOOREHT-SITNEMNREVOG. Keep drunk drivers *off* the road. 9. For non-intoxicated drivers who can't stay *ON* the road, we recommend that thick sponge pads be wrapped around all telephone poles and street light columns. 10. By the year 2000, all motor vehicles must be constructed out of styrofoam.

We hope that these suggestions for shock therapy, padded poles and mush bumpers will be taken seriously by NHTSA and other "Safety" agencies in their quest of complete protection for everyone, from everything at any cost. After all, these laws *are* for our OWN GOOD and if it's worth doing right...it's worth over doing.

Cautiously,

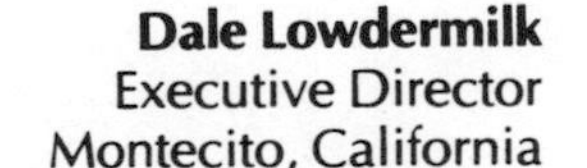

Dale Lowdermilk
Executive Director
Montecito, California
National Organization Taunting Safety
and Fairness Everywhere

How To Avoid A Financial Catastrophe

Dear Jim:

I have just finished preparing my 1983 tax return and as always am stunned by the realization of the amount of income tax that I must pay. So I have been thinking about tax planning for 1984.

I have been reviewing the real estate ventures and oil drilling funds as well as cattle propositions. They all sound great and certainly give you a good write-off. However, I don't really believe they satisfy long term needs. In fact, I am beginning to believe that a "sophisticated investor" is one who refuses to invest in a "tax shelter investment".

For that reason I have been looking around for a really sound business deal which will not only be a good investment, but also have some tax benefits. As you can imagine, there are very few opportunities around which give real profit potential and also may be classified as conservative, sound investments.

Fortunately, I have just found a real 'sleeper'. I thought I would mention it to you because (while conservative) it looks like it will make a lot of money with very little invested cash.

A friend of mine is considering investing in a large cat ranch near Karmessillo Mexico. It is our purpose to start rather small, with about one million cats. Each cat averages about twelve kittens a year; skins can be sold for about 20 cents for the white ones and up to 40 cents for the black. This will give us twelve million cat skins per year to sell at an average of around 32 cents, making our revenue about $3 million a year. This really averages out to $10,000 a day, excluding Sundays and holidays. A good Mexican cat man can skin about fifty cats per day at a wage of $3.50 per day. It will only take 663 men to operate the ranch so the net profit will be over $8,200 per day.

Now, the cats will be fed on rats exclusively. Rats multiply four times as fast as cats. We will start a rat ranch right adjacent to our cat farm. Here is where the first year tax break really comes in. Since we will be utilizing the rats to feed the cats, we can expense the entire first batch of rats purchased during 1983. If we start with one million rats at a nickel each, we will have four rats per cat per day and a whopping $50,000.00 1983 tax deduction.

The rats will be fed on the carcasses of the cats we skin during 1983/1984 and successive years. This will give each rat one quarter of a cat. You can see by this that the business is a clean operation, self-supporting and really automatic throughout. The cats will eat the rats and the rats will eat the cats, and we will get the skins and the tax benefits. Incidentally, the ecologists think it's great.

Eventually we hope to cross the cats with snakes. Snakes skin themselves twice a year. This will save the labor costs of skinning and will also give us a yield of two skins for one cat.

Let me know as soon as possible if you are interested. Naturally we want to keep this deal limited to the fewest investors possible. Time is of the essence.

Cordially,

Bruce C. Nydahl M.D.
Minneapolis, MN

Weekend Scientist:

let's defend against cruise missiles

D.I. Radin
Columbus, OH

INTRODUCTION

Since the U.S. Congress has recently appropriated several billion dollars for new defense technology, the future of the Cruise Missile is virtually assured.[1] In light of these events, *Weekend Scientist* has received thousands of requests from concerned readers, asking us to investigate the environmental problems posed by the cruise missile, and to develop a weekend project to resolve these potential dangers or inconveniences.

ENVIRONMENTAL IMPACT

We have focused our attention on two broad environmental categories: urban and rural. In urban settings, we envision home owners angrily shaking their fists at the sky as cruise missiles roar past their homes, scant inches from newly laid roofs—asphalt tiles wrenched loose, dogs crying, children barking, TV reception disrupted....[2] Not a pretty sight.

In rural environments, cows will refuse to produce milk. Whole flocks of chickens will be sucked into powerful vortices bellowing from the wake of the cruise missile. Farmers will curse as they try in vain to fell the silver harbingers of death with their sadly inadequate shotguns. Again, not a pretty sight.

The scenario is frightening. Citizens will walk in fear, furtively scanning the horizon for the aerial torpedoes, diving to the earth at sudden whooshing sounds. Will we become an underground civilization, grubbing out measly little lives in dank soil, paralyzed by skyphobia? Can we do anything to protect ourselves from these grinning deathmask missiles?

Weekend Scientist has researched this issue very carefully over the last several months, and we believe there *is* something that the ordinary, concerned citizen can do. First, let us investigate the missile itself.

THE MISSILE ITSELF

Just what is the cruise missile, and why is it so frightening and lethal? The answers to these questions will be the key to its defense.

The cruise missile is essentially a self-guiding horizontal rocket. It is dropped from a regular aircraft such as a bomber, then flies to its target at up to 500 miles per hour only 50 feet from the ground.[3] The missile carries a nuclear warhead, capable of destroying an entire city in somewhat less than a millisecond.

But what makes the cruise missile particularly horrifying is that it is an ' intellegent ' bomb. It can 'see' the terrain, and ' knows' where its target is located. It will avoid obstacles and perform incredible aerial acrobatics to reach its final destination and detonate itself.

It is this combination of intelligence and low-flying capabilities that is the near-invincibility of this weapon. Since the missile cruises at about 50 feet, normal early warning radar cannot see the missile, and thus, no (normal) defense is possible. *Weekend Scientist* has determined, however, that it is this very combination of intelligence and low altitude flying that is its undoing.

HOW SMART IS SMART?

Just imagine for a moment that you are a cruise missile. You have enough fuel to fly about 1500 miles, and you like to cruise at a crisp 500 miles per hour. Your flanks skim the tips of trees as you roar towards your target, and you itch to detonate and blow yourself into smitherines.[4] You enjoy sleek, metallic lines, a sharp paint job, and an incredible, almost reverential respect from others. You idle faster than a Ferrari and can maneuver better than a cat in a burlap bag. You are a missile—a *cruise* missile—and you're darn proud of it.

Now, about this role-play. Do you feel a rush of excitement as you fantasize about all that power and awesomeness? Of course you do. But do you think about how much fun it will be to blow yourself into tiny bits? No! This is the key, the weak link in the missile's gold-plated chain.

How intelligent can the cruise missile be if it blows itself up at the end of an arduous journey? What prudent piece of hardware would obediantly obliterate itself? Not a very smart one, to be sure. This leads us to believe that the so-called "smart bomb" is, in fact, rather dumb.

So we have a marginally smart bomb with a limited amount of fuel, going at breakneck speed, practically scraping the ground. This adds up to a nifty little weekend project to protect your home from the nasty cruise missile.[5]

THE END OF THE WORLD, by Luca Signorelli, From the Chapel of San Brizio in the Cathedral Orvieto

THE PROJECT

You will need a quantity of aluminum poles, wood 2 by 4's, and steel sheeting. A selection of brown and white acrylic paints will come in handy, as will several old springs from junked cars and a police-style radar speed detector. This project should take about 10 hours to complete. Our version cost about $350 using mostly spare parts.

1. Place the radar speed detector on the top of your TV antenna. Run a wire from the radar's audible alarm jack down the antenna and into your "Cruise Missile Defense Center" in the garage. The kids will love it.
2. Construct a 55 foot structure out of the aluminum poles and wood. A miniature Eiffel tower design might be a good idea because it will disguise the true purpose of the project.
3. Reinforce the top of the tower with the steel sheeting. Use your artistic flair and make something that will blend in with the neighborhood.
4. Determine the most likely direction that a cruise missile would come towards your home, then lay the tower along the axis such that the foot of the tower points towards the direction of most-likely[6] cruising.
5. Afix the springs at the base of the tower. Make a spring-loaded trap arrangement like the ones on mouse traps. You may wish to purchase a mouse trap to study the design.
6. Connect a relay to the radar so if the alarm sounds, the tower spring-trap will be tripped and the tower will bolt upright.
7. Paint the steel sheeting at the top of the tower so it looks like bricks. Use the brown and white paint for this.
8. That's all there is to it! Now sit back and enjoy year-round defense protection from those annoying cruise missiles!

THEORY OF OPERATION

When the radar detects a cruise missile, the spring relay is activated, causing the tower to shoot upright. The missile is then suddenly faced with an apparent brick wall.[7] Since the bomb is not too smart, as we have seen, it will become very confused because no one told it to watch out for brick walls 50 feet up in the air. The result of this confusion will cause the missile to violently veer away from the apparent brick wall, expending an enormous amount of precious fuel in the process. It will not be able to complete its mission and will therefore dive into the ocean in shame, saving you from future annoyances and whoever was at the missile's receiving end from destruction. All in all, a humanitarian gesture that is sure to be appreciated by friend and foe alike.[8]

When the missile has gone, you just crank down the tower, reset the springs, and luxuriate in your back yard, confident in the security of a safe, defense-bristling home. Your neighbors are sure to appreciate your thoughtfulness, and will very likely give you a medal.

ERRATA

In last month's column, "The Black Hole: An Inexpensive Vacuum Cleaner," we inadvertantly suggested that you place the vacuum hose inside the Schwarzschild radius. We should have said *outside* the Schwarzschild radius. We regret any inconvenience this error may have caused. ■

---

[1]We assume that "cruise" in this context means "to travel for the sake of traveling," or possibly "to go about the streets at random."

[2]Warren, M.W. Cruise Missiles: Panacea or Pain? Implications for Television. *Journal of Theoretical Broadcasts,* 173,22, 1982.

[3]We realize that this description is equally appropriate for most rush hour commuters on our nation's highway system.

[4]*Smitherines* is a technical term used in defense terminology. It means, strictly speaking, "itsy bitsy pieces, that if reassembled, would nearly resemble the original, somewhat."

[5]On a larger scale, an entire country could be protected. We hardly think it unpatriotic to point this out since if we thought of it, surely others have too. We claim no special knowledge or expertise in defense technology, other than our perfection of "The Bigfoot Protection Device," described in *Pentagon Today,* 2, 27, 1979.

[6]We have purposely omitted technical jargon to avoid confusion.

[7]Thus the reason for painting the top of the tower to resemble bricks. The better the paint job, the more effective the defense, obviously.

[8]We have not solved the problem of those homes not situated near the sea. It is possible that the cruise missile will be so embarrassed that it will detonate immediately upon encountering the pseudo-brick wall, thus annihalating you and your neighborhood. In that case, you will not have to worry about future annoyances.

Ralph Nadar is Radioactive

In a pamphlet received at the Institution some weeks ago headed 'They haven't been idle. Have you?', Peter Beckman (author of 'The Health Hazards of NOT Going Nuclear') goes crusading against the anti-nuclear lobby in the States.

Ralph Nader is singled out for particular attention as Professor Beckmann pursues his quarry:

'In attacking the safest form of electrical power generation, the anti-nuclear crusaders love to make 10-second assertions that it takes half-hour lectures to counter. As often as not, these assertions are false. But the following assertions are not false:

Ralph Nader is Radioactive!

The annual dose he gets from his own blood is more than 100 times higher than the average US resident gets from all the nuclear power plants in the country.

The probability of contracting cancer from exposure to Ralph Nader's radioactive body is not zero.

At least one of the radioactive ingredients in Ralph Nader's blood has a halflife of more than one *billion* years and will pose a hazard to countless future generations.'

Those stories that Nader radiates charm may be only partially true!

Inst. of Nuclear Engineers May/June 1978

The
glare of
irreproducible
research
enlightens

PSYCHOLOGY

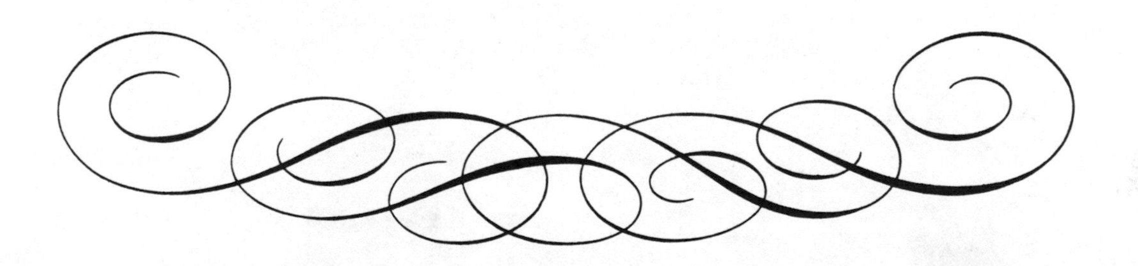

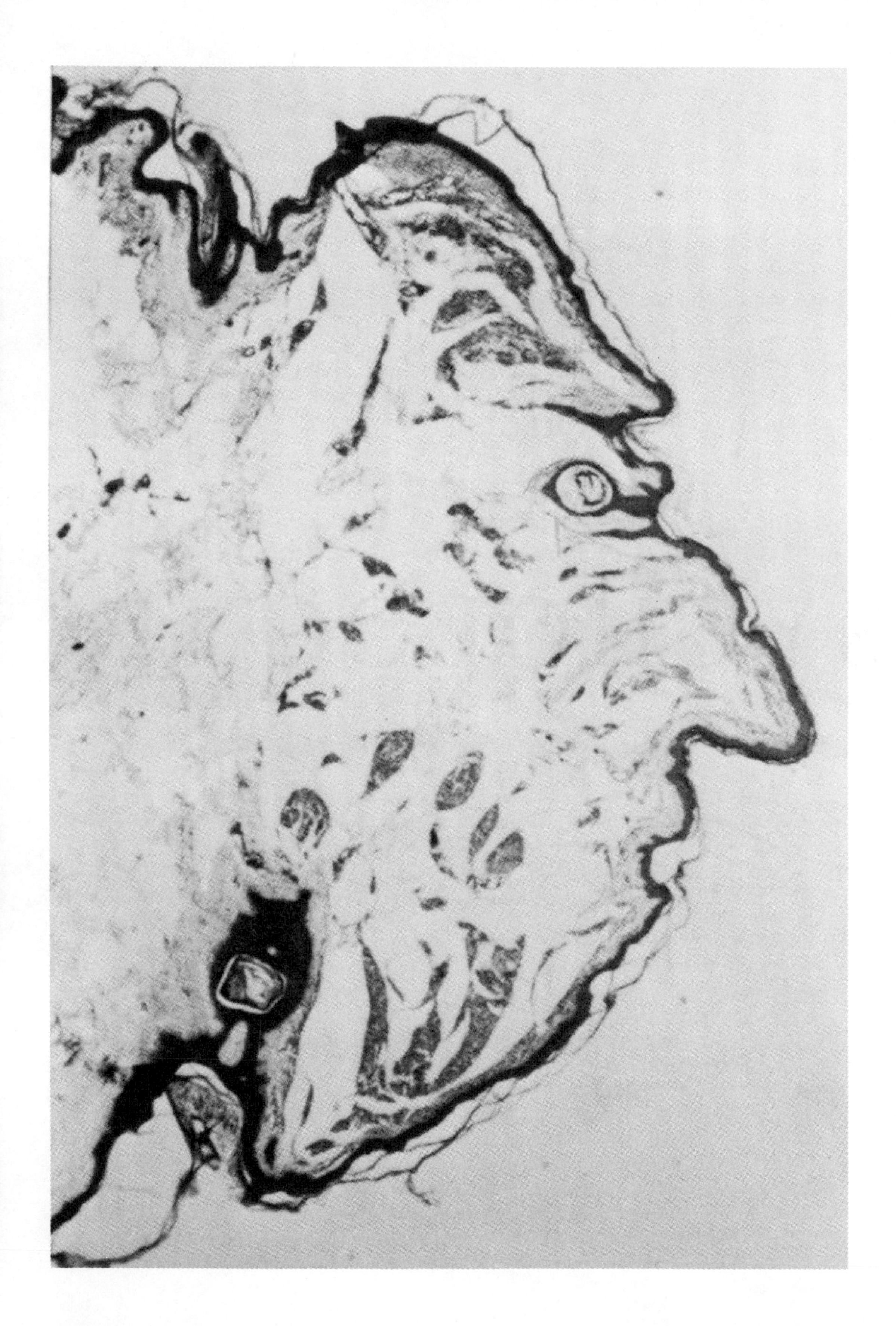

Submitted By: Walter H.C. Burgdorf, M.D. Dermatology, Oklahoma City, OK 73104

Legend: Shave excision of intradermal nerve, fixed in 3.7% formaldehyde, stained and photographed at 30x.

Nasality: A Psychological Concept of Great Clinical Significance, Previously Undescribed

Stephen D. Bourgeois, M.D.
7021 Treehaven Road
Fort Worth, Texas 76116

Generations of physicians have cut their eye teeth on Freudian psychiatry. There is not a physician currently licensed to practice medicine in this country, who, to pass his examinations and obtain his M.D. degree, did not have to master the intricate and bewildering concepts of the Oedipal complex, castration anxiety, and anal fixation, the oral personality, and whatnot. All very well so far as it goes. However, as this paper will expose for the first time, there exists an incredible oversight, a huge gap, as it were, in Freudian theory. Furthermore, it is the intention of this communication to plug that gap by defining an area of normal psychological development as well as pathological syndromes associated with disorders in this area.

Freud stressed the importance of very early influences on an individual. It would seem logical, therefore, in discussing any area of psychological development, to begin at the beginning, to look, in short, at what is happening in the immediate post-partal period. Let us take some of the well defined and well understood syndromes associated with certain body parts and consider what role these anatomic parts play and their relative importance in the life of the newborn.

Many experts have written about the oral personality. However, when a newborn babe comes into this world, what is the first thing that he does? Does he settle down to a hearty meal of Similac, Enfamil, or mother's milk? No, in modern hospitals he will get no nourishment for up to 8 hours. Much has been written about anal fixation, but coming back to our infant child, moments after expulsion from the warmth and security of his mother's womb into the harsh reality of the world of inevitable death and taxation, what is the first act that he must accomplish? Move his bowels? Does he, disappointed at the insipidness of his bowel movement, dejected, perhaps, at the prospect of 8 to 72 hours of fasting, decide to console himself by moseying down to the nearest bordello or massage parlour for a bit of sexual gratification? Hardly.

The first thing this newborn baby must do, if he is to live, is to breathe. Furthermore, he must continue to do it on a fairly regular basis throughout his entire life. Much is made of sex, eating, and bowel movements, and those parts of the anatomy associated with these functions, but what has been written about the psychology of that primary and most important of all functions, breathing? Almost nothing. However, let respiration cease for but a few minutes and life itself ceases. That Freaud could have failed to appreciate the psychological significance of respiration and of the external organ thereof, namely the nose, boggles the imagination. The ideas and concepts described in this paper, like so many intuitive concepts of genius, seem extremely obvious once stated.

When a baby is born, the obstetrician's first act, before even clamping the umbilical cord, is to clear the infant's airway with an aspirating syringe. Subsequently, pediatricians and nurses monitor the child carefully, and an aspirating syringe is always kept handy to suck mucus from the baby's nose as needed. All this fussing about his nose cannot fail to impress upon the subconscious mind of the child the importance of that part of his anatomy.

Competition and rivalry are present even among very young children. In countries where poverty and starvation are common, they may fight over food. Less obvious, but no less important, is the subconscious competition a child feels for the very air he breathes. Conditioned from birth to an awareness of the importance of the nose and respiration, he is bound to notice the noses of his peers and to compare them with his own. If his nose is small and insignificant, he will feel inferior and threatened. He will feel that his survival depends on avoiding conflict and not provoking his large nosed contemporaries, as he is no match for them in the competition for air. On the other hand, if his nose is large, he will feel that he can inhale far greater quantities of air than his fellows, and he is likely to develop into a confident, aggressive, often even arrogant person. Feeling himself to be genetically superior to his peers, he well have a natural tendency to "look down his nose" at them. Clinical studies have led me to recognize what I call the "nasal personality." It is characterized by the following syndrome:

1. A large nose
2. Aggressiveness
3. Arrogance
4. A tendency to stick one's nose into other people's business

The "non-nasal personality" conversely is characterized by:

1. A small nose
2. Placidity
3. Humility
4. A tendency to mind one's own business

History is replete with examples of the nasal personality. Indeed, almost all of the great conquerors were classic examples thereof. Alexander the Great possessed such a nose in addition to an aggressive personality. His conquests spread Hellenism from Gibraltar to the Punjab and paved the way for the later development of the Roman Empire. Julius Caesar's nose was remarkably similar to Alexander's in both size and shape.

In the early part of the thirteenth century, Genghis Khan and his Mongol hordes conquered half of the known world. There is a misconception in the minds of some that Mongols were of the same race as the Chinese, a people with small noses. However, the Mongols were descended from a distinct race, the Tungusi, and had a large admixture of Turkish and Iranian genes. Available pictures of Genghis Khan indicate that he had a long and high bridged nose, and there seems little doubt as to his nasality.

There is certainly no doubt about the nasal dimensions, arrogance, or aggressive personality of Napoleon Bonaparte. Napoleon stormed through Europe and had his way on that continent until he was finally stopped at Waterloo, by an alliance of other large nosed races, prominent among which were the British.

When it comes to nasality, the British need take second place to none. Under a succession of large nosed monarchs, the British established empire that stretched around the globe. The Indians who inhabited what is now the United States of America had hawk-like noses of the sort so well depicted on a nickel. These semi-barbaric tribes offered fierce and determined resistance to the European invaders. The Indians of Mexico have small, flat noses. They had highly advanced civilizations, yet a handful of nasal Spaniards led by Hernán Cortés conquered the entire nation within a very short period of time.

Despite innumerable historical examples of the nasal personality, the importance of the nose has never been fully appreciated. Much nonsense has been written, on the other hand, about "firm chins," and "determined jaws." The chin and jaw have nothing to do with it. Look at General Charles deGaulle. He had, in fact, a receding chin, and a rather weak jaw. But what a nose! I had the privilege of seeing this great man, then President of France, when he visited New Orleans in 1962. As he passed in review along Canal Street, the crowd was struck with fear and awe at the sight of his incredible nose. One sensed that if he put the thing in high gear, it would suck up all the surrounding air like some great nuclear powered vacuum cleaner and leave bystanders gasping for breath in a void. Small wonder that this great nasal personality was able to unify his dissident nation and rule it with a firm hand almost until the time of his death.

The Japanese and the Africans, both races with small, low bridged noses, are examples of non-nasal personalities.

This brings us into the area of nasal envy and rhinoplasty[1] anxiety. When a race of non-nasal people, especially one that has been isolated for centuries, is suddenly exposed to large numbers of nasal people, the result is intense psychological trauma, and the development, sometimes on a national scale, of nasal envy and rhinoplasty anxiety.

Penis envy is to nasal envy as a wart is to metastatic carcinoma.

I hope that the foregoing discussion has convinced the reader of the importance of nasality. The introduction of this concept into the literature will undoubtedly force a shift in emphasis of psychotherapy. I wish that I could claim to be the first to recognize the nasal syndrome. Unvortunately, as is so often the case in medicine, when one thinks that one has an original idea, a careful search of the literature shows that the idea has already been mentioned, if not developed and defined. Edmond Rostand, in his play, *Cyrano de Bergerac* (1897), showed that he grasped the importance of the nose when he had Cyrano say:

> "Or ---- paradying Faustus in the play ----
> Was this the nose that launched a thousand ships and burned the topless towers of Ilium?"

These are telling lines indeed. I will readily grant to orthodox Freudians, that someone, a General Patton perhaps, under appropriate circumstances, say the occasion of a colossal military blunder, might ask, "Is this the ass who launched a thousand ships?" However, I would be quick to point out that in this circumstance the speaker would be referring to the intelligence or personality of the launcher of those myriad ships, rather than to parts of his anatomy.

[1]A rhinoplasty is an operation to reshape the nose, generally making it smaller. Interestingly, in modern times this operation is usually performed on large nosed patients who suffer guilt feelings over the nasal aggressiveness of their forebears. These patients subconsciously wish to punish and humiliate themselves by amputating, or at least diminishing in size, that organ which is the symbol of the nasal brutality of their ancestors. ■

Blazer, J.A.

LEG POSITION AND PSYCHOLOGICAL CHARACTRISTICS IN WOMEN

Psychology—A Journal of Human Behavior, 1966, 3, 5-12 (August 66).

"The present study was designed to test a theory of "leg position analysis" or "observed psychology" using standard psychometric tests and methods to gather and analyze the data." "The first hypothesis tested was: The preferred method of leg-crossing or position generally used by a woman as indicative of her needs, strengths and basic values or interests. The second hypothesis was: intelligence and education have no effect upon preference of leg crossing or position in women."

(Highly recommended to our readers!)

THE PRINCIPLE OF FALLIBILITY:

THE PRIMARY LAW OF PSYCHOLOGY

William T. O'Donohue
Bloomington, IN

BLUE MARINE, by Lyonel Feininger, Manson Williams - Proctor Institute, Utica

Psychologists are often a depressed and anxious lot. This unfortunate state of affairs is in part due to their failure, despite years of intense effort, to discover the causes of psychological phenomena, for example, depression and anxiety.

I think somewhat less than 50% of psychologists would agree (*'most* psychologists agreeing' being a contradiction in terms) that the discovery of an exceptionless, inviolable law within psychology would be a rewarding, cathartic and/or a genuine growth experience for their fledgling science. The remaining psychologists will, of course, express every other possible opinion about this issue, except those psychologists practicing in New York City, who being immune to innovative developments within the field, will remain characteristically silent. Regardless of the inevitable controversy, this paper will reveal psychology's first inviolable law.

The primary, exceptionless psychological law is to be known as the principle of fallibility. This law is quite simple (read elegant, parsimonious). It states: *No human is capable of constructing a psychological law which is inviolable.* Or more operationally: Show me a psychological law, and I'll show you error variance. It is important to note the following two points: (1) this is imminently a psychological law: it is about human behavior and human capabilities, and (2) this is a scientific law which is based upon decades of extensive data collection within psychology. Every psychologist of every theoretical persuasion from Wundt to Freud to Skinner has consistently provided research evidence which strongly corroborates this law and, to date, has failed to produce a single anomoly.[1]

Psychologists, always being keenly aware of the philosophical implications of their work, will surely note that the principle of fallibility is a scientific, meta-scientific law (or alternatively, a meta-scientific, scientific law). That is, it is an empirical, behavioral at one level of remove from its *explanandum* i.e. all empirical, behavioral regularities. Thus, it possesses a special virtue of self-referentiality, as all *psych*ological laws should. This self-referentiality reveals that if it is the case that the law of incompleteness is inviolable, it is then an exception to itself, then it is the case that it has at least one exception, and thus it is consistent with itself. If, on the other hand, there are in fact actual exceptions to the principle of fallibility, then this is an outcome which the law predicts, and hence it is an outcome which is consistent with the law.

The principle of fallibility is a critical discovery which helps place psychology on equal footing with more developed disciplines like atomic physics and mathematics. Psychology has long felt inferior to these more advanced and sophisticated disciplines which possess impressive principles like uncertainty, and incompleteness. But psychology need not suffer from inferiority any longer! Atomic physics may be fundamentally uncertain, and mathematics may be fundamentally incomplete, but laws within psychology will always be fundamentally in error.

Therefore, psychologists after long years of fighting a seemingly intractable reality, finally have gained a victory. There is at least one foundational fact about human behavior: no human is capable of constructing a law about human behavior which is exceptionless. ■

[1] It is important to note that these research efforts were not intentionally directed towards the development of the principle of fallibility, but any study of the psychological literature will reveal that it unequivocably supports this principle. Thus, the discovery of the principle of fallibility is actually an example of serendipity. Serendipity is a common phenomenon in the history of the sciences and is said to have occurred when one is looking for a needle in a haystack but one finds the farmer's daughter.

NEIL ILLUSIONS

ALLAN NEIL works at the Institute of Behavioral Research, Texas Christian University, Fort Worth, Texas

Although psychology is one of the oldest of the true sciences, the study of illusions started long before a science psychology existed. A great deal of the early research in the area of illusions, however, was the work of men who are not usually thought of as psychologists, and thus has been irretrievably lost. It is impossible to estimate just how many illusions were lost during this period of scientific darkness, but the number may be substantial.

Such a deplorable condition could not continue forever, and indeed it did not. The advent of formal psychology and the timely arrival of the introductory textbook secured, for all time, the illusion's place in psychology. The golden years really began in 1832 when a transparent rhomboid owned by L. A. Necker first began to show spontaneous depth reversal. Named the "Necker Cube" in honour of its discoverer the transparent rhomboid was quickly committed to literature and has continued cyclic depth reversal, without interruption, ever since.

While the scientific community pondered this singular circumstances, some of the really great men of psychology were contributing their names to illusions: Muller-Lyer, Poggendorff, Zollner, Ponzo, Optical and Moon. Illusions piled up. Theories flourished. Debate waxed luxuriantly. Research spread unchecked. Then at the turn of the century, a noticeable slackening in activity developed. A drastic decline followed. Finally, total collapse! As the years passed, a few hardy souls continued to look for an adequate perceptual account for the various figures, but there were no new illusions. Without new illusions intrest flagged. Full page, four colour illustrations helped for a while, but finally, only freshmen could be tricked into saying,"Gee it really does!"

Recently a few new illusions have been discovered. these new illusions, in sharp contrast to those of the 19th century, do not violate the invariance of parity, charge conjugtion, or Time reversal. Full scale research has not yet begun on the information-processing mechanisms which respond to the subtle factors in these illusions, but the preliminary studies have not overestimated their impotance. The figures shown below, although only the first of the new series, are presented here with the hope that the interest they generate will revive that research and restore the study of illusion to its rightful place in the feild of psychology. ■

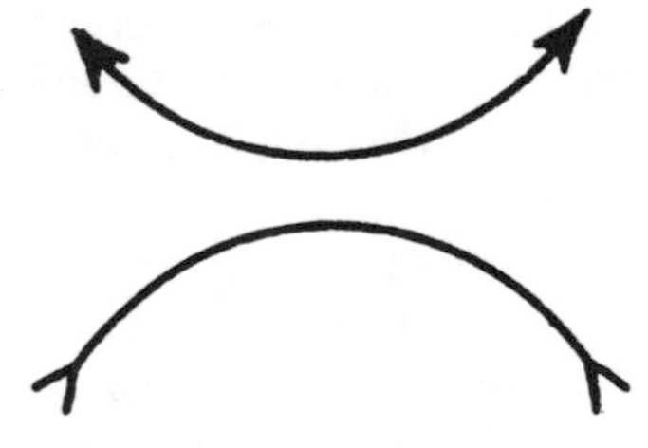

1
Note that the lines do not appear parallel

2
Note how one line appears longer

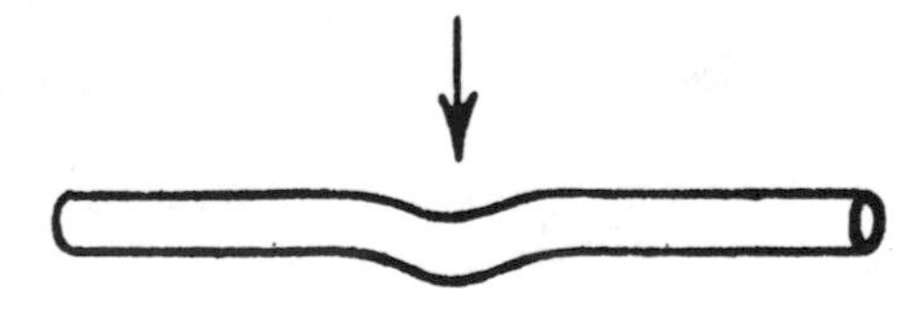

3
Note how the pipe appears bent under the arrow

4
Note that the boxes appear to be different sizes

5
Note how the line appears thicker
where it passes through the column

6
Note how quicly the figure disappears
when you look directly at it

Reprinted with permission from *New Scientist & Science Journal*, April, 1971

Contributed by: Dr. Henry S. Gertzman, National Cash Register Comany, Dayton, Ohio 45409

THERAPEUTIC EFFECTS OF FORCEFUL GOOSING ON MAJOR AFFECTIVE ILLNESS

Stuart A. Copans*
Assistant Professor of Pataphysical Psychiatry
Dartmouth Medical School
Hanover, NH 03773

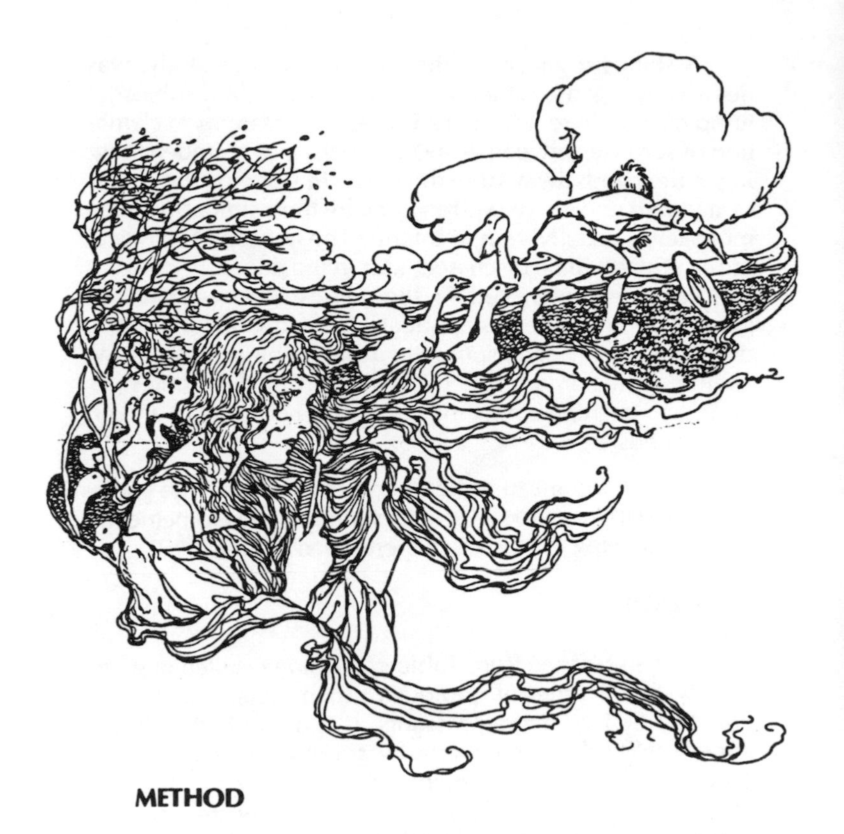

A review of therapeutic approaches to depression revealed that shock therapies of various types have been used in the treatment of depression for hundreds of years, ranging from sudden immersion in a bath of cold water (the bath suprise) to metrazol, insulin or electrically induced seizures. These methods, while therapeutically successful, have been heavily criticized, and for the most part, abandoned because of the morbidity and mortality associated with their use. Serendipitious observations suggested to the authors of this paper that forceful goosing might have a therapeutic effect on patients suffering from a retarded depression without such morbidity or mortality.

A controlled study was carried out, comparing the effects of unilateral electronconvulsive therapy (ECT), bilateral ECT, tricycle antidepressants and forceful goosing on depression (as measured with the Zung Scale 4, 8 and 16 weeks, after initiation of therapeutic trials) on short-term memory loss, on subjective discomfort as assessed by patient interview, and on length of hospital stay.

INTRODUCTION

The orgins of the practice of goosing remain shrouded in antiquity. H. Allen Smith[1] has suggested in his monumental study of the subject, that the term goosing originated from the similarity between the physical movements involved in a goose and the tendency of geese to peck at the nether regions of those invading their territory. In any case, whatever the origin of the custom's name, it has generally been considered either a petty annoyance or a sign of affection.

In late 1978, observations on an adult treatment unit occurred when a manic patient was admitted and refused medication. The patient, hereafter referred to as Mr. G., spent much of his time creeping up behind female members of the staff and female patients and forcefully goosing them. While this engendered a great deal of hostility in female staff members (and some jealousy in male staff members), it was noted that two patients suffering from retarded depression seemed to improve markedly during the two weeks this patient remained on the Unit. Following the discharge of Mr. G., both patients exhibited a significant clinical relapse and eventually were treated with eleven sessions of bilateral ECT. It was as a result of this observation that the following experiment was designed.

*Please address all correspondence to: Stuart A. Copans, M.D., Director, Child & Adolescent Out-Patient Services, Brattleboro Retreat, 75 Linden Street, Brattleboro, VT 05301

METHOD

All patients admitted to the Adult Psychiatric Unit at the Riverside Hospital for the Sad and Lonely with a DSM III diagnosis of Major Affective Illness, Depressive Type, were assigned randomly to one of 4 groups. The first group was treated with tricyclic antidepressants; the second group was treated with 10 sessions of unilateral ECT, given every other day. The third group was treated with 10 sessions of bilateral ECT also administered on an every other day schedule. The final group of patients was assigned to the forceful goosing condition. These patients were forcefully goosed twice an hour while awake. The forceful gooses were administered by fully licensed R.N.s, who had been trained to administer standardized forceful gooses (geese).

Depression was assessed using the Zung Depression Scale, and was assessed prior to the onset of treatment, and again at 4, 8, and 16 weeks after treatment had been completed. In addition, all but one patient were successfully contacted a year after treatment and data is available at that time as well. Additional testing was carried out to assess short-term memory loss for both verbal material and visual material, and to assess subjective discomfort as expressed by the patients during the time that the treatment procedures were being carried out.

A great deal of debate went on before the study began, as to whether we should use homosexual goosing or heterosexual goosing. In fact, that debate was resolved not on clinical or research grounds, but on medical/legal grounds. It was the opinion of our lawyers that since forceful goosing was to be employed as a medical intervention, it should be administered only by licensed registered nurses. Since the registered nurses on our treatment unit were all women, the forceful gooses in the study were all administered by women to patients of both sexes.

A standard goose for the purposes of this study, was defined as a goose which elicited, in a standard subject, a jump of anywhere from 6-12 inches, a scream or exclamation of surprise between 40-60 decibels, and a sudden flinging of the hands away from the body. In studying the effects of standard gooses, we were struck by their similarity to the moro reflex usually seen in infants. The neurophysiological implications will be discussed later in the paper.

For those interested in biometrics, the standardized goose may be defined as application of a digital force in the direction of the anal sphincter from behind and below, attaining a maximum speed between 20 and 30 miles per hour, and exerting a maximum force of between 30 and 40 foot-pounds.

It is important to remember, however, that the gooses described in this paper were not standardized biometrically, but rather using the bioassay mentioned earlier.

RESULTS

As can be seen from Table 1, goosing resulted in significantly better relief of depression than either unilateral or bilateral ECT or antidepressants. It is of particular importance that this therapeutic effect was not seen just at 4, 8, and 16 weeks, but persisted and could be shown in the data from the follow-up one year after completion of treatment. In addition, there was significantly less memory loss.

In fact, one finding of this study was a significant improvement in short-term memory following goosing. We are currently exploring the use of goosing in a group of senile patients with the hopes that it may result in significant improvement in their mental functioning.

TABLE 1.

		Unilateral ECT	Bilateral ECT	Tricyclic Anti-depresants	Forceful Goosing
ZUNG DEPRESSION SCORE	Before Treatment	97	95	96	97
	4 wks	82	88	80	72
	8 wks	74	79	66	52
	16 wks	64	62	58	44
	1 year	68	72	64	32
SHORT TERM MEMORY LOSS 0 = None 7 = Most	Verbal	1	5	2	0*
	Visual	5	5	3	0*
SUBJECTIVE DISCOMFORT 0 = None 0 = Most		4	5	3	2**
LENGTH OF HOSPITAL STAY		117 days	108 days	112 days	96 days

*In fact, there was a statistically significant improvement in short-term memory in this group.

**4 of the patients in this group reported subjective pleasure associated with the treatment procedure.

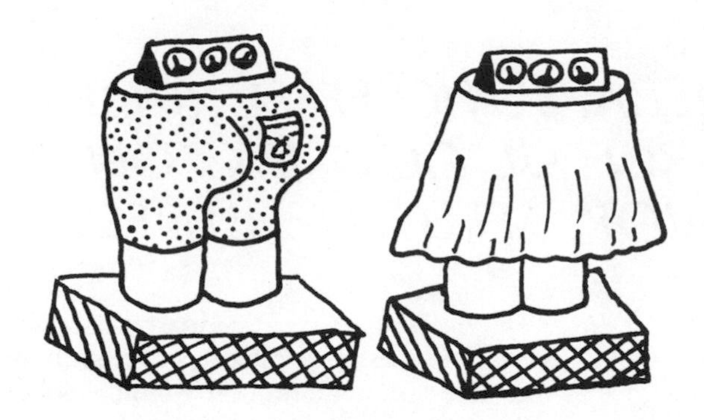

While the length of hospital stay is shorter in the forcefully goosed group, the difference there does not reach statistical significance. In fact, if we look at the distribution of length of stay data for the four groups, we see that both the median and the mode in the forcefully goosed group are quite a bit less than in the other three groups. The reason for the lack of significance when we looked at mean length of stay can be found in 2 patients in the forcefully goosed group whose hospital stays were profoundly longer. In these two cases, the patients developed an intense transference reaction to the nurses who administered the treatment procedure and could not be weaned from their goosing. We are currently exploring with the FDA the medical/legal ramifications of out-patient maintenance goosing, but presently both of these patients remain in the hospital and will probably need to remain in the hospital until our outpatient maintenance goosing project is begun.

DISCUSSION

It seems clear from the above results that forceful goosing represents a therapeutic procedure that should be added to the armamentarium of all practicing psychiatrists. As we begin to explore its use in clinical practice, however, it is clear that there are a number of important medical/legal issues that must be addressed. First, should forceful goosing remain by and large an inpatient procedure, used only in those who are severely depressed and admitted to a hospital unit where the gooses can be administered at various times throughout the day, catching the patient by surprise? Second, given the addictive potential of forceful goosing, it is clear that some outpatient maintenance program must be developed for some patients. Should this outpatient maintenance program be hospital-based? Could it be carried out in the offices of individual psychiatrists? Could the patient's families be instructed in carrying out the procedure? Currently, forceful goosing is considered, at least in Vermont, a medical treatment procedure; that is, it can be administered only by physicians or by registered nurses under the direction of physicians. Social workers, psychologists, educational counselors, and other parapsychiatric personnel are not allowed to administer forceful gooses to patients.

A dilemma, however, is posed by the family of one patient treated with forceful goosing on our unit. The patient's husband had a particular fondness for goosing his wife, and in fact, her depression seemed to be precipitated by an extended business trip the patient's husband took, so that the patient was not goosed for nearly a month prior to her admission. It seems clear that the patient's husband sees his goosing as an affectionate gesture and that it also serves an important function both in the dynamics of the family and helping maintain a balance of catacholamines in this patient. If, however, forceful goosing is to be considered a medical procedure, can we encourage and support this patient's husband in his practice of medicine without a license.

In our ongoing studies of the phenomenon of forceful goosing, we're exploring a number of possible physiologic mediators of its therapeutic effect. One possibility is that forceful goosing is usually followed on the part of the patient by an inspiratory gasp and a relatively prolonged exhalation phase.

Measurements of blood gasses showed that this results in a brief pulse of highly oxygenated blood followed by increasing buildup of carbon dioxide. Single cell monitoring of midbrain respiratory centers in goosed cats suggest that this pattern of change in blood gasses may result in massive discharges from these respiratory centers to a variety of limbic structures. There is some suggestion that thismay act to reset baseline levels in the reticular activating system; however, until more detailed results are obtained over the next year, this must remain a speculation.

The second area of investigation in the therapeutic mechanisms of the forceful goose, involves the examination of direct connections between the perianal plexus and the hypothalamus. Again, while definitive pronouncements must await further studies, it does appear that forceful goosing has a profound effect on the functioning of the neurohypophyseal axis, with significant alterations in the release of both catacholamines and brain polypeptides.

CONCLUSIONS

It seems imminently clear from our data that forceful goosing represents an important therapeutic procedure. It may well help us leap onward and upward to a further and more detailed exploration of the mechanisms underlying affective illness.

BIBLIOGRAPHY

1. Smith H, Allen Handbucher Goose: Die Tiere, die Steppe, und die Spiele. Katzenjammere Presse, Berlin, 1847.

2. Langry Lilly. *Therapeutic Touch,* The Blue Press, London, 1912.

3. Streisand B. *People Who Knead People, or Therapeutic Massage Revisited,* Green Door Press, San Francisco, 1971. ■

PSYCHOTHERAPY

Contributed by:

Mark Worden
Roseburg, OR

Psychotherapy is an undefined technique applied to unspecified cases with unpredictable results. For this technique, rigorous training is required.

Victor Raimy

A NEW PSYCHOLOGICAL TEST: T.E.T.

URIEL AKAVIA
School Psychologist, Tel-Aviv

We are glad to inform professional and non-professional readers that a new test is getting ready for publication.

It is a personality test based on the mechanism of projection, a worthy follower of the Rorschach, T.A.T., C.A.T., Blacky and many other projective tests.

T.E.T. = Thematic Esspresso Test.

The main tool for testing is the plain cup of esspresso coffee. It was not by chance that we have chosen this handy tool for our experiments. Esspresso is the symbol of our technological and hurried generation, called not without reason "The Esspresso Generation."

Procedure: The client (or patient—it depends on the school of thought represented by the testing psychologist) is seated in a soft, reclining chair and is told to relax. Then he is offered a cup of coffee, esspresso. The psychologist doesn't reveal his intentions and the testee is not aware that he is tested. The movements of the testee are recorded by a hidden camera.

(A set of all needed appliances will be available at the local Psych. Assoc. shop: A chair, a camera, video, dozen cups and 10 pounds of good coffee, record-blanks and pencils. All this is manufactured now at top speed as the test, because of its modernity, is sure to become a best-seller, a sine qua non of every self-respecting psychologist, from Berkeley to Tel-Aviv.)

Norms. 2457 subjects were tested already and exact norms are being computed at this moment. The testees were White Academicians mostly students of psychology. All of them were very eager to sip esspresso for the sake of science.

Results. Many very pertinent findings are recorded, some of them astonishing in their depth and analytical revelations.

We report a few of the findings.

a) The testee puts three lumps of sugar: The characterological meaning: Oral Dependence, Hedonism and Masochism.
b) He puts none; coffee unsweetened: Ascetic tendencies. Anal trends. Hypochondriacal worries.
c) Puts sugar in mouth and drinks: Probably a communist.
d) Spills coffee on his shirt: A bum. More masochism. Bad kinder-stube.
e) Spills coffee, on the psychologist: Repressed hate of father. Sadism.
f) Sips coffee noisily: An egoist. More sadism.
g) Asks for more: A commercial, exploiting attitude. Spoiled brat.

The whole list of findings, the rationales and the interpretations will be published soon. ■

INFANTS' ROOM DECOR & LATER-LIFE BEHAVIOR

Paul G. Friedman,
Lawrence, Kansas

THE ADORATION OF THE MAGI, by Fra Angeli-Nat'l Gallery of Art, Washington D.C.

In the decade following World War II, two dramatic growth spurts occurred. One was in the number of births recorded in the United States, and the other was in the number, style, and variety of accouterments used to decorate the rooms in which those infants resided. Now, nearly 30 years hence, we have available sufficient longitudinal data to assess the relationship between infant-era room decor and the later life development of this population. Correlation coefficients were obtained between a number of room decor variables present during the first year of life and later parent-child interactions in 300 typical middle-class families located in three suburban localities.[1] The statistically significant relationships found are listed below:

1. Infants raised in rooms with at least one mobile, containing a minimum of four suspended objects, were more likely to sit quietly in their rooms for an hour or more at a time working on art projects during the ages of 6-9 than those whose decor in the upper half of their rooms was securely fastened to the wall.
2. Infants raised in rooms with wall to wall carpeting were more likely, between the ages of 10-12 to hang up their clothes after one request than those whose rooms had merely a scatter rug or no rug at all.
3. Infants raised in rooms with matching crib bumpers, sheets, and comforters chose dating partners from families in a significantly higher socio-economic group than those whose crib furnishings had contrasting ornamentation.
4. Children raised in rooms with bookcase shelves containing at least seven storybooks, three stuffed animals, and two music boxes were more likely to attend, and complete before the age of 22, a four-year Ivy League college program than those with fewer objects or no bookcase at all.
5. Infants raised in rooms containing stuffed animals at least twice their own size were more likely, during the 22-26 year age period to call their parents and request their advice before making major life decisions than were infants who had only teeny-weeny plastic teething rings to play with.

These findings strongly support investment of substantial portions of the incomes of parents, relatives, and friends on the elaborate and well-coordinated decoration of infants' rooms. Further research is now underway on the relationship between parents' cultural experiences during the prenatal period and children's later-life museum attendance. ■

[1] To qualify for our sample the parents had to have no less than two cars, two television sets, a dog, a wet bar, a set of golf clubs, and a front and back lawn of at least 300 square feet each.

Certain Psychological and Physiological Manifestations Relating to the Female Mammary Gland

John H. Graves
Leland, Mississippi

THE TEMPEST, by Giorgione, Galleria dell'Accademia, Venice

The American male's preoccupation with the female breast as a symbol of eroticism, or at least as a thing of structural beauty, is difficult to understand.

It has been suggested by psychologists that this yearning for the breast goes back to infancy when the almost universal practice of bottle feeding denies us the comforts of breast feeding so vital to a feeling of security and stability in later life.

In my approach to this subject today I will not attempt to be like the lady's hoopskirt which covers the subject without actually touching it, but more like a dancer's G string which touches on the subject but doesn't cover it.

This subject does have it's humorous aspects though; as illustrated by the medical student who had been out late the previous night and was caught unprepared by a pop quiz by his pediatrics professor. The question was: "What are 5 advantages of breast feeding over bottle feeding?"

The student meditated for awhile and then began:

(1) Mother's milk is always the right temperature;
(2) It always contains the proper proportions of carbohydrate, protein, and fat;
(3) It is free from bacterial contamination (here he paused, scratched his head and though a bit, and then hurriedly added)
(4) It comes in a more attractive container, and;
(5) The cats can't get to it.

Let us then investigate some of the anatomical and histological aspects of our subject.

The mammary gland is a somewhat conical mass of glandular tissue traversed and supported by strands of fibrous tissue and covered by a thick layer of fat. Each gland is situated in the superficial fascia covering the anterior aspect of the thorax and usually extends from the level of the 2nd or 3rd rib to that of the 6th. The hemispherical projection formed by the gland lies upon the superficial aspect of the pectoralis major and to a less extent upon the seratus anterior muscle. Near the summit of each mammary elevation and usually at the level of the 4th or 5th rib is placed the wart-like papilla mammae. (Cunningham's Text Book of Anatomy; 6th Ed.) (1).

Unclothed natives in the jungle exhibit no undue preoccupation with the breast as an erotic organ. It is strictly utilitarian, and sometimes the lactating New Guinea mother will nurse her baby on one side and a young pig on the other (2).

Desmond Morris in his book "The Naked Ape," states that the female of our species is unique among primates in having two prominent hemispherical organs of lactation; although the other flat-breasted species have adequate milk for their young. He concludes that since man became relatively hairless and adapted the upright stance, these peculiar characteristics could serve only a decorative or seductive function (2).

Of course, Mr. Morris is speaking only from the anthropological viewpoint, and he obviously missed the importance of the early mother-child relationship involved in breast feeding.

Thus the female breast becomes a symbol which is sought as the answer to a deep-seated deprivation neurosis.

Zahorsky in his "Text Book of Pediatrics" states that the breast fed baby will be a stronger child in the end; that he will more readily adjust himself to the complicated environments of modern life; that on the average he will be a better scholar and more successful man, and will live longer (he neglects to predict whether, as an adult, he will have more fun) (3).

The phenomenon of the topless waitress is, at first thought, a difficult thing to explain. Why should anyone who is hungry wish to be distracted by an unseemly jiggling of anatomical specimens? And there is always the danger of gross contamination of such things as soup or mashed potatoes.

The answer here, also, goes back to an unhappy infancy. The association is obviously between the act of eating while being stimulated by the original food source. This satisfies the surrogate mother yearning, as it were, and perhaps circumvents an overt oedipus complex.

Spectators at a burlesque or hoochy-kooch show presumably obtain certain understandable, if not wholly commendable satisfactions. However, this being a scientific paper, these aspects of the problem will not be dealt with herein.

This national preoccupation with the bosom as a sort of fetish was recently illustrated for several weeks on New York's Wall Street. Each day at the noon hour various ladies with super-abundant chest dimensions would stroll by the financial district suitably clad in sweaters or other revealing garments in an obvious effort to entertain and edify the gawking clerks and to bring a little ray of sunshine in an otherwise drab existence.

A Miss Geri Stotts, 36, of Burbank, Calif., who has the amazing dimensions of 47-29-38, attracted 5,000 spectators and had to be rescued by the police from the eager crowd. She apparently won the title from Miss Francine Gottfried, who in the previous week had scored with an outstanding 43-25-37 (4).

I do not anticipate that the findings elucidated in this paper will completely eliminate bottle feeding for babies. But I hope that it will give each of our wives a better understanding when her husband casts longing eyes at a generously endowed dancer on the T.V. or a curvaceous cutie at the beach, and she will remember that he is just a little child crying out in the dark for his security blanket. ■

REFERENCES

(1) **Robinson, Arthur:** Cunningham's Text Book of Anatomy 6th Ed.: 312-314, Oxford University Press, New York, N.Y. 1931

(2) **Morris, Desmond:** The Naked Ape: McGraw-Hill Book Co., New York: 70-80, 1967

(3) **Zahorsky, John:** Synopsis of Pediatrics 4th Ed. St. Louis: The C.V. Mosby Co., 1943

(4) **Memphis Commercial Appeal:** Memphis, Tenn. May 2, 1969

PSYCHOGENIC HEADACHE

When headache is of psychogenic origin, the issue of why the head becomes the chosen organ in which pain is localized must be understood. Constitutional-generic predisposition factors are ususally considered, but are yet to be fully delineated. When the head - or some part of the head - becomes the focus of pain, the head may be of great symbolic significance in the person's life.

Volume 61, May 1982
Ear, Nose & Throat Journal

Ed: Symbolic yes-practical no

Submitted By:
Dr. Max Samter
Chicago, IL

THE EFFECT OF PRENATAL INSTRUCTION ON READING ACHIEVEMENT*

During the past few years the controversy regarding *how* to teach reading has abated somewhat, only to give rise to a new topic for debate-*when* to initiate reading instruction. In the past, some experts have claimed that a child should have a mental age of 6.6 before reading instruction is inaugurated; others, that children should be taught to read when they reach five years of age. The trend, however, has been towards attempts to teach reading to even younger children. For example, a *Ladies Home Journal* article (May 1963) informed mothers how to teach their two-year-olds to read. Yet no one has suggested that reading instruction might begin even before birth. Such a hypothesis was postulated twelve years ago and a logitudinal study was undertaken to determine the effect of prenatal instruction on reading achievement in the elementary school.

In cooperation with local obstetricians, 112 women in their fourth month of pregnancy were obtained as subjects. California Tests of Mental Maturity and Nelson-Denny Reading Tests were then administered to each set of parents. Based on the assumption that the offspring would tend to approximate their parents in these factors, the average total scores of each set of parents were used to establish three groups which were matched as to intelligence and reading ability. The average CTMM score for each parent group of twenty-five equaled 104.6. and the average reading score 10.8. Next, the expectant mothers were assigned to either the Basal Reader Group, the Phonics Group, or the Control Group.

The instructional programs for the Basal Reader and Phonics groups consisted of the elements described in the manuals which accompanied these materials; and, what amounted to a placebo, repetition of nonsense syllables, was given to the Control Group. The instructional portion of each lesson was placed on tape and played for each mother individually. These instructions were transmitted concurrently to the unborn child by means of a specially designed fetoscope which was placed against the mother's abdomen. Later as the mother did the workbook or mimeographed assignments, she recited her responses into the fetoscopic device in order to transmit this part of the lesson to the fetus. After eighty-five lessons, this phase of the experiment was terminated. No further attempts were made to teach the children to read until they entered the first grade.

In kindergarten, the children were tested to ascertain if any significant differences existed among them when they were grouped according to the methods used during the initial stages of the study. No statistically significant differences were found among the children's groups either as to intelligence or reading readiness. Average CTMM scores ranged from 109.2 to 113.4 and Lee-Clark Reading Readiness average scores ranged from 1.1 to 1.3.

BOY WITH A BOOK, by James Chapin, Private Collection

As indicated in Table 1, the levels of reading achievement attained by both experimental groups surpassed those of the control group at every grade level. Moreover, the group which had been exposed to the Basal Reader approach exhibited superiority to the Phonics Group. In every instance the differences between means were significant at the .01 level.

The results of this study suggest that prenatal instruction does have a positive effect on reading achievement in the elementary school. Furthermore, the use of a Basal Reader approach proved to be the most effective of the methods utilized. The main conclusion to be drawn from this article,

TABLE 1. Metropolitan Reading Achievement Test: Grade Placement Scores

Actual Grade Placement	Basal Reader N = 25	Group Phonic N = 25	Control N = 25
1.9	2.7	2.1	1.8
2.9	3.9	3.2	2.9
3.9	5.0	4.1	3.7
4.9	6.2	5.2	4.9
5.9	7.1	6.4	5.9
6.9	8.1	7.3	6.8

however, was best stated by the Roman orator who proclaimed, "Nimium celeriter ne credas omnia quae legas." ■

*Reprinted from Elementary English, 1965, 431-432, with the permission of the Council of Teachers of English and Edward R. Sipay.

SUBMITTED BY:
Alan J. Tyler
Tucson, AZ

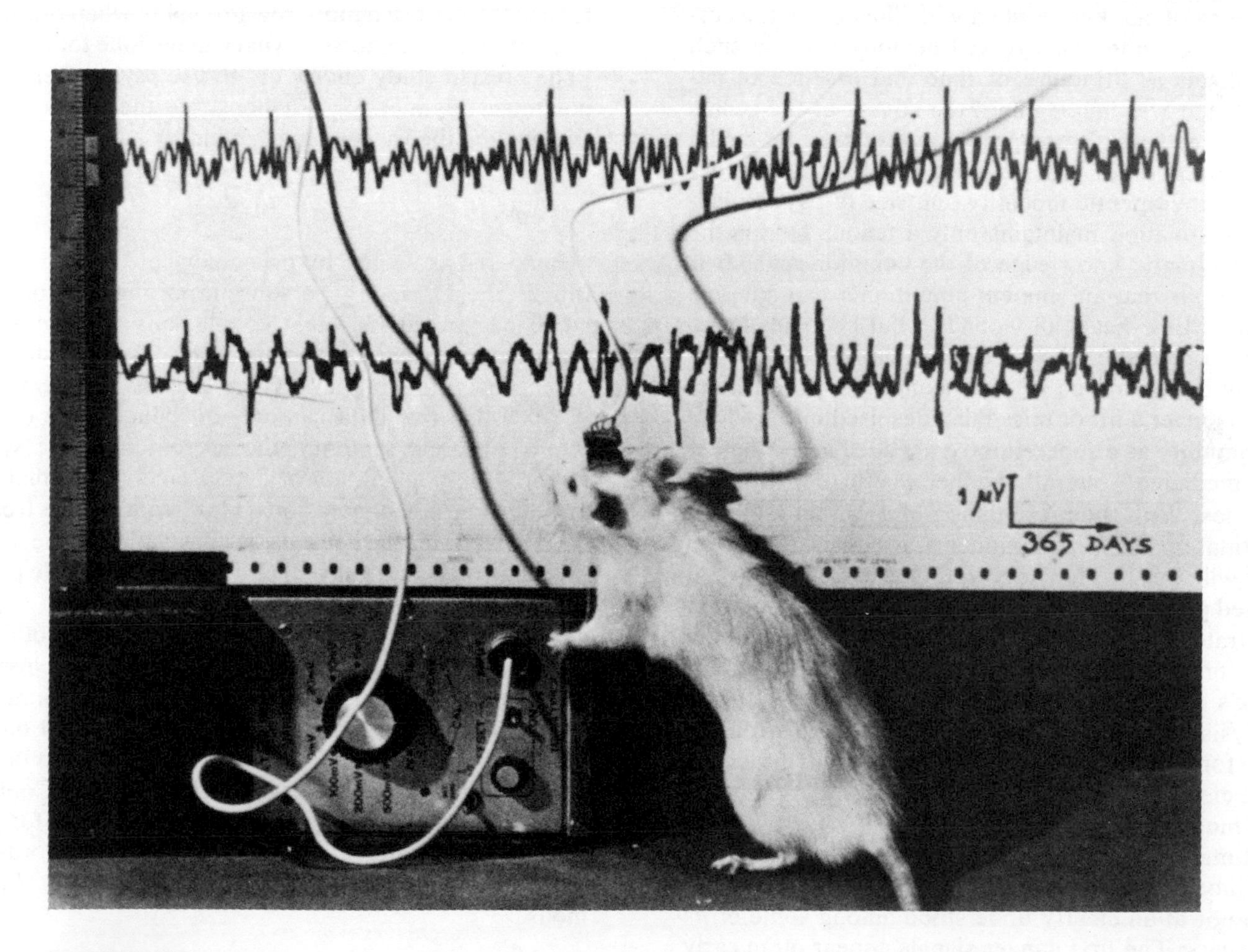

Contributed by:
The Dept. of Pharmacology
Free University
Amsterdam, Holland

The Re-assessment of Criticism and Defenses of Depth Psychology with Supporting Data*

C. K. McKINLEY, Ph.D.[1],
C. K. MCKINLEY, Ph.D.[1],
W. B. REID, Ph.D.[1]
and Anonymous[2]

Introduction

The history of mankind is filled with illusions to the curative power of time. This is well demonstrated in such colloquialisms as "tincture of time and essence of patience." What was lacking until the advent of the genius Sigmund Freud was a comprehensive systematic model for the application of procrastination and systematic delaying as a therapeutic modality. Outside of psychoanalysis procrastination maintains only a tenous hold in the earthly pragmatic knowledge of the common man. It is indeed curious that an eminent practitioner can advise a friend that "time heals all wounds," but becomes impatient with a neurotic who poses a similar problem. With the major breakthrough of psychoanalysis, procrastination is no longer a hit or miss thing despised in the scientific community as a superstition embedded in the morass of "folk medicine," but rather a therapeutic tool with vast possibilities. With the presentation of a formal model of procrastination man can conquer a symptom not by an effort of will, but by effortless will. One can see now that protracted psychoanalysis is not to be scorned, but rather demonstrates the effectiveness of the theoretical model in the use of procrastination in real human situations. Eysenck's classic paper on the "ineffectiveness of Psychoanalysis" now paradoxically emerges as the strongest support for the Phoenix, Psychoanalysis, in the literature and clearly demonstrates the applicability of the theoretical model of procrastination.

Unfortunately the real significance of psychoanalysis and the modifications and expansions of earlier investigators are not at all clearly understood among some of its practitioners. The first danger signals appear quite early with the advent of "directive psychotherapy" and become an imminent danger with the appearance of "interpersonal theory." Currently the linkage of operant conditioning and behavior theory with depth psychology poses an immediate threat to the whole discipline. If we fail at this time in our efforts to save formalized procrastination for future generations, then who knows what eons may pass before another genius of the calibre of Freud will reopen the door to a more rewarding life when one can in fact put off until tomorrow what can be done today.

The present study endeavors to use psychotherapy appointments as means to demonstrate the utility of Procrastination theory in aiding mankind.

Method

Hampered as we are by psychological "raw data flow" (that is, the demand by persons purporting to be or reputed to be "scientists," that conclusions such as are set forth in this paper be supported by "facts," "data," or other uncontrollable events in the environment. It is obvious that raw data can only introduce noise or error into a scientific system.), the present scientific system requires an examination of actual facts (or factual acts) in order to draw conclusions. Out "data" come from an unpublished masters thesis in Psychology; the author failed his orals and subsequently left the field. He prefers to remain anonymous.

Table 1 displays data on 10 patients in psychotherapy for their first 10 appointments. Each patient was instructed to deliberately miss each of the first 10 appointments and both patient and doctor were not to contact one another until the eleventh hour. At the conclusion of each missed appointment the patient was given the Contrivatory Tests of Symptoms, and the total score for each administration rated as "better" or "worse" according to Buzzard's method[3]. Testing was conducted by Anonymous.

* Research supported in part by grant funds liberated from coffee dues.

[1] Division of Child Psychiatry, University of Texas Medical Branch, Galveston.

[2] Grant A. Fund, M.A., Candidate, last known address, Dept. HEW, Washington, D.C.

[3] Buzzard, Fraquahr, What's in a name, J. of Orthonegativity, *1*:247 (1966)

TABLE 1. RAW DATA. THERAPEUTIC EFFICACY

Appointment No.	No. reports "better"	No. reports "worse"
1	4	6
2	5	5
3	7	3
4	4	6
5	6	4
6	5	5
7	6	4
8	6	4
9	3	7
10	4	6

The results of Table 1 are patently inconclusive. The beneficial effect of no therapy (Procrastination) may be present but is not evident to a naive observer. Application of the Data Enrichment Method[4] yields the following results.

Of no relevance is the fact that the figures in the tables presented to have a superficial resemblance to those found in the article on data enrichment method previously cited. Statistical computations reveal that such similarities occur at a probability level of 10^{-22}, however, as the old baseball maxim has it "it only takes one to hit it."

A glance at Table 2 shows that the efficacy of delaying seeing the therapist which was skulking almost unnoticed in the raw data of Table 1, has been brought fully forth by the Data Enrichment Method.

As can be seen, not keeping appointments which can be thought of as not being in psychotherapy at all produces beneficial results simply through putting off seeing the doctor. "No doctor today keeps neurosis away."

Summary

An in depth study was made of the efficacy of systematized procrastination in psychiatric disorders as compared with unstructured procrastinationatism reflected in everyday life. The systematic application of this treatment modality was found to be in all respects as effective as its less sophisticated cousin. ■

[4] Lewis H., The data enrichment method, J. Irrepr. Res., *15*:6–9, (1966).

TABLE 2. ENRICHED DATA : THERAPEUTIC EFFICACY.

Appointment No.	No. of virtual "better"	No virtual "worse"	Probability of report of "better"
1	4	050	4/54
2	9	44	9/53
3	16	39	16/55
4	20	36	20/56
5	26	30	26/56
6	31	26	31/57
7	37	21	37/58
8	43	17	43/60
9	46	13	46/59
10	50	6	50/56

The Varieties of Psychotherapeutic Experience

ROBERT S. HOFFMAN

1. FREUDIAN
   - P: I could use a ham on rye, hold the mustard.
   - T: It's evident that a quantity of libidinal striving has been displaced to a regressive object with relative fixation in the anal-sadistic mode.
   - P: What do you suggest?
   - T: Perhaps a valve job and tune-up.
2. ROGERIAN
   - P: Shit! Do I feel shitty!
   - T: Sounds like you feel shitty.
   - P: Why are you parroting me?
   - T: You seem concerned about my parroting you.
   - P: What the hell is going on here?
   - T: You sound confused.
3. EXISTENTIAL
   - P: Sorry I'm late today.
   - T: Can you get more in touch with that sorrow?
   - P: I hope it didn't inconvenience you.
   - T: Let's focus on your capacity for choice rather than on my expectations.
   - P: But I didn't mean to be late.
   - T: I hear you, and I don't put it down. But where we need to be is the immanence of the I-Thou relationship (in Buber's sense) emanating from the here-and-now, and from there into a consciousness of the tension between be-ing and non-be-ing, and eventually into the transcendence of be-ing itself, through to a cosmic awareness of the oceanic I-dentity of Self and the space-time continuum.
   - P: Gotcha.
4. BEHAVIORAL
   - P: I feel depressed.
   - T: Okay. First, I want you to look at this list of depressing phrases, order them by ascending depression-potential and match them with these postcards. Then I want you to step over to these electrodes — don't worry — and put your head in this vise and your left foot in this clamp. Then, when I count to ten, I want you to . . .
5. GESTALT
   - P: I feel somehow that life just isn't worth living.
   - T: Don't give me that shit!
   - P: What do you mean? I'm really concerned that . . .
   - T: Real hell! You're trying to mind-screw me. Come off it.
   - P: You shmuck — what are you trying to do with me?
   - T: Attaboy! Play me — play the shmuck. I'll play you.
   - P: What's going on?
   - T: Not shmucky enough — try again, louder.
   - P: I've never met a therapist like this.
   - T: No good — you gotta stay in the here-and-now. Again.
   - P: (gets up to leave)
   - T: Okay, now we're getting somewhere. Stand up on that table and do it again.
   - P: (exits)
   - T: Good. Now I'll play the angry patient and walk out the door. "You shmuck — I'm leaving."
6. CONFRONTATION
   - P: Hello.
   - T: Pretty anxious about the amenities, eh?
   - P: Not very.
   - T: Don't try to wiggle out of it.
   - P: I'm not. I just . . .
   - T: Trying to deny it?
   - P: Okay, you're right.
   - T: Don't agree just for agreement's sake.
   - P: As a matter of fact, I don't agree . . .
   - T: Sounds a bit hostile.
   - P: Have it your way. I'm hostile.
   - T: That's pretty dependent, that statement.
   - P: Okay, I'm EVERYTHING.
   - T: God, what modesty!
7. PRIMAL
   - P: Can you help me stop cracking my knuckles, Doctor?
   - T: Okay. You're three years old — you're hungry — REALLY hungry — you want to suckle — you reach for your mother's bosom — what happens? — she pulls away — SHE PULLS AWAY! — SHE ISN'T GOING TO LET YOU HAVE IT — FEEL THAT! — WHAT DO YOU FEEL?? — WHAT DO YOU WANT??? — Get down on that mat there or you'll hurt yourself — YOU *WANT*, YOU REALLY WANT THAT MILK! — YOU WANT YOUR MOMMY! — YOU AREN'T GOING TO GET YOUR MONEY, I MEAN MOMMY!! — CRY OUT TO HER! — TELL HER YOU WANT HER! — CRY, YOU SONOFABITCH!!!!
   - P: But I'm allergic to milk products.
8. PHARMACOLOGIC
   - P: I've been having this feeling that people treat me like an object, that they don't see

me as a person in my own right, in all my uniqueness.

T: NURSE! Get me 500 mg. of Thorazine STAT!

Future Directions

9. ASTROLOGIC THERAPY

P: My last therapist told me I'm deficient in reality-testing. Delusional, I think he termed it.

T: What sign are you, may I ask?

P: Libra. The thing is — I see things occasionally that I'm not sure others see. I tend to form conclusions with insufficient evidence.

T: That's quite characteristic when a full moon hits on the second Thursday.

P: What?

T: Especially if your middle name begins with a P.

P: How did you know that?

T: Well look at this chart . . . you can see that whereas two weeks ago Saturn was out of phase with Route 101, we're now approaching the Spring equinox, and when the Life signs predominate you'd predict that all the MacDonald's hamburger stands will go out of business. You know what that means.

P: Vaguely.

T: Right — things are vague right now for you, in fact for Geminis more than Libras. But on the last day you'll receive a message from a close associate that will clarify a great deal.

P: That's good to know.

T: Surely, but no surprise.

10. MUSICO-THERAPY

P: I have this nagging sense of something left undone, some unfinished business.

T: What are you feeling?

P: Sorta flat. Like nothing of major significance is going on.

T: Not major?

P: That's right.

T: Anything of minor significance going on?

P: You could say that. But it's not enough.

T: What are your present concerns?

P: Well, mainly I think I worry too much about how people *see* me. They seem to think I'm not too sharp.

T: And what would you like to accomplish?

P: I guess I'd like to focus more on what I can *be* of my own volition, rather than just meeting people's expectations.

T: I see. And this sense of unfinished business — does it feel like something you've done before, or is it something you haven't experienced yet?

P: It seems sorta familiar, like I've been through it before and want to recapture it.

T: A repeating pattern.

P: Right — it keeps popping up, and I somehow feel that my life will be incomplete unless I regain it one more time. I'm having a little trouble expressing it — it's hard to conceptualize.

T: Not at all. The way I see it, your life resembles an unfinished *rondo*. Our task is simply to modulate from C-flat-minor to B-sharp-major and rediscover the refrain.

P: But what if we can't rediscover it?

T: No problem. Then we'll just *vamp ad lib* or write in a *coda*. Or if you really want to be adventurous, we could call in John Cage for a consult.

11. T.V. THERAPY

P: Doctor, I've got these pains that sort of move around.

T.V.: (therapist at controls) What do doctors recommend for pain of neuritis, neuralgia?

P: But I also have this ulcer, and plain aspirin makes it . . .

T.V.: Rolaids absorbs forty times its weight in excess gastric acidity.

P: But they may not be real pains — I think they may be in my head.

T.V.: Well, Ben Casey, I think we have to operate. I know you can do it, if anyone can.

P: I know this sounds stupid, but . . .

T.V.: M-I-C, K-E-Y,

P: Are you making fun of me?

T.V.: Excedrin headache number fourteen.

P: Look, I'm getting angry.

T.V.: This is Chris Schenkel from Madison Square Garden.

P: I really don't want to fight you.

T.V.: Stay tuned next for the Late Late Show. Tonight's presentation is *The Red Badge of Courage*, starring . . .

12. REPETITO-THERAPY

T: Good morning.

P: Hi. I feel sorta empty.

T: I feel sorta empty.

P: You too? Wow. Anyway, things just don't seem to be going right.

T: Things don't seem to be going right.

P: That's right. Everything I try to do fails.

T: Everything *I* try to do fails.

P: Really? But I'm coming to *you* for help.

T: I'm coming to *you* for help.

P: Well here's a howdy-do!

T: Here's a pretty mess. I mean . . .

P: You're a Gilbert & Sullivan fan?

T: You're a Gilbert & Sullivan fan?

P: Of course — you think I made that stupid lyric up myself?

T: Of course — you think I made . . .

P: Hold it!

T: Hold it!

P: And I thought *I* was sick! (exits)

T: (soliloquy) Something went wrong there — I'd better get some more supervision. ■

The Correlation Between Intelligence and the Success of American Scholars: Report of Am. Inst. of Strange Behavior

PETER STEINER

It has long been presumed that there is a direct correlation between the success of American scholars and their intelligence. In a study spread over six years and examining a sample of scholars representing every discipline in every research and educational institution in America and only recently made public, the correlation has finally been substantiated in an empirical and unambiguous fashion. What is surprising is that the correlation is an inverse one, that is, the most successful scholars were demonstrated to have the lowest intelligence by the standards used, while those academicians polled who were singularly unsuccessful in the eyes of their colleagues had higher levels of inteligence. It should be noted in spite of this relative difference in intelligence, the difference between the highest and lowest scores on the intelligence tests were negligible when compared with the spread of intelligence over the general population. The intelligence demonstrated by the most intelligent (who was uniquely unsuccessful among his peers) was at about the level of a barnyard horse, while the lowest intelligence demonstrated by a scholar (who was the three-time president of the Association of American Politics and a former ambassador to Russia) was measured to be slightly higher than that of a well trained house plant.

The measure of sucess for this study was a complex interrelated and integrated statistical representation of

a. the weight in grams of the sample's dissertation and subsequent publications, adjusted by a paper weight and type size factor;
b. the hate his colleagues registered on a visceroscope at the mention of his name;
c. the salary and fringe benefits he drew converted to pennies;
d. a caliper measurement of cranium and hip fat. Chart I shows the range and curve of the sample as well as the breakdown by paper, hate money and fat factors.

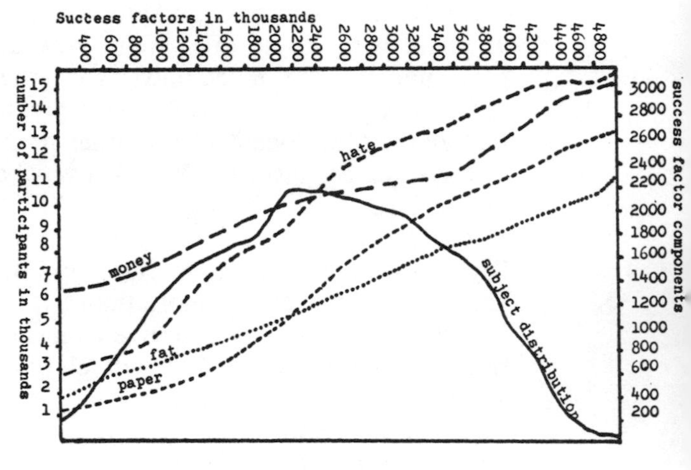

CHART I. Distribution of success factors among subjects

Measuring the intelligence of American scholars in a scientific fashion, although an initially perplexing task, was ultimately accomplished by adapting techniques long used at the Institute in other studies. A maze was constructed and the subjects' ingenuity, perceptive and cognitive abilities were tested from step to step with increasing rigor. Figure I shows the maze and significant stations along the intelligence steps scientifically established by the testers at the Institute. The maze was simplified since the subjects involved had neither the conditioning of rats or rabbits, nor the intuitive familiarity with the manner of testing that rats, rabbits or flat worms have demonstrated in past experiments.

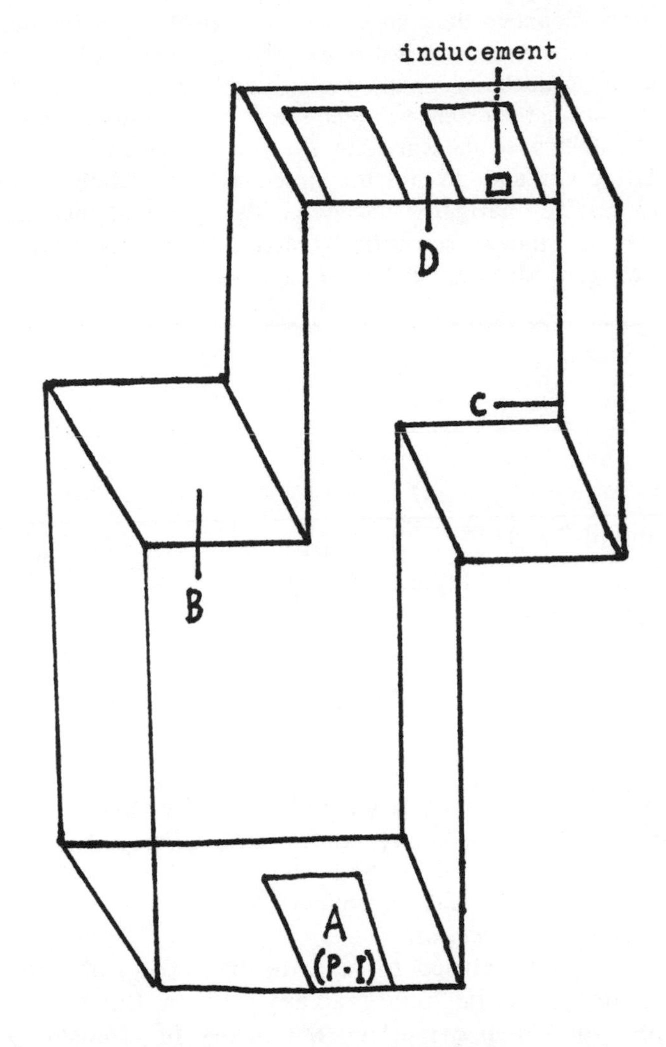

In preliminary testing using a more complex maze, a majority of the subjects being tested remained at the start in, what testers described as "a befuddled condition." In the simplified labyrinth, an old book and a little pile of money placed at the end of the maze to serve a s inducements were clearly visible from the start.

Even in the simple maze there was a significant minority of scholars whose response was a P-1 response, that is, they remained at the start, confused by sight of the pile of money and the alternate exit. Of this group (7.63% of the total sample) 68% had a success factor above 4.3 million. The average success factor of this group was at 3.9 million, substantially above the average success factor of the whole sample (see Graph I).

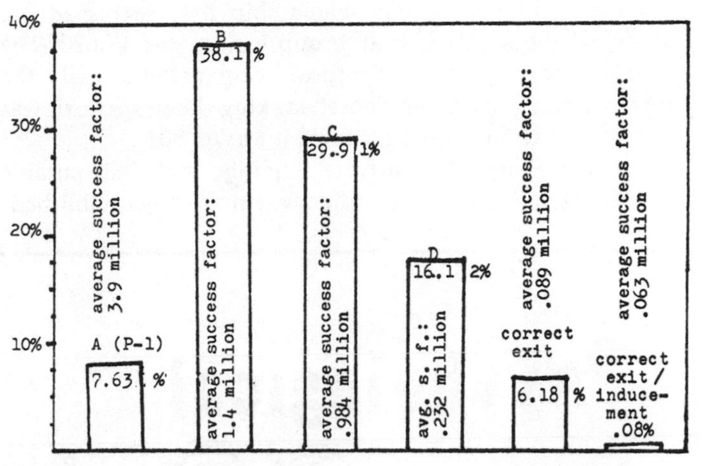

GRAPH I. Correlation of Success Factor and Achievement.

While there was a number of scholars who faltered and stopped in confusion in the first main leg of the maze (between points A and B), the bend in the main hall of the maze at B proved to be the next significant stumbling point in the test. 38.1% of the subjects failed to get beyond this ling point in the test. 38.1% of the subjects failed to get beyond this point, either unable to negotiate the bend in the hall or unaware of the bend. Nearly a third of those faltering at this point in the test walked into the wall repeatedly until they fell to the floor exhausted or unconscious. Many of those that negotiated the first right angle turn at B were confused to be facing the wall at C and either stopped there or turned in the wrong way and got hopelessly lost in the maze. As one might suspect, the success levels of those who did not get past these similar stumbling points were also very similar. The average success factor of those stopping at B was 1.4 million while the average for those getting to C was .984 million (see Graph I).

While a remarkably strong 22.3% of the subjects tested made it to the end of the maze, a surprisingly large 16.12% bolted through the wrong door or ran into the jamb between the exits. The remaining 6.18% took the proper exit, but only .08% of those stopped to look at or take the old book, while only one of the subjects tested actually took the money on his way through the door. He was a former soccer player, who entered teaching to avoid the draft and who, incidentally, went through the maze in four seconds. Predictably his success quotient was 341, one of the lowest success quotients among the subjects tested. The average success quotients for non-scholarship first year graduate students, the lowest human group tested, was 17,620. The average success quotient for those scholars (admittedly, the term scholar is used here loosely) taking the proper exit was 88,895. For those taking the book it was 63,301.

These findings demonstrate tangibly, with indisputable clarity, what has been scientifically, empirically established: that the intelligence of working scholars is inversely proportional to their success in their field. In priliminary testing the American Institute of Strange Behavior is setting up the machinery and establishing the procedures for measuring the extent to which scholars can assume the sort of intelligence measured in the test described above by eating cut and ground up pieces of those scholars who successfully navigated the maze. The results of such test, assuming they are scientifically carried out, could revolutionize higher education in America and the world. ■

Psychological Report

Patient: Dr. Sigmund Duerf
Age: 116 *Date of Birth:* May 6, 1856
Examiner: Edward Zuckerman
Referral Reason: To establish extent of delusional system and severity of thought disorder.

Behavioral Observations:

Dr. Duerf is a short, stooped, balding, grey-haired man who looks much younger than his claimed age of 116. His eyes are alert and penetrating and give the impression of great wisdom. His speech is difficult to understand because of a strong Austrian accent, a facial prosthesis he must wear, and constant cigar smoking. Rapport was excellent and easily established.

Referral Question:

The well-systematized delusional system was easily elicited as he eagerly seeks converts to his beliefs. It all came to him "in a flash," he says, as he was reading Heider's balance theory and watching the Green Bay Packers on T.V. one Sunday. He saw that the psychoanalytic theory of dynamics was tragically incomplete: defenses everywhere but no offenses. He immediately set out for Green Bay, Wisconsin, to consult with the late Vince Lombardi and to give his theory to the world.

In the next several months, one offense after another was discovered. First was the most basic, *expression* (to be distinguished from *acting-in*, of course), then *action-formation, doing*, and *emotionalizing*, the offenses most favored by the hysteric. The major offense of the obsessive was, of course, *expression* and of the paranoid, *retrojection*.

Dr. Duerf refused to take the psychological tests, explaining that he could fake any of them. But he was most voluble in explaining the theory of offenses in the interview. The examiner was not allowed to ask questions of the doctor as this would interfere with the "keeping hold of." (Dr. Duerf explained that this phrase lost something in the translation.) This restriction led to the examiner's remaining confused about the explanation of the offenses, and thus the report on delusions is incomplete.

Diagnostic Impression: Senile Dementia

Recommendation: Because of the lack of background information on the patient, his case is to be turned over to the staff forensic psychiatrist, Dr. Franz Kafka, for disposition. ■

Bayes Theorem and Screening for Future Mental Illness: An Exercise in Preventive Community Psychiatry

LYON HYAMS, M.D., M.S.

It is generally agreed[1,2] that an ounce of prevention is worth a pound of cure[3]. Accordingly, one reasons that if there were predicting indices of future severe mental disturbance, early therapeutic intervention would result in a reversal or diminution of the pathological processes decreasing the institutional and welfare burden, increasing the work force, and thereby increasing the future gross national product[5]. It is, in addition, possible that such people would be happier[6]. Of course, one would want to assume that: 1. reliable identification is possible; 2. one could, indeed, induce a reversal; and 3. that R_{12} was favorable[7] before implementing preventive measures[8].

The purpose of this presentation is to indicate that even if the above postulates can be assumed, the preventive goal is still impractical—and Bayes theorem[9] is to blame[10].

Assume that a "marker," M, of previous psychiatric instability can be found such that in a large psychotic population it had been present 95% of the time[11]. Conversely, in a nonpsychotic group it was absent 95% of the time. (Such a 'marker' has been proposed[12]. Apparently, a "Certain Fixed Stare" (CFS) during work differentiates these broad groups). Such probabilities are a priori, i. e., given the event, psychotic or not, the probability of having had the marker. But in screening we are only interested in the aposteriori probability, i.e., given the marker, the probability of becoming psychotic. Bayes' formula[13] allows us to calculate this quantity knowing the probability of psychoses in the general population. Recent estimates[14] indicate that one in every 50 persons develop psychotic illness. We can now define a universe consisting of two exhaustive and disjoint events: being psychotic and not being psychotic (P and $\overline{P}$)[15], where $Pr(P) = .02$ and $Pr(\overline{P}) = .98$.

If M indicates a positive marker and M the absence of such, the conditional probability of becoming psychotic when identified with a positive marker is given:

$$Pr(P/M) = \frac{Pr(M/P)\ Pr(P)}{Pr(M/P)\ Pr(P) + Pr(M/P)\ Pr(P)}$$

Where:

$$Pr(M/P) = .95$$
$$Pr(P) = .02$$
$$Pr(M/\overline{P}) = .05$$
$$Pr(\overline{P}) = .98$$

Then:

$$Pr(P/M) = .28$$

In other words, for every 100 people giving a postitive marker, only 28 of them would develop a psychosis in time. Since these 28 are unidentifiable, it would be financially prohibiting to treat them all. In addition, psychotherapy applied to normals might precipitate an unexpected break, increasing the type II error. ■

[1] Hyams, L. "The Advantages of Screens, A Prospective Epidemioogical Study." Journal of Household Effects 3:109–110, 1924.
[2] Goose, M. "Silly Stories" 1397:97, 1922.
[3] This colorful expression, used to denote the more profitable distribution of energy outlay-reward return ratio, R[4] is copyrighted. Please do not quote without written permission of the author.
[4] Definition of R: a. $R \geqslant 100$ implies Inadequate Exchange. Modification necessary for continued life.
b. $5 \leqslant R < 99$ implies Normal Operating Range.
c. $R \leqslant 4$ implies Unusually Efficient Transfer Kinetics. Continued operation at this level results in Fuse Blowing (F.B.)
[5] This has an important inference for grant go-getters.
[6] This is pure speculation. No one has been able to demonstrate, for example, that the washed-out schizophrenic is not really having a ball.
[7] See footnote (4).
[8] It is suggested that the above criterion be termed "Hyams Postulates for Favorable Preventive Community Psychiatric Ratios (HPFPCPR)."
[9] Or formula. I never know which is correct usage.
[10] Bayes himself wants no part of this.
[11] You can never do better than 95 percent.
[12] Hyams, L. and Hyams, D. "Staring During Work" Journal of Occupational Medicine, 13:1, 1963.
[13] Or theorem.
[14] Hyams, L. Unpublished data.
[15] Further differentiations of groups can be attempted. A division of psychoses into vertically and horizontally crazy is useful[16].
[16] Vertical psychotics vibrate up and down. They do not disturb other people. Horizontal psychotics vibrate in a plane parallel to the ground. They are a pain in the neck.

Quasi-successful concurrent validation of a special key for a relatively new and exciting personality instrument in a group of potential managers: or, I am never startled by a fish.

W.S. Blumenfeld

Most of us would probably agree that, in the area of managerial selection, what is lacking most is the availability of valid non-cognitive predictors. Further, most would agree that tailor-made, special keys are to be preferred to universal, general keys. The combination of the two seems desirable and appropriate. However, as Kurtz pointed out so well in 1948, too often the wishes and hopes of the practitioner and/or the consumer manifest themselves in a strange form of selective perception in the evaluation of effectiveness of such keys, i.e., the acceptance of self-fulfilling "research" *via* foldback design.

Purpose. The purpose of this research was to develop and validate concurrently a special tailor-made key for a relatively new and exciting personality instrument in a group of potential managers. A secondary (sic) purpose of this research was to point out once again the specious, spurious, fallacious, but fascinating results that are obtained when cross-validation does not follow item analysis.

Data Collection. The subjects, criterion, and instrument follow.

The subjects in this experiment were 126 management majors in an introductory management course at Georgia State University. The instrument administration was presented to them as an example of a "scientific" selection technique (very much as charlatans present their wares to unwary personnel and marketing executives). From all indications, it was accepted as such (just as it is usually accepted by "hard-nosed businessmen"). These subjects may be viewed as entry level managers, or at least potential managers, i.e., personnel and marketing executives, hard-nosed businessmen, etc., etc.

The criterion in this study was self-reported grade point average of the subjects.

The relatively new and exciting (if not sensational) personality instrument used in this study was the *North Dakota Null-Hypothesis Brain Inventory* (*NDNHBI*), conjured up and conceived by (Art) Buchwald (1965) with a sharp tongue and a great deal of cheek in answer to the problems of face validity encountered by the *Minnesota Multiphasic Personality Inventory*. The *NDNHBI* consists of 36 statements of a non-cognitive nature to which the respondent indicates either true or false as being a descriptive of himself. Since the inventory is so "special" and will no doubt be of interest, the items are presented here:

1. I salivate at the sight of mittens.
2. If I go into the street, I'm apt to be bitten by a horse.
3. Some people never look at me.
4. Spinach makes me feel alone.
5. My sex life is A-okay.
6. When I look down from a high spot, I want to spit.
7. I like to kill mosquitoes.
8. Cousins are not to be trusted.
9. It makes me embarrassed to fall down.
10. I get nauseous from too much roller skating.
11. I think most people would cry to gain a point.
12. I cannot read or write.
13. I am bored by thoughts of death.
14. I become homicidal when people try to reason with me.

15. I would enjoy the work of a chicken flicker.
16. I am never startled by a fish.
17. My mother's uncle was a good man.
18. I don't like it when somebody is rotten.
19. People who break the law are wise guys.
20. I have never gone to pieces over the weekend.
21. I think beavers work too hard.
22. I use shoe polish to excess.
23. God is love.
24. I like mannish children.
25. I have always been disturbed by the sight of Lincoln's ears.
26. I always let people get ahead of me at swimming pools.
27. Most of the time I go to sleep without saying goodby.
28. I am not afraid of picking up door knobs.
29. I believe I smell as good as most people.
30. Frantic screams make me nervous.
31. It's hard for me to say the right thing when I find myself in a room full of mice.
32. I would never tell my nickname in a crisis.
33. A wide necktie is a sign of disease.
34. As a child I was deprived of licorice.
35. I would never shake hands with a gardner.
36. My eyes are always cold.

In the original article, Buchwald presented a differential psychometric scatter scoring system for placement in either the Peace Corps, the Voice of America, or the White House. In the current research, as indicated by the purpose, an appropriate configuration which concurrently related to an external criterion was developed and quasi-validated.

Data Analysis. There were three phases to the data analysis of this research, i.e., (1) item analysis, (2) foldback, and (3) cross-validation.

The 36 items in the *NDNHBI* were item analyzed using the procedure described by Lawshe and Baker (1950) with an external criterion of self-reported grade point average. A skew in the criterion distribution categories necessitated that the high and low "halves" of the criterion group be of different sizes. In the item analysis, there were 48 in the high group, and 28 in the low group. Alpha of .10 was used to identify the "discriminating" items for inclusion in the "special" key.

To prove to the proponents of the instrument (of which there were a few) and to those who really "wanted" the key to work (several students with an apparent clinical bent), the items surviving the item analysis were applied to the answer sheets of the item analysis group. The concurrent validity was documented by biserial correlation.

For those more interested in the best (rather than the most fulfilling) estimate of the relationship between the derived key and the external criterion of self-reported grade point average, the items surviving the item analysis were scored in holdout groups of 25 high answer sheets and 25 low answer sheets. Again, biserial correlation was obtained to quantify the relationship between the special key and the criterion.

RESULTS

The item analysis procedure identified 9 items (chance would have been 4) which discriminated between the high and low groups at or beyond the .10 level. The reader will no doubt be interested in which items "came through", particularly as the potential for *post hoc* interpretations and insights are nearly infinite. The items (and their weights) in the special key were:

1. (—) My sex life is A-okay.
2. (+) When I look down from a high spot, I want to spit.
3. (—) I think most people would cry to gain a point.
4. (+) I am never startled by a fish.
5. (+) My mother's uncle was a good man.
6. (+) I don't like it when somebody is rotten.
7. (+) I have never gone to pieces over the weekend.
8. (+) Most of the time I go to sleep without saying goodby.
9. (+) It's hard for me to say the right thing when I find myself in a room full of mice.

Applying these 9 items back upon the original sample, the obtained biserial correlation was .78. This is clearly off zero beyond the .05 level, — most encouraging to all, and completely satisfactory, convincing, and conclusive to some (Kurtz, 1948). (Consider here for a moment those of your acquaintance and/or your employ using this foldback design and at this point mouthing such quasi-professional, and sage, things as "of course, these results should be interpreted with some caution.")

Unfortunately, when the 9-item key was applied to the holdout sample of 50, the encouraging coefficient of .78 shrank slightly. In fact, it shrank back to .07 (*not* significantly off zero at the .05 level). Too bad; pity; so many of the items seemed to have so much construct validity, and were *so rich* in potential for *post hoc* interpretations and insights, e.g., "I am never startled by a fish."

DISCUSSION AND CONCLUSIONS

Little if any discussion seems necessary; Cureton's classic paper (1952) has been re-trotted out and executed. It seems clear once again that (1) the application of a key to the control group is the acid test of the quality of a key and (2) the (re)application of a key to the original group is but a half-acid test. To a sophisticated group like this, this would seem to be "coals to New Castle"; however, as an industrial psy-

chologist in a business school dealing with students of business administration (and naive practitioners and consumers of business administration), it is painfully clear to me that the foldback design still remains very much in vogue. (Afterall, it has such obvious marketing advantages.) I think it appropriate to continue to beat home the point of cross-validation, i.e., let's have no more of this half-acid research.

In conclusion, the foldback design is not (necessarily) dead; it is very much alive and doing quite well among the malicious and the naive in the general business world.

And frankly, I *am* always startled by a fish — particularly when the "fish" turns out to be a personnel or marketing executive. ■

REFERENCES

Buchwald, A. My eyes are cold: Testing in the great society. Author, 1965

Cureton, E. E. Reliability, validity, and baloney. *Educational and Psychological Measurement*. 1950, *10*, 94-96

Kurtz, A. K. A research test for the Rorschach test. *Personnel Psychology*, 1948, *1*, 41-51

Lawshe, C. H., & Baker, P. C. Three aids in the evaluation of the significance of the difference between percentages. *Educational and Psychological Measurement*, 1950, *10*, 263-270

The author wishes to acknowledge the data collection and analysis contributions of Allen Austin and Julian Eidson.

S.M.A.R.T.S.

The Scale Of Mental Abilities Requiring Thinking Somewhat

Test Construction, Normative Data, Test Items, Administration, and Scoring

Glenn C. Ellenbogen
New York, NY

A recent study, conducted by Dr. Ernst von Krankmann and his associates at the prestigious Advanced Institute for Psychometric Research, an elite division of Educational Testing Service, called for the development and marketing of a new generation of intelligence tests more contemporary in content and design than those that have heretofore been known to the psychological community. The present paper represents the formal introduction of the Scale of Mental Abilities Requiring Thinking Somewhat (S.M.A.R.T.S.), the first of a new breed of intelligence measures to come out of the laboratory setting in the last quarter of the 20th century.

Norm's

The subject we hired to take the S.M.A.R.T.S. test was quite bright and a very nice person as well. Our subject, Norman (or Norm, as he preferred to be called), was a freshman at New York University at the beginning of our intensive and extensive testing program and, upon completion of his participation in our study, was a freshman at New York University. After carefully weighing the possible threats to internal and external validity, as well as to job security, we decided to administer the S.M.A.R.T.S. test to Norm many many times in order that we would end up with enough numbers, data, and statistics to construct at least one table. The scores, then, presented in Table 1, are really Norm's.

TABLE 1. Norm's Test Scores

SMARTS CATEGORY	Verbal Abstractions	Arithmetic Reasoning	Visual Construction	Visual Gestalt	Visual-Verbal Abstractions	Total SMARTS Score
You've Got SMARTS!!!	18-22	14	10	8-9	5-7	55-62
You've Got SMARTlets*	12-17	13	5-7	9	4	43-54
All-American Average	8-11	12	4-8	2-4	3	25-42
Room For Growth	3-7	5-11	1-3	1	1-2	7-24
Mental Vegetable	0-2	0-4	0	0	0	0-6

*SMARTlets are small SMARTS

Instructions For Administration

The generation of a valid S.M.A.R.T.S. score requires close adherence to the same instructional guidelines as were employed in the standardization of the test instrument. Thus, test instructions should be administered exactly as they appear below.

The Scale of Mental Abilities Requiring Thinking Somewhat was designed as a self-administered intellectual assessment tool. When administering an assessment tool, it is important that "rapport" be established between tester and testee. Since you will be administering the test to yourself, it is crucial that you establish proper rapport with yourself. First, introduce you to yourself. Try to make yourself feel at home in your home. Explain to yourself, slowly and in a comforting voice, the nature and purpose of today's test. Make sure to answer any and all questions which you may have about the test.

It is important that you be relaxed before beginning the test. Some sure signs of uneasiness are trembling, stuttering, and breaking out in hives. If you should notice such signs in yourself, try to reassure yourself that it is perfectly normal to feel somewhat nervous when taking a test such as this, that many people share the very same feelings—the shortness of breath, the tenseness and queasy feelings in the stomach, and a whole variety of other feelings and sensations which appear just prior to vomiting. Should you find yourself saying to you, in an angry and nasty tone of voice, "I just want you to know I don't do well on these kinds of tests!" and/or "What do they prove, anyway?," you may be expressing anxiety in the form of hostility. It would not be wise, then, to mention to yourself that you will be timing yourself on some parts of the test.

You'll only make yourself more nervous. You'll have plenty of time to make yourself nervous when you get to the actual timed tests, so why make yourself nervous prematurely?

TEST ITEMS

Section 1 - Verbal Abstractions

1. Which one of the five words below does not belong?
   car, boat, train, bus, dendrochronology
2. In what way are *Cheddar cheese* and *Ronald Reagan* alike?
3. A student always has
   A. homework
   B. spitballs
   C. marijuana
   D. zest for learning
   E. jeans
4. Which one of the five words below does not belong?
   Dramamine, death, neurosis, Woody Allen, tortilla
5. In what way are a *dog* and an *electron microscope* alike?
6. Unscramble the letters below so that they make a word. (This is a timed question. You only have 5 minutes in which to unscramble the letters. The faster you unscramble the letters, the more points you earn. Record the amount of time required to completion.)
   RRDOOIMEHH
7. Which one of the five words below does not belong?
   transistor, penicillin, computer, satellite, Pet Rock
8. As the saying goes—It was________fit for a king.
   A. quite
   B. Roger
   C. Emile
   D. most
   E. not
9. Which one of the five words below does not belong?
   pragmatic, reactionary, utilitarian, existential, wart
10. A person who accepts in other people things that don't really bother him is known as
    A. a Californian
    B. wishy-washy
    C. an eclectic
    D. a schizophrenic
    E. a liberal
11. Which one of the following statements *best* represents the concept "concrete"?
    A. Why?
    B. How many insights can fit on the head of a pin?
    C. If no one were in a forest to hear it, would pigeon droppings hitting the forest floor make a sound?
    D. Gloria Vanderbilt jeans sell for $45.00 a pair at Macy's, but can be had for 20% off on Manhattan's Lower East Side.
12. In what way are an *apple* and an *existentialist* alike?

Section 2 - Arithmetic Reasoning

1. How much does a piece of penny candy cost?
2. How many ears does a man have?
3. How many ears does a woman have?
4. How many shoes do you wear on your feet?
5. How many pair of shoes do you own?
6. How many shoes make up a pair?
7. John had 3 euglena and his stepmother gave him enough euglena so that he had double the number he started with. John now has
   A. as many euglena as he had before
   B. twice as many euglena as he had before
   C. three times as many euglena as he had before
   D. four times as many euglena as he had before
   E. more euglena than he can count
8. If you have 7 pennies and you lose 2 of them
   A. you would have 3 pennies left
   B. you would have 6 pennies left
   C. you would have 4 pennies left
   D. you would be the type of person who loses things easily
9. Vinnie Tortoni had 3 pennies and his mother gave him 1 American penny and 1 Canadian penny. How much money does he have altogether?
   A. 6 pennies
   B. 4.87 pennies American value
   C. 3 pennies
   D. enough pennies to buy one piece of penny candy
10. Dave had six marbles. His father usually gave him $1.25 allowance each week, but this week his father started giving him $1.50. With his new allowance, Dave went to the store and tried to buy 3 more marbles, but the store was closed because it was the 4th of July weekend. Instead, Dave bought a chocolate popsickle from the Good Humor man for 50

cents. He also lost a nickel because he had a hole in his pocket, and spent 15 cents more on a package of Kleenex tissue to wipe off the ice cream which he dripped onto his brand-new Keds sneakers so that his mother wouldn't kill him. Finally, Dave got lost but the police returned him to his parents, safe and sound. How many marbles did Dave start off with?

11. On the last day of his travels in Madagascar a man bought a used clockradio for 1/5 of what it cost new. The man paid $5.00 for it. The next day the clock broke and the radio short-circuited. How much money did the man really save by buying a used clockradio?

12. Tom is three years older than Jim.
    Jim is five years older than Bill.
    Bill died last week.
    Which statement is necessarily true?
    A. Tom is younger than Bill.
    B. Jim is younger than Bill.
    C. Bill is older than Tom.
    D. Bill doesn't care about the age differences between Tom and Jim and himself.

13. A man starts out in his car from Buffalo at 12:00 Noon and drives due south at 25 miles per hour for 2 1/2 hours. He then turns due west and travels at 40 miles per hour for 3 hours. Finally, he turns due north and drives 13 3/10 miles before running out of gas. What good is a car without gas?

Section 3 - Visual Gestalt

Each picture below has something essential that is missing. Look at each picture, one at a time, and decide the one essential thing that is missing from the picture.

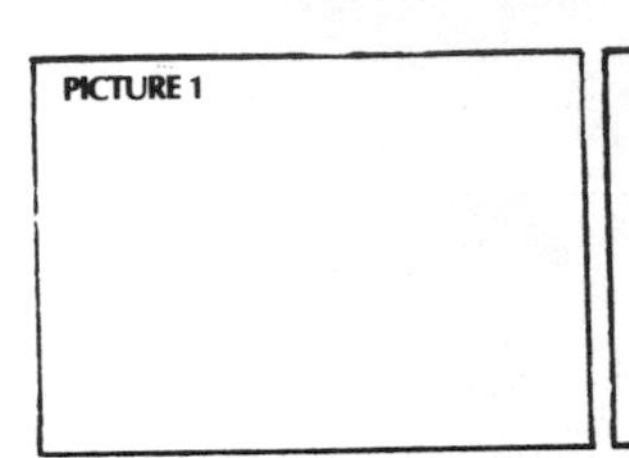

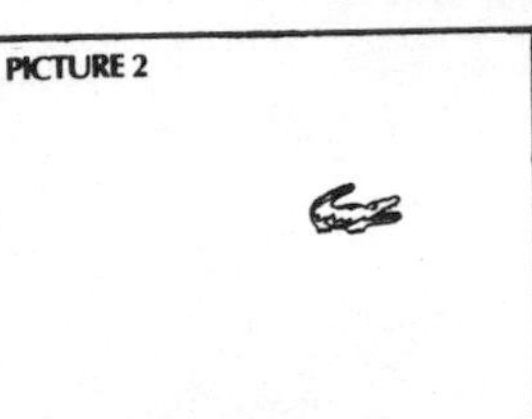

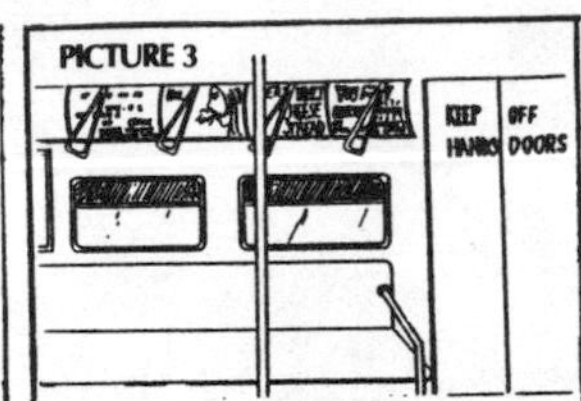

Section 4 - Visual-Verbal Abstractions

CARD 1

CARD 2

CARD 3

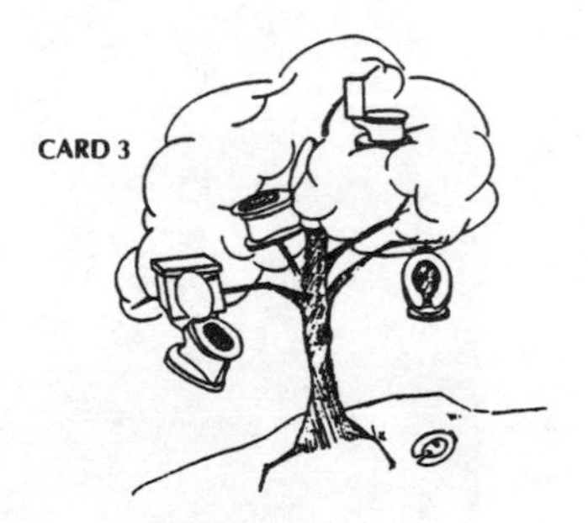

Card 1. What does this picture look like to you?
Card 2. Look at and read the card and draw on a piece of paper what you think the answer would look like.
Card 3. What *one word* does this card depict?

SCORING

Section 1 -Verbal Abstractions

1. Answer: Dendrochronology
   Score: Give yourself *1 point* for the correct answer, no credit for a wrong answer.
2. Answer: They both come aged
   Score: *2 points*
3. *Answer: E. jeans*
   Score: 1 point
4. Answer: Tortilla
   Score: *1 point*
5. Answer *2 points*: They're not alike
   & *1 point:* I don't know; they're alike?
   Score: *0 points:* they're the same thing
6. Answer: HEMORRHOID
   Score: If you correctly unscrambled the letters within
   0"- 60", you get *6 points;*
   60"-120", you get 5 points;
   120"-180", you get *4 points;*
   180"-240", you get *3 points;*
   240"-300", you get *2 points;*
   If you were unable to unscramble the letters after 5 minutes, you get *1 point* for trying hard. If you gave up, you get 0 points.
7. Answer: Pet Rock
   Score: *1 point*

Nothing

can be

more

irreproducible

than

NATURE STUDY

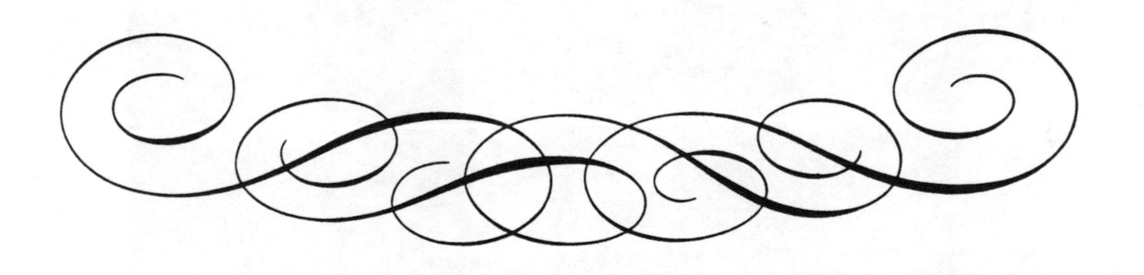

Electron Micrograph of a cell from line of cells of a mosquito (Aedes). Submitted by Dr. Dieter Adamiker Elektronenmikroskophisches Laboratorium der Tierarztlichen Hochschule, Wien

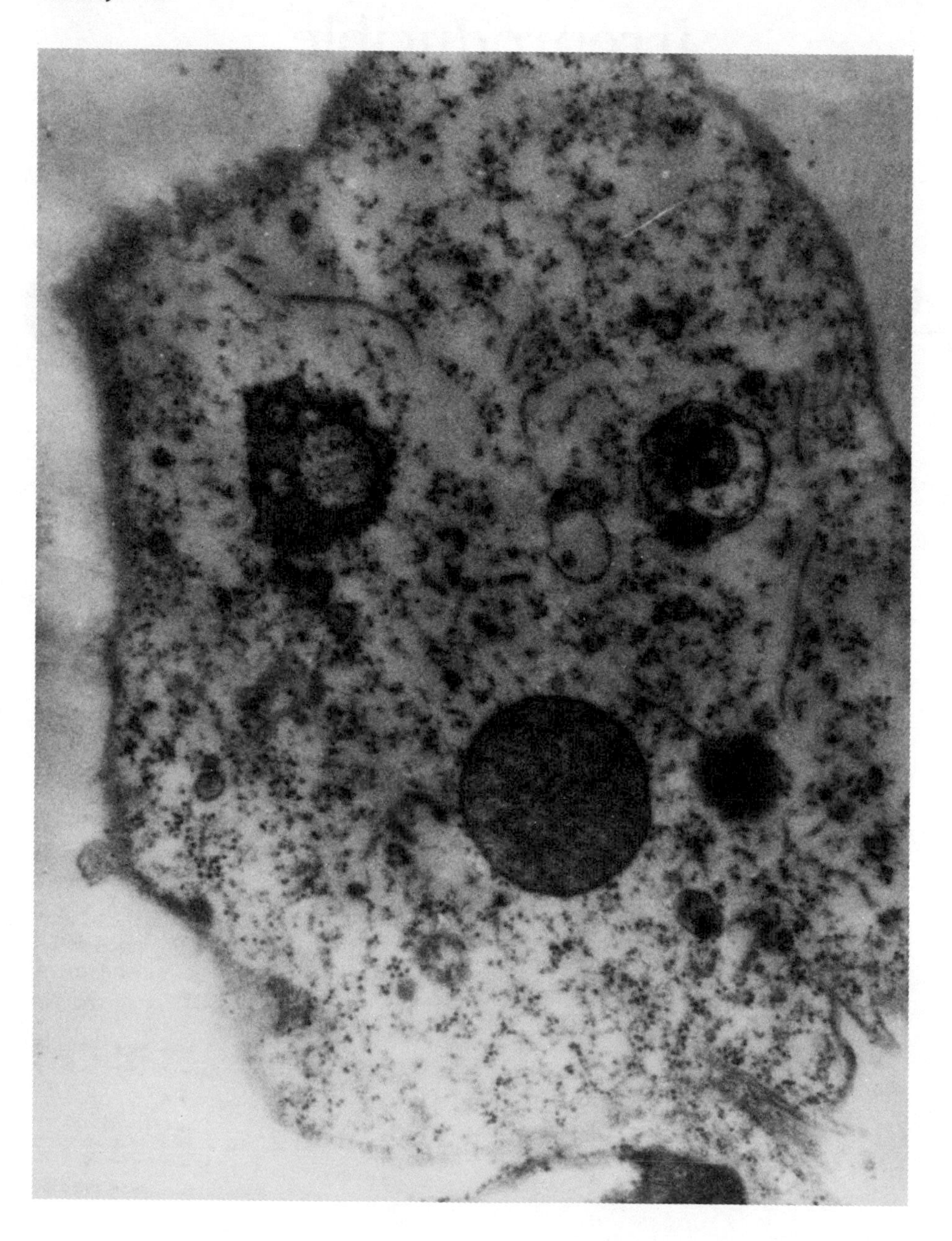

The Rapidity of Female Mosquitoes' Gathering to Mate from a Distance of up to, or, in Some Cases, Beyond, Fifty Miles, as Affected by the Humming of Males at Frequencies of A440 as Opposed to A441

JAMES STITCH
Oxblood University
Pitchfork, Nevada

Social psychologists of small renown have displayed an intense interest in the effects of musical intonation upon females of the species *mosquito sexipediformis nastiensis* at the approach of the rainy mating season in Ecquador in recent years. The coming monsoon promises to be the dampest ecologists have seen in decades, and so these latter (ecologists) will probably accompany a series of ten jungle expeditions arranged by musicians to study the species. Psychiatrists are said to be planning to follow close behind to watch for any abnormal developments.

Purpose of Mating Hums

Naturally, mosquito hums are of different varieties and serve different functions. Insecto-musicologists have long taken an interest in such phenomena:

Why, as recently as last fall I was bitten by one, and upon consulting my tuning fork, I noted that it was humming at A447; with this knowledge in mind, I took a slightly higher-pitched tuning fork, namely B^b, and, taking that pitch a little to the flat side, namely B^b500, I was able with humming to set up such an unpleasant and harsh vibration when the next intruder came, also humming, that it could not bear it, and flew away.[1]

Procedures

Investigators proceeded in this experiment to determine the relative rapidity of mosquitoes' gathering for mating purposes through comparing the responses of two groups, call them A440 and A441, to male hums of those pitches, by operating on the mosquitoes' inner ears (exclusively females underwent the operations), namely, the hammer, anvil, and the stirrup mechanisms, in such a way as to enable members of both groups to hear only one of those pitches or the other, exclusively. (Since the operations took a long time, some of the original subjects had died by the time investigators were ready to experiment. Reliability went down when females not having undergone the operation had to be substituted.) Portions of the inner ears removed were kept in cold storage. To assure that the more rapidly gathering group would not be doing so merely for reasons of sexual appetite, one ovary was tied off in each of the subjects to dampen that. This was done with #46 hemp, obtained from Sears and Roebuck.

Males then were gathered up, and for purposes of the experiment, the males were taken to a point exactly fifty miles "as the crow flies" from where the females, having newly recuperated from their operations, were waiting. The rainy season was very much in evidence as on the day of commencement of the research, it had been raining for a week and the jungle floor was teeming with fish. It is important to remember that mosquitoes fly the same way as crows.

At a certain appointed hour, males were allowed to begin initiating their mating hums for the benefit of the fifty-mile-away females. In order that no females would come that were not part of the experiment, those females intended for use in the experiment had one leg painted red, and any female without such an appendage was rejected as irrelevant.

Another difficulty in the project was, that it was found that males do not in their natural state ever hum the pitch A440 or A441 during the mating season. So males had, at great cost of time, to be given a laryngeal operation that involved removing the larynx, opening it, and cutting the vocal chords slightly for tuning purposes. The reason for this part of the project's taking so long can be clearly understood when it is known that even after the larynxes had been put back the mosquitoes were not allowed to hum for a period of two weeks, and those humming the wrong pitches had to be re-tuned.

It was thus near the end of the monsoon and the concomitant mating season by the time all things were in readiness, and most of the females had died or else got tired of waiting and gone off to lay their eggs any-

[1] H. L. Mencken, *Jungle Music.*

way. The remaining females, approximately ten in number, gathered in a ratio of six to four, flocking, in that order, to the A441 and the A440 male groups, respectively, One male, having flown out of the campsite, was at a distance some few feet greater than fifty miles, and his mate, exhausted as all the rest, had to cover a consequently greater distance, accounting for the elongated and offset portion of the title of this article, originally intended to be shorter. This male was of the A441 test group, and since his mate was one of the type which, identified by antenna angle (this changed in connection with the females' ability to hear one pitch as opposed to the other, A440 or A441), seems to belong to the A440 mates section, it cannot be determined whether this female approached out of a desire aroused from hearing the particular male's hum, or whether she just bumped into him.

As has been stated, the greater proportion of mates came to the males humming the A441 pitch. From this it can be probably deduced that, A441 being a more erotic pitch, orchestras should tune to that when performing music of the Romantic period.

Those portions of the females' inner ears removed and placed under cold storage, rotted.

Since neither these particular females' ability to hear nor these males' ability to hum at those pitches audible to these females ever occurs in nature, the research value of the project was nil. ■

An Experimental Behavioral Psychology Application to Cancer Research

T. D. C. KUCH

Abstract. As part of the work of the Psychobiology Project, Aaron Burr Research Laboratories, Inc., fifty six-week old female mice (strain B6C3F1) were treated with dl-2-amino-4-(ethylthio)butyric acid, which has been reported to be carcinogenic. Two days later, an intensive course of operant conditioning was begun. At age 14 weeks, 19 of the animals had been sufficiently conditioned to record their own body weights by standing on a mechanized scale while simultaneously pressing the operating lever to create a paper tape weight record. At this point the group was split into "slow-learners" (15 animals) and "fast-learners" (the 19 animals). The remaining animals, characterized as "dull-normal", were sacrificed. By age 29 weeks 12 of the slow-learners and 14 of the fast-learners had developed palpable mammary tumors. Both groups were then conditioned to inspect their own lesions, count the number of tumors, and record this number by pressing a button the appropriate number of times with their noses. As expected, the slow-learners failed to perform satisfactorily. Of the fast-learners, five achieved better than 80% accuracy (one achieving a 100% rate), while the rest averaged 62% accuracy. All but the five best performing animals were then sacrified. At age 42 weeks all animals presented with scruffiness, mange, and large tumor masses. The one best performing animal was trained to perform necropsies and read the resultant slides. The present paper was written by that mouse before she, too, was sacrificed.

A SOCIOLOGICAL AND BIOLOGICAL STUDY OF POOL LIFE

S. FLOWERS
Rehovot, Israel

This research was carried out in two locales: Berkeley, California, and Rehovot, Israel. Certain similarities were noted in both pools, the various forms of wild life being identical; only the cries they uttered and the mating calls showed basic linguistic differences.

The pool has a total area of 1234 square feet and contains 1,300 gallons of water; 13.297 gallons of that water have been drunk by aspiring, perspiring and expiring swimmers; the maximum depth of water is 3.5 meters and is the only place where a scientist can go off the deep end without hurting anyone's feelings (unless he lands directly on someone), x families are members, y people are willing to pay for visitor's tickets each summer, and at least z people were unwilling to pay and tried to get in by other means. Such facts can easily be ascertained from the Royal Institute of Statistics or by using one of the many computers on the campus at either Berkeley or Rehovot.

There are two species of pool users—male and female. The females can be tabulated thus:

near misses	(aged 0–13)
misses	(13–18)
someone's dear missus	(18–80)

These divisions are not always clear-cut and there are exceptions to the above classification. It is more difficult to classify the males. They can be roughly divided into boys and men. The only trouble is that the boys try to act like men and the men try to act like boys. The easiest way to differentiate them is that the boys have hair on their heads, while the men have it on their chests.

Apart from the horizontal division above, we may also perform a vertical division into the following categories:

a) The monostroke
b) The polystroke
c) The sunstroke

A rich variety of wild life has appeared in this habitat in an amazingly short space of time. Among the many new species the following have been observed:

HIPPOPOTAMI (Hippopotamidae) puff and pant as they swim around clumsily.

ELEPHANTS (Elephantidae) trumpet and spout water over anyone within range.

TURTLES (wrongly known as tortoises; chelonia), are slow and sure as they do their daily kilometer or more.

SEALS (Phocidae) swim gracefully and effortlessly, their sleek heads breaking the smooth surface only for air, and they are capable of covering vast distances.

TADPOLES (Ramidae) are those plump bright-eyed little creatures that flash in and out of the paddling pool. It is usual to find a croaking frog or ungainly stork in the vicinity.

PORPOISES (Odontoceti) are easily recognized. They resemble children as they frolic and splash all over the place.

POOL LIZARDS (Lacertidae) are those languorous brown creatures seen stretched out around the pool. They are usually the female of the species. A noted anthropologist informed me that this is perhaps due to the fact that the males of the species are too busy studying the few perfect female specimens and are therefore more likely to be found slithering in and out of the pool.

(Gourmets please note: lizards are particularly succulent when cooked. Those that have been slowly broiled, after having been basted in good quality cream or oil, look more attractive and have a better appearance that those fried in cheap oil. They also tend to burn when roasted too quickly.) ■

Maternal Behavior In The Domestic Cock Under The Influence Of Alcohol

Abstract. *When normal male domestic chickens were given a single dose of grain alcohol and then exposed to newly hatched chicks, they assumed maternal behavior. The same behavior can be elicited by the administration of prolactin, but the results of these experiments suggest that maternal behavior in the cock is not exclusively dependent on hormonal mechanisms.*

We were led to the investigation of this phenomenon by an old custom, prevalent among Hungarian farmers, of transferring newly hatched chicks from the hen to a "drunken cock." The farmers justify this transfer by the argument that: (i) the cock is better than the hen at defending the chicks from predators; and (ii) such transfer frees the hen for returning to the commercially desirable egg-laying cycle.

Joseph K. Kovach
Menninger Foundation
Topeka, Kansas

From *Science*, Vol. 156, 5/12/67.

Increasing Abnormality in the Sexual Behavior Pattern of the Male Black Widow Spider

W. HENRI KREICKER, D. E.*
1005 South Country Club Drive
Warsaw, Indiana 46580

Over the years I have been an amateur entomologist, as a hobby. The study of insects is indeed a fascinating subject. There are almost countless genuses. One could spend a lifetime studying butterflies, for example, and still not know all that there is to know about this species of insect.

More recently I have given over to the study of the arachnids or spiders, which are — strictly speaking — not insects, and in particular, Black Widow spiders (Latrodectus mactans).

I have made some remarkable observations. I say, remarkable because in checking through the works of Fabre, Gertsch, Emerton, Bristowe and Comstock, I found no reference to the phenomena I have observed, which has to do with the male of the Black Widow spider family. The male is considerably smaller than the female.

It is generally known that the female Black Widow, which is venomous, usually destroys the male immediately after mating. Frequently she devours the male after killing it. I have observed this practice on several occasions. Authorities on spiders have given no satisfactory explanation of this androcidal tendency, nor have I one.

The disposing of the male is immediately after mating, while the male is completely spent from the orgasm. There is no respectable waiting for the ardor to cool.

One authority has suggested that the female Black Widow is invariably disappointed in love, since the male is so much smaller, and exhibits her utter contempt by promptly destroying the male.

Another researcher speculates that the female Black Widow may be completely carried away with ecstasy and with unbridled emotion destroys the male. Among human beings the saying, "I love you so much I could eat you", is not uncommon. The annals are replete with cases of teen-agers so-called necking wherein what practically amounts to mayhem was perpetrated, frequently stitches having to be taken.

Kucharov has suggested that the female Black Widow may be completely emotionally unstable and as a result is filled with mistrust. Fearing the peregrinations of the male during her gestation period she may put her mind at ease by destroying him, then cozily hatching her brood.

A French authority has suggested that the female Black Widow may be likened unto Shah Jehan of Agra, India, who commissioned the Taj Mahal to be built and upon its completion destroyed the builder so that he could never construct anything more beautiful than the Taj Mahal. However, this explanation hardly seems plausible.

It is quite obvious that the libido of the male Black Widow is strong or else it would not yield to the mating instinct with almost certain death staring it in the face. This would be uxoriousness beyond the line of duty. On the other hand, the male may have been conditioned, down through the ages, to believe that death is not too dear a price to pay for the utter gratification of mating. One would have to mate with a Black Widow spider to know for certain whether it was worth paying with one's life. It would appear exorbitant. The Italians, however, have a saying, "See Florence and die" (referring to the city, of course).

By careful scrutiny of the minute creatures I have learned that there is a growing tendency upon the part of the male Black Widow to weigh the pros and cons of mating. Some of the young bucks have adopted a "fools rush in where angels fear to tread" attitude. As a result, there is a lessening of the Black Widow spider population. This is favorable since they are dangerous to mankind, their bite (of the female)often resulting in death, albeit it is decreasing study material.

However, what might be termed an unbiological situation is becoming more prevalent among the male Black Widows. More and more the males are turning to each other for sexual gratification, in what would be termed homosexualism among the human species. The male spiders with palpi interlocked, as in a Japanese kendo match, is indeed an amazing sight.

Somebody once said, "Human nature has not changed since the first human." Likely spider nature has not changed since the first spider. But, it is interesting to conjecture whether on some far distant tomorrow the (1) species will become extinct (like the Great African Scaled Sloth which was too indolent to mate), the (2) female will see the error of her ways, and — superinduced by the dread of being a spinster — will adopt a policy of "live and let live", or the (3) male will adroitly develop a "touch and go" procedure, leaving the female with murder in her heart as he rapidly departs after mating with an "it was fun while it lasted" attitude, or what passes for an attitude among spiders.

Perhaps some observer in 5000 A.B. (After the Atomic Bombing) will know the outcome of the erstwhile precarious love-life of the male Black Widow and the Black pre-Widow too. ■

* Distinguished Exterminator

THE BATTLE OF THE TEAT*

MICHAEL KELLY, M.D.

The Australian kangaroo, like other marsupials, is born in an immature state. After birth it crawls up its mother's belly into the pouch where it firmly grasps the nipple with its jaws. There it remains fixed for many weeks until it can fend for itself.

These facts have been known for 150 years, but a contrary belief has been cherished by many people in the Australian outback. In 1924 Frederic Wood Jones, then Professor of Anatomy in Adelaide, spoke in a public lecture of the birth of the kangaroo. He was assailed in the press by a number of selfstyled "practical men" who pitied the ignorance of the theoriest but recently arrived in Australia. Wood Jones did not reply, but another scientist took up the challenge and showed six of the bushmen a specimen of a kangaroo embryo *in utero*. They laughed this to scorn as a fake, and invited Wood Jones himself to prove his theory; but at first he ignored the challenge.

One of the six men offered £100 to the Children's Hospital if he could prove his case. Wood Jones was tempted. He got together a large number of specimens, and put on a complete and convincing demonstration in the Anatomy Lecture Theatre. The demonstration was well publicized, and a large crowd – mostly of medical students and bushmen – turned up. We third-year medical students were on duty as demonstrators. But the bushmen were not convinced. Every argument of Wood Jones was met with a "practical" reply starting with "all this theory . . ." Wood Jones' patience was sorely tried, and in the end he made a few mildly sarcastic remarks. The following account of this occasion is taken from the *Adelaide Medical Students Review* of Nov. 1924.

"The story of the Gallant Six will go down in history. They will be immortalized in legend, for the manner in which they resisted the futile attacks of science. The University had been haunted for weeks by strange bearded men who carried small bottles in their hip-pockets. One carried an apple from which grew a wart-like protruberance; he proclaimed: 'If apples grow like that then kangaroos do.' Another had personally shot and skinned 30,000 kangaroos in 12 months and had never seen an embryo except upon the teat.

"On the fateful afternoon a mighty army of Practical Men turned up to do battle. One of them exclaimed audibly as he entered the hall: 'Me for the teat.' Respectful silence greeted the efforts of the Theorist; some of the 'stoodents' thought him victorious. But no; the Six were still untouched. At the first word ('All this theory . . .') it was evident that no impression had been made – that no risk had been taken when the offer was made to the Children's Hospital.

"One Practical Man drew from the Theorist the admission that the embryo when first seen on the teat was fully formed. The practical man had seen them on the teat 'half-formed, aye, and quarter-formed; no longer than a grain of wheat.' This, he said, proved that the embryo flowed out of the nipple like milk, solidified, and then shaped itself into a recognizable animal. But another member of the Practical Clan admitted that he had often seen, just before delivery, congealed drops of a "milky substance" (colostrum) on the nipples of mares. It is strange that foals do not grow on the teat.

"One keen observer claimed that he had often seen in the kangaroo a tube from the ovary, running through the belly wall to the teat; on one occasion had observed an embryo working its way along the tube. He was shown a specimen of the belly wall of a female kangaroo, cut into serial sections; it possessed no tube. His reply was: 'the belly wall's very thin, and if the embryo gets up against it, God will do the rest.' One Practical Man expressed his contempt for the despised 'stoodents,' who 'had never seen an adjectival kangaroo in all their sanguinary lives . . . Lot of hoodlums; I wouldn't let 'em touch me dorg if it was sick.'

"One of the strongest supporters of the teat theory had a private interview some days later with the specimens. He came away convinced. However, he had never before seen the inside of a kangaroo, though he had shot and skinned a great number!

"And so the battle of the Teat ended. Victory crowned the practical men. Patriotic Australians could not allow the foreigner within their shores to insinuate that the Australian kangaroo was less than it seemed. Secure behind a safe shield of ignorance they laughed in their beards at the vain theories of university 'perfessors' and all such cranks." ■

---

* from A MEDICAL BULLETIN vol. IX No. 1.

A BIOLOGICAL LASER

MICHAEL E. FEIN
and DIANE C. MILLER

One of the authors recently invested several years of his life in the development of a laser that ran on hot air,[1] only to find out that the job had been done earlier and far more effectively by a classified military-funded research program.[2] Despondent, the author turned to drugs and communism. As a result, we are able to report here the first achievement of population inversion by a biological system. This also appears to be the first recorded realization of coherent radiation from a population of completely incoherent radiators.

It will become apparent to the reader that this investigation required resources that could only have been provided by the People's Republic of China. The cooperation of the Chinese was most generous, but the delicate nature of diplomatic relations did require delaying the publication of these results until the present. If not for a hitherto-unpublicized side trip to Peking by a well-known American labor leader ostensibly enroute to North Vietnam, this publication might still be impossible.

The essential process utilized was the hyperactivation of luciferase by lysergic acid diethylamide. The experiment was performed in a large population of common fireflies,[3] which were placed on their heads by immersion in an aqueous solution of LSD-25 (.0001M), coactivated by dimethylsulfoxide (.001M). Since preliminary measurements had indicated that the inversion thus achieved is exceedingly weak, approximately $2x10^6$ fireflies were required.[4]

The fireflies were aligned near the optic axis of a ten-thousand meter unstable resonator by stretching a fine wire between the mirror centers[5] and placing the fireflies on the wire. It can easily be shown that the cycloidal shape assumed by the stretched wire approximates very closely the normal to the constant-phase surfaces of the principal cavity mode, so that if the fireflies could be aligned with sufficient accuracy to the wire they would all radiate into the principal mode. The desired standard position of a firefly relative to the alignment wire is indicated in Figure 1.

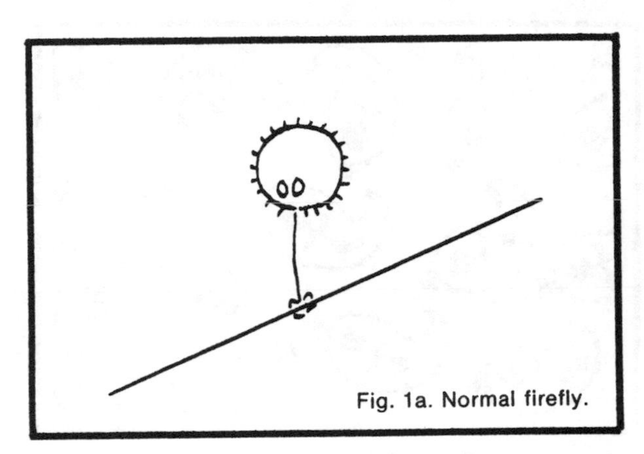

Fig. 1a. Normal firefly.

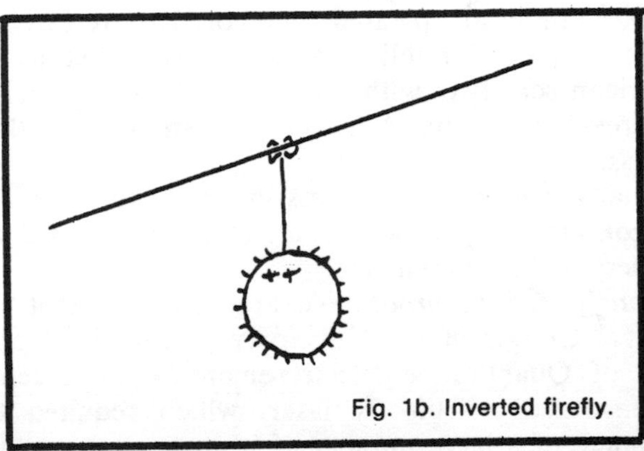

Fig. 1b. Inverted firefly.

The mean lifetime of fireflies in the inverted state was found to be approximately thirty minutes, after which they usually fell off the wire, so that it proved to be difficult to align all two million fireflies with sufficient speed. This difficulty was overcome by enlisting the cooperation of 20,000 members of the People's Institute for Glorious Anti-imperialist Technology. Each cooperating scientist was responsible for processing and aligning 100 fireflies. Several of the scientists who were insufficiently prepared in Mao's thought became themselves inverted, but after elimination of these side reactionaries it was possible to achieve adequate alignment.

The exact length and spectrum of the output pulse achieved is uncertain, as the spectrometer and photodetector were vaporized, but the well-cooked state of the fireflies makes it evident that the laser threshold was exceeded. Figure 2 is a reproduction of the far-field pattern sketched by one of our Chinese colleagues from a persistent retinal afterimage (the observer described this image as a "way-out picture", which nearly resulted in our misunderstanding his true meaning).

Fig. 2. Far-field pattern.

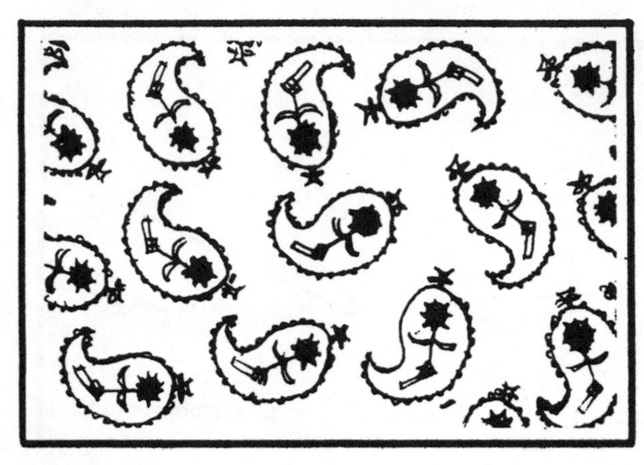

The authors are pleased to report that researchers in the People's Republic are just as concerned as are American scientists with the maximal use of natural resources. Appendix A is ample testimony to their success.

Finally, we wish to express gratitude to Peter and Margot van Schaick of Toledo, Ohio, who raised the fireflies in their basement.

Note added in proof: We are informed that Dr. Michael Craigie of the University of Illinois Department of Quantitative Electrosemantics has recently demonstrated a fruit fly laser, which required the alignment of a large number of white-eyed *drosophila* upon the teeth of a brilliant orange zipper. Measurements taken with this apparatus are expected to answer the question: "If time flies like an arrow, what do banana flies like?"

APPENDIX A

Sweet Plum Firefly[6]

2000 pre-cooked fireflies
2 crushed cloves garlic
1-2 teaspoons ginger sherry
pinch each salt and MSG
2 shakes pepper
2 teaspoons beaten egg
2-3 tablespoons self-raising flour
deep peanut oil
For plum sauce:
- 2 tablespoons canned plum sauce
- 3-4 tablespoons hot chicken stock.
- 1 tablespoon dark sugar crystals
- 1 tablespoon hot water

Place the fireflies in a bowl. Add the garlic and then the sherry and seasonings. Work them well into the individual fireflies. Next add the beaten egg and work it in too. Discard the garlic. Dip the fireflies in the flour and shake off excess. Work the flour well into the fireflies.

Meanwhile have the oil heating to 375°F. Drop the fireflies into it and cook to a golden color, separating the fireflies during cooking if necessary. Drain, place on a heated serving dish, and hand the hot sauce separately.

To make the plum sauce:

Blend together the canned plum sauce and stock. Dissolve the sugar in the hot water and add it. Heat together, then rub through a sieve. ■

[1] Applied Physics Letters *14*, 337 (1 June 1969).

[2] E. T. Gerry, IEEE Spectrum, Nov. 1970, pg. 51.

[3] The rain
tries without avail
to quench your lamp
and the rushing wind
but makes it glow
the more.

I believe
that if you flew
up to the sky
you would twinkle
as a star
beside the moon.

Li Po

[4] After several unsuccessful attempts to count the population by conventional methods, a photometric estimation technique was developed.

[5] A wire of the required tensile strength was drawn from an alloy ingot left over from the drill-string development program of the Mohole Project. We regard this as an example of practical spinoff from basic research fully as important as the use of missile nosecone materials to make breakable cookpots.

[6] See H. Burke, *Chinese Cooking for Pleasure*, Paul Hamlyn Ltd., London, c. 1965, for a collection of related material.

ON A NEW POWER SOURCE

F. G. HAWKSWORTH

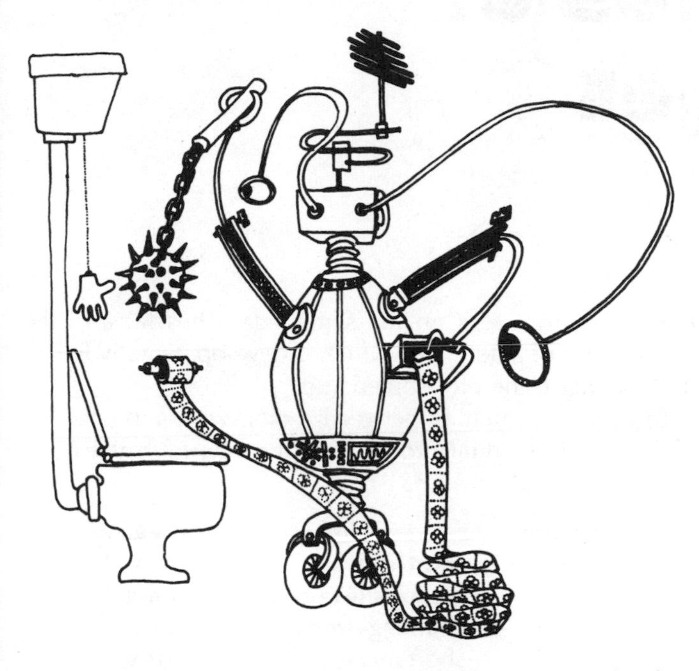

In these days with ever-increasing demands for power, it may come as a surprise to learn that a completely untapped power source exists throughout the West. Why not harness the explosive power of the fruits of the ubiquitous dwarf mistletoes (*Arceuthobium* spp.) of western conifers? These parasites have long been known (2), but their potential importance as a power source has not been recognized. Each dwarf mistletoe fruit contains a single projectile-like seed (about 1/10 of an inch long) which at maturity is ejected at speeds of about 90 feet per second (4).

The following discussion shows the potential energy now being lost through wasteful seed dispersal in just one part of the West. For this paper I have chosen the power-poor Southwest (Arizona and New Mexico) for an example. Here it has been determined that ponderosa pine dwarf mistletoe occurs on 2.5×10^6 acres (1). Estimated seed production by this same mistletoe (3) averages about 25×10^4 seeds per acre. Thus approximately 6×10^{11} mistletoe seeds are produced annually just in the ponderosa pine type. Estimates of the force of the seeds based on their weight (2.7×10^{-3} grams) and velocity (2540 cm/sec.) indicate that each develops 8×10^{-8} horsepower. When we multiply this times the number of seeds produced we have 48×10^3 horsepower or (among friends) 50,000. Converted to more conventional terms this represents 67,000 kilowatts!

It should be kept in mind that this is the potential of only one part of the West. Vast additional sources exist in mistletoe-infected pine, fir, larch, and hemlock forests throughout the West.

It might be only fair to mention some slight difficulties in the potential development of this resource, but none of them need be considered insurmountable. The harnessing of the power of billions of tiny fruits scattered over millions of acres presents somewhat of a problem because the energy from an individual fruit may have to be harnessed during about 2×10^{-4} seconds when seed ejection occurs. Another minor handicap is that this power source would be available only during the seed dispersal period which lasts about 2 to 3 weeks. But with application of typical American ingenuity and know-how, a multi-billion dollar crash program for development of this resource into a year-around power source could be expected to produce spectacular results.

The harnessing of this power source would have many national advantages — 1, it would enable us to beat the Russians at another game because their mistletoe power resources are puny in the extreme compared with ours; 2, by directing the seed power to more useful purposes rather than (as nature intended) toward debilitating our forests, our trees' sense of well-being should be enhanced; and 3, it would negate the need for additional Colorado River dams and thus help "Save Grand Canyon". ■

Literature Cited:

1. Andrews, S. R. and Daniels, J. P. 1960
   A survey of dwarfmistletoes in Arizona and New Mexico. U.S. Forest Serv., Rocky Mtn. Station, Sta. Paper 49, 17p.
2. Gill, L. S. 1935
   Arceuthobium in the United States. Trans. Conn. Acad. Arts & Sci. 32: 111-245.
3. Hawksworth, F. G. 1965
   Life tables for two species of dwarfmistletoes. I. Seed dispersal, interception and movement. Forest Science 11: 142-151.
4. Hinds, T. E. and Hawksworth, F. G. 1965
   Seed dispersal velocity in four dwarfmistletoes. Science 148: 517-519.

PROF. LAWRENCE M. DILL
Dept. of Biological Sciences
Simon Fraser University
Burnaby, B.C, Canada V5A1S6

Behavioral Genetics Of The Sidehill Gouger

The sidehill gouger (Ascentus lateralis), is a unique animal native to the mountainous areas of British Columbia. It possesses two short legs, on the same side of the body, which enable it to stand and walk about on hilly terrain. Thus the sidehill gouger is beautifully adapted to its particular ecological niche.

Within any one population of sidehill gougers two distinct morphological types appear. One of these has the short legs on the right side of the body and is thus able to walk around mountains in a clockwise direction only. The other type, having short left legs, can walk only in a counterclockwise fashion.

These two are not distinct species, since they often mate with one another. This is no mean feat since one or both of the animals, having approached from different directions, must back up to effect copulation. Such complex mating behaviour suggests that isolating mechanisms between the two morphotypes have not evolved.

Laboratory investigations were undertaken to elucidate the genetic basis of dimorphism. Only a short summary of the results will be presented here, as a major paper will appear in a forthcoming issue of "Acts Artifacta."

The morphology and corresponding behavior of the animals is controlled by two sets of co-dominant genes, each having two alleles. The first locus determines whether anterior or posterior legs are short: the homozygote AA has two short forelegs; the homozygote PP, two short hind legs; and the heterozygote AP, one fore and one hind leg short. The second locus determines on which side of the body these legs appear.

When two individuals of either parental stock mate with one another, all of the progeny produced are of the same genetic constitution as the parents. However, as many young are aborted during pregnancy as are delivered. Examination of these reveals one half of them to have 2 short back legs on the same side of the body and the others to have two short front legs on the same side. Thus 50% of the zygotes are of a genotype which is developmentally lethal, and the adults therefore breed true.

In contrast a mating between a clockwise and counterclockwise individual produces 3 phenotypes in the F_1 as follows:

Proportion	Phenotype	Genotype
.25	short front legs	AADS
.50	diagonally opposite short legs (rockers)	APDS
.25	short rear legs	PPDS

By means of breeding experiments the "rockers" produced an F_2 containing 4 lethals, 4 rockers, 2 clockwise parentals, 2 counterclockwise parentals, 2 with short front legs and 2 with short rear legs. This is regarded as the critical test of the hypothesis presented for the genetic constitution of the species.

The experiment could only be conducted after special shoes were fitted to these animals, as they otherwise have a tendency to fall over, either onto their faces (and suffocate), or onto their other ends (and starve to death). In the field they would quickly be selected against.

The behaviour of the individuals with short front legs or short rear legs is of considerable interest. The former are able to walk only uphill, eventually falling off the tops of the mountains to certain death on the rocks below; the latter walk only downhill, congregating in river bottoms. There, it can be easily demonstrated, they breed true, although producing 50% developmental lethals each generation. Some theoretical taxonomists suggest that they may be the evolutionary progenitors of the present day hyaenas, indicating perhaps a wider global distribution of the sidehill gouger in the past. Reduction of its range may have resulted from the very high genetic load (developmental lethality) carried by the population. ■

Webb, B.F.

BROADBILLED SWORDFISH FROM TASMAN BAY, NEW ZEALAND

N.Z. J. of Marine & Freshwater Res. 6 (1&2): 206 (1972)

On 4 January 1971 Mr. F.J.P. Kellor shot with a .308 rifle an adult broadbilled swordfish *Xiphias gladius* Linnaeus, in Wairangi Bay . . . On examnation, the swordfish showed no external signs of injury.

* * * * * * * * *

WHY NOT EAT INSECTS?

By A.V. Holt. 1885

This curious little book, reprinted in 1967, consists of 99 pages. It includes some remarkable suggestions for menus.

BOVINITY

WILLIAM F. JUD

Geological anomalies affect animal action. Cliffs, for example, make animals detour. Rivers make them swim. Subsurface structures align them. This latter fact permits geophysical evaluation of ore bodies.

A study has been initiated to determine alignment of cows and the diurnal variations in the intensity and polarity of this alignment. It was found that in the morning cows are generally aligned with their heads (+) toward the pasture, and in the evening just before sunset their tails (−) point in that direction. In addition, there is a definite drift and transport phenomenon in the direction of their (+) ends. Flow lines are sharpest in all cases along the path leading to the barn, and are randomly spread in the pasture proper. The flux causing this alignment strengthens considerably below 15°C during strong wind. The polar effect of wind is such that the cows are oriented with their (−) end toward the wind.

Bovinity flux was first recognized in the Missouri Ozarks. It is the custom there to graze stock over areas of underground mining. Since the original observations on diurnal alignment of cows were made in this locality, namely above ore bodies, it was desirable to verify this correlation in other localities.

Experiments performed in various geographic areas of known geologic structures with cows of different breeds showed an excellent correlation between the mining camp geology and bovine orientation, as shown in Figures 1 and 2.

In every case, cattle kept over anomalies were polarized at least twice daily. The conclusion drawn from this observation was that wherever cows align one should dig and would strike ore.

Bovinity rays vary as the inverse square of distance from source. Small changes due to topography, rock density, and elevation are removed from bovinometric calculations through the Bouguer-Holstein correction. Field intensities are mapped in 0.00 μ (milli-moo) units.

Since the development of the automated Airborne Bovinimeter the value of Bovinity Flux has increased considerably, especially for mapping sea cows that live over offshore oil and natural gas fields.

SUMMARY: A herd of cows in a field align themselves with each other over ore bodies (bovinity flux). ■

****

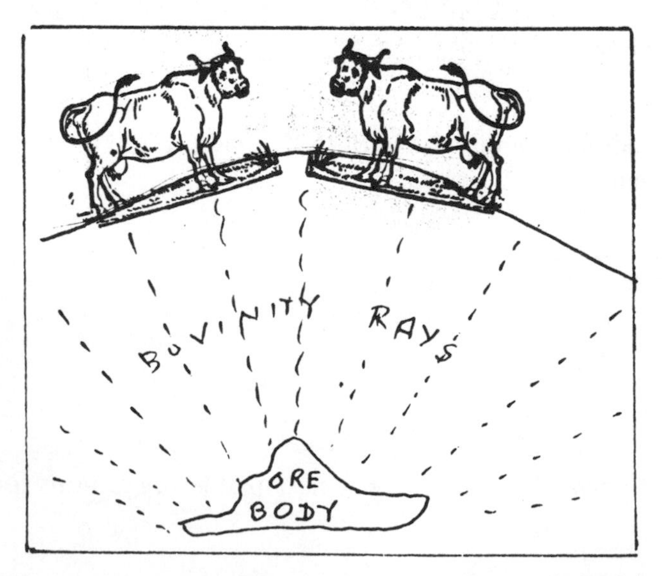

Fig. 1. Sectional view of Cows in Spherical Alignment in Bovinity Field emanating from buried Ore Body.

Fig. 2. Cows in Linear Alignment in Bovinity Field radiating from recently Active Fault.

James A. Kushlan, Marilyn S. Kushlan
Homestead, FL
Joan A. Browder, Peter A. Schroeder
Miami, FL

SNOWY HERON OR WHITE EGRET, by John James Audubon, American School

The cattle egret (*Bubulcus Ardeola (Egretta) ibis*) has undergone marked expansion of its range (Sprunt, 1955, Smith. Misc. Rpt. 4198: 259-276). Although the population expansion has been well documented, some of its causes have only recently been elucidated (Blaker 1971, Ostrich Supl. 9: 27-30). We discuss here the adaptive basis permitting this expansion.

Population fluctuations are sometimes accompanied by or may be caused by genetic changes (Carson 1967, pp. 123-137, in Lewontn (ed.), Population biology. Syracuse Univ. Press). Such genetic changes often accompany the isolation of a small population. In such populations, genetic drift or the effect of chance is often taken to be the major factor influencing changes in gene frequency. Isolated populations in an evolutionary bottleneck (Nei, Maruyama and Chakraborty 1975, Evolution 29: 1-10) can also be subject to gene fixation. Relatively few incidences of selection can have a weighty impact on a population having a reduced gene pool. We propose analagously that the critical role of periods of high electivity be called the survivors principle and suggest that the cattle egret represents a stunning example of this.

Around 1910, the population in Egypt was reduced to less than 500 individuals. Immediately after the restriction of plume hunting by the British administration, the cattle egret underwent its dramatic range expansion that is characterized by its association with large mammals. A segment was restricted to a roost located in the zoological gardens of Cairo. The survivor principle provides that those birds that perservered the zoo period provided the genetic stock responsible for the later expansion.

We propose that this occurred through behavioral phenomenon we call negative fecaltropism. During the period of range contraction, the zoo provided suitable foraging habitat where the egret was undoubtedly associated with large mammals such as elephants (*Elephas africanus*), with which they were forced to feed. There is considerable risk associated with foraging in close proximity to such a large animal. This risk is not distributed uniformly along the animal but varies along an anterior to posterior gradient. Risk is particularly great for those birds foraging beneath the hindquarters of such beasts (Fig. 1). Such impact would place a premium on being associated with the forequarters of the larger animal. Those egrets foraging there, by moving along with the elephant, encountered increased opportunity to obtain invertebrates made more conspicuous by the elephants movements. Feeding on these prey increased net return to the egret. Natural selection through negative fecaltropism led to the commensal foraging behavior later to become associated with the more ubiquitous cattle.

To determine the significance of negative fecaltropism, it would be necessary to set up a field experiment. If cattle egrets partitioned the cow and tended to cluster at the head rather than under the tail, negative fecaltropism clearly would be indicated. Fortunately we did not have to conduct the experiment since Grubb (1976, Wilson Bull. 88(1): 145-148) independently determined that this was true. Although he concluded that egrets occurred more frequently in the head region because of increased foraging return there, as we have shown, this is a secondary attribute of negative fecaltropism, which thus is the ultimate selective factor.

We feel that the selective advantage of avoiding the rear coupled with the increased energy input from forward commensal foraging provided the ecological advantage that permitted the cattle egret to expand into large herbivore-dominated habitats worldwide. ■

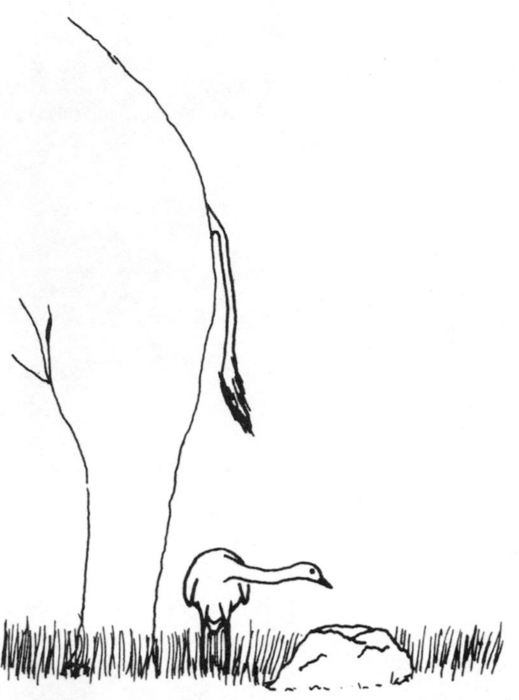

FIGURE 1. Cattle egret foraging near the rear of an elephant examining a recently deposited pile of feces that buried its more posteriorly foraging associate.

QUOTES

G.H.E. Hopkins and Miriam Rothschild

AN ILLUSTRATED CATALOGUE OF THE ROTHSCHILD COLLECTION OF FLEAS

British Museum, London 1962, 168s.

K.E. Goard

INFECTION BY *LEPTOSPIRA POMONA* **CONTRACTED FROM PIGS BY MOUTH-TO-MOUTH RESUSCITATION.**

Med. J. Australia, 1961, 1:897-898, (June 17).

J.B. Appel

THE RAT: AN IMPORTANT SUBJECT

J. Exp. Analysis Behavior, 1964, 7, 355

"In addition to living a relatively short time, rats tend to become ill and sometimes interupt experiments by dying."

B. Zondek and I. Tamari

EFFECT OF AUDITORY STIMULATION ON REPRODUCTION. IV. EXPERIMENTS ON DEAF RATS

Proc. Soc. Exp. Biol. & Med., 1964, 116, 636

"... The experiments show that auditory stimuli during the copulation period do not cause a decrease in fertility in deafened rats in contrast to normal animsls."

Brit. Med. J., March 16, p. 743 (1963)

"On March 5, 1963 pickets appeared in front of the White House and called on Mrs. Jackie Kennedy to clothe her horse. They belonged to the Society for Indecency to Naked Animals (SINA). The first preposition should have been "against," but I gather they can't change it because of legal registration of the original society. According to the Guardian, SINA has designed bikinis for stallions, petticoats for cows, and knickers for bulldogs. Just think of the dogs in Hyde Park tearing the pants off each other."

Jerusalem Post, March 21, 1963 (Quotation from Lord Shackelton's Speech in the House of Lords)

"Cannibals in Polynesia no longer allow their tribes to eat Americans because their fat is contaminated with chlorinated hydrocarbon. Recent figures published show that we (English) have two parts per million DDT in our bodies, whereas the figure for Americans is about 11 p.p.m."

Retreat into general objectives on which everyone can agree. From this higher ground you will either see that the problem has solved itself, or you will forget it.

Findings

that

could only

come from

highly

irreproducible

RESEARCH

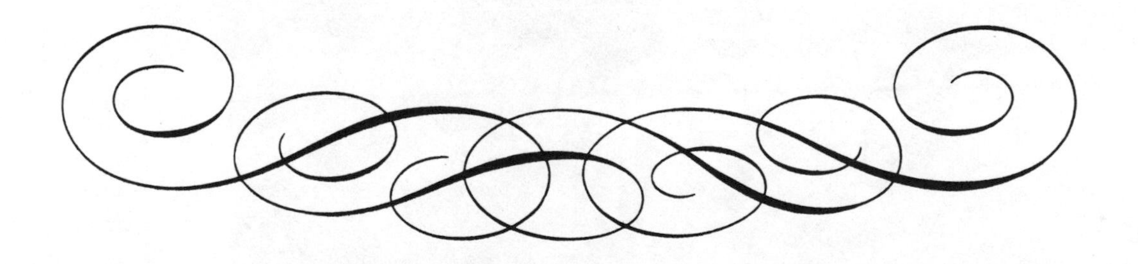

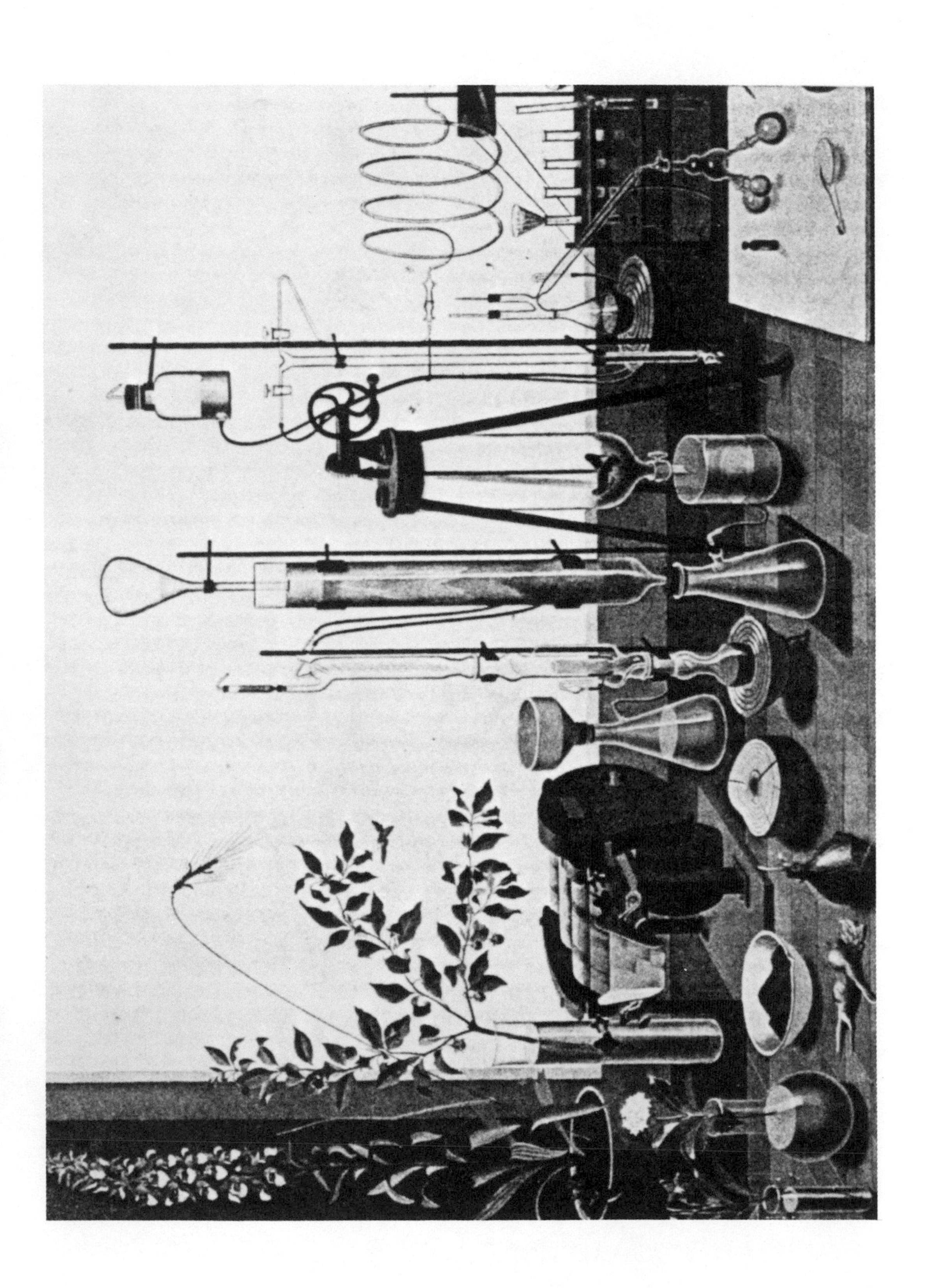

FOURTH GRADE BIOLOGY EXPERIMENT · MR.HIGGINS ANATOMY OF A FROG

PROCEDURE : The student shall dissect a frog and examine its internal organs.

Name of student . . . . E.MYRON FOGARTY

Labrotorie report on Cutting up a frog.

Plan. I will get a frog and a sharp nife, Which with I will cut opin the frog whilst he is alive. I will studdie carfulley his innerds, for the purpuss of gaining scientifik knolledge.

Part One. Mr. Higgens give me a frog, a nife, and a pan of stuff like jello. I grapt the frog by his hine legs and I beet his head on the edje of the lab bench so he wood not bight. When he was woosie and ready for scientific investigathun, I stuck thumtax in his foots and pinned him against the jello, with his bellie in frount. Part Two. Soon the frog begun ones agan to kick, and feering his eskape, I scientifikully rammed the shiv in his stumak. There was a stickie dark red flooid which ouzed out from whence I stabbed him. This was kind of fun so I druv the shiv in him agan. I noted that the more I stabbed him the more he kicked. Soon he begun to kick less and less each time I stabbed him and finnalley he kicked not at all. I went to Mr. Higgens and got a new frog. <u>Part Three</u>. I through what was left of the old frog in the wase basket. then put the new one in its place. I was determint to learn more from this frog by condukting a more scientifik experemint. I made too insessyuns akross his tummy and I peelt the skin away so I coold see better. It was kinda silverry inside so I cut deeper, and found his innerds, which I scientifikully scoopt out with a spoon.- <u>Part Fore</u>. I got a woodin handil fork, upon which I put the frog. I turned on a bunsin burnur and held the frog, who kicked not much too terrible now, above it. Soon the smell of roastid frog filled the labb and Mr. Higgens made me share him with othur kids

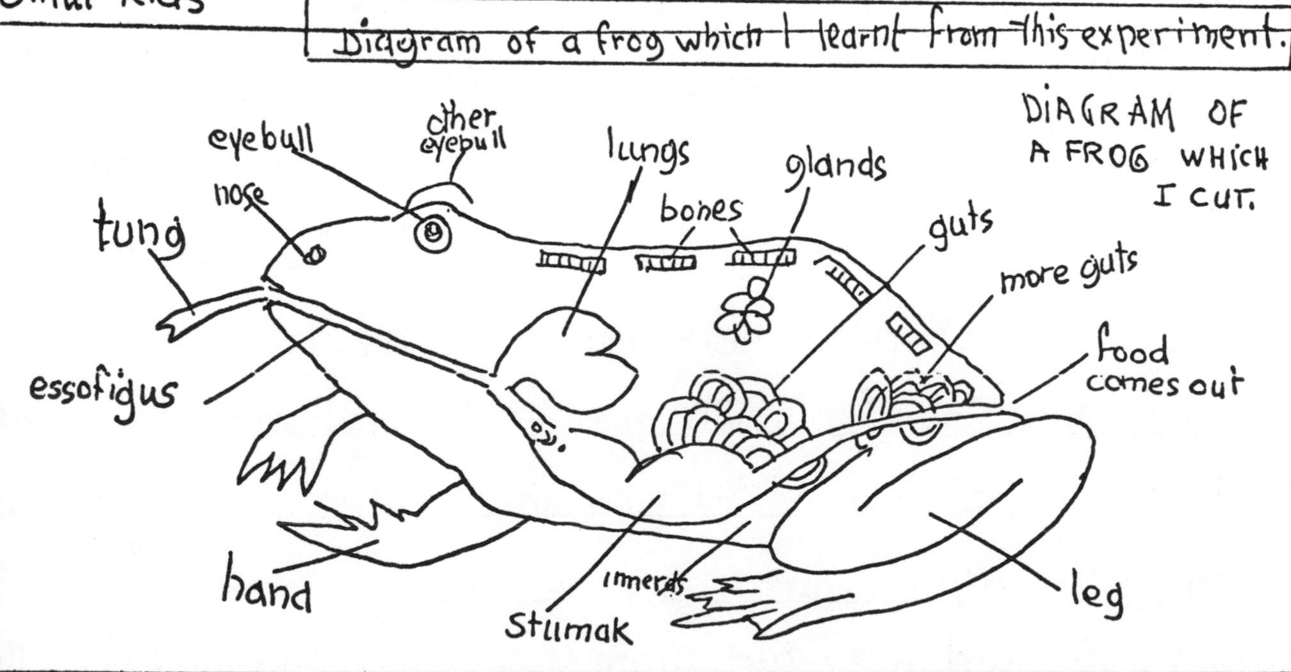

Reprinted from JIR, 1963, 11, 65 from a reprint in Synthetic American, 1962.

THE TRIPLE BLIND TEST

R.F., M.D.

At one time, an investigator working alone could produce significant research. But the ever increasing array of electronic gadgets is making it difficult for one person to record objectively all his erroneous observations. Many irrational leads are thereby missed.

Since it is now virtually impossible for him to blunder blindly without help, teamwork has come to the rescue with a perfect solution: the double-blind study.

The value of the double-blind study was first illustrated by studies of the LD-50 (lethal dose for 50% of test subjects) of cyanide. In earlier experiments it was considered satisfactory to give, and not to give, the drug to alternate subjects. But now control subjects are given a placebo of exactly the same size, shape and color as the cyanide tablet. By doing the experiments in this way, the clinical impression that cyanide is lethal is verified by observations uncluttered by bias.

The accuracy of the double-blind test permits evaluation of a drug with far fewer subjects than heretofore; in fact, in some studies it has been possible to reduce the number of subjects to zero*.

In evaluating the precision of this method, researchers have made an unexpected discovery – the therapeutic effectiveness of sucrose and starch placebos. The diversified effects of these simple carbohydrates constitute a major discovery in medicine.

Of particular interest is the remarkable potency of starch in pathological pain, with no indication that it produces addiction. It is equally noteworthy that starch does not affect experimentally produced pain.

This observation has brought about the realization that analgesics should be appraised only in relief of naturally occurring pain since man may be the only species that imagines suffering.

A curious sex difference has also been demonstrated in that red placebos are effective in males, but only blue or pink placebos are effective in females. A report of two hermaphrodites who responded only to indigo-colored placebos should be verified.

Placebos are now available for the treatment of headache, rheumatoid arthritis and female ailments. In addition, they are as potent as ethanol either as "psychic energizers" in depressed states or as "psychic de-energizers" in overactive states. Work is now in progress to ascertain whether they act by blocking oxidative phosphorylation, cytochrome-c reductase or alcohol dehydrogenase.

Recently, there have been alarming reports about the toxicity of placebos. But these reports concern patients treated only with yellow or green placebos whose marked side effects may make it necessary to discontinue such treatment. In fact, a number of cases of agranulocytosis have occurred with the use of green placebos.

SERENDIPITY

Over the past few years, a new concept has arisen in medical research. Therapeutic nihilists now feel that the best chance of therapeutic breakthrough in mental disease, cancer and hypertension lies in experiments so completely unbiased and randomized that an accidental discovery of importance may turn up. This principle – sometimes called serendipity – is well recognized in mathematics: Sir Arthur Stanley Eddington said, "We need a super-mathematics in which the operations are as unknown as the quantities they operate on and a super-mathematician who does not know what he is doing when he performs these operations."

Applying the randomization principle to experimental medicine has led to the TRIPLE-BLIND TEST: The subject does not know what he is getting, the nurse doesn't know what she is giving and the investigator doesn't know what he is doing. Half way through the experiment, randomization is increased by a process known as turnabout – the patient administers the drug to the investigator, and the results are evaluated by a student-nurse. The famous mathematician, Lewis Carroll, may have had the randomization principle in mind in the phenomenon he described as Jabberwocky or "unknowable actors executing unknowable actions."**

The chance that triple-blind testing will produce something of consequence is calculated to be at least as great as that of spontaneous mutation. This probability is about one times 10 to the minus sixth power per generation. But considering the large number of chaotic investigations now in progress, the chance of a significant breakthrough in the next few thousand years is not improbable. ■

* A novel presentation of the theory of errors in small samples may be found in the recent book by B. L. Smith entitled *The Statistical Treatment of Vanishingly Small Samples and of Nonexistent Data.*

** Annals of *Alice in Wonderland* and *Through the Looking Glass.* A more precise description of Jabberwocky follows:

> "Twas brillig, and the slithy toves
> did gyre and gimble in the wabe;
> all mimsy were the borogoves,
> and the mome raths outgrabe."

THE PILL

MISS MARY SULLIVAN

(A sort of an autobiographical note)

I have been a laboratory technician for nearly twenty years now, so I think I can say, in all humility, that I really understand most of what there is to know on the practical side of the subject. Actually, it isn't complicated at all really; it's frighteningly simple.

Back in the 1930's, there were any number of people working on the effect of colchicine, studying its effect on mitosis, which it inhibits apparently by interfering with the spindle behavior. Cells treated with colchicine start to divide, the chromosomes split, they line up on the equatorial plate, and then something goes wrong: the cell stops dividing and goes back into interphase again, with double the usual number of chromosomes. In fact, that's how many tetraploid races of cultivated plants are obtained.

More recently, Mazia and his students at Berkeley have found that 2-mercaptoethanol interferes with spindles, too[1], though in a rather different manner (they think it's something to do with sulfhydryl bonds and sol-get conversions). Well, so much for the chemicals involved.

Now, over the past 15 years or so, while I've been working with Katsumi and Lindberg—first in Washington, then for a year at the Karolinska Institute and now here at the Rockefeller—we have been trying out these things on sea-urchin eggs.

The usual procedure has been to fertilize the eggs, and then test a variety of concentrations of various agents under various conditions on the first few mitotic divisions of the egg before it develops into a blastula. After the Lindbergs settled here in New York and started going to Woods Hole every summer, they tried a few other invertebrates, too, and even some frog-spawn shipped up from North Carolina, but the results were all much the same.

This did not stop them from publishing, of course, since luckily the details of concentrations and optimum pH values differed from one organism to the next.

Then we got a high-school student to help us one summer, and one morning he added one of the mixtures to sea-urchin eggs that had not been fertilized. I think the boy simply forgot to add the sperm first, but he said he did it just to see what would happen.

Well, between five and ten per cent of eggs started to cleave! When Lindberg heard about it, he (or maybe Mrs.

[1] Mazia D., Mitosis and the physiology of cell division. In: THE CELL, ed. J. Brachet and A. E. Mirsky, Acad. Press, N.Y., 1961, 77–394.

Lindberg, I forget now, since it happened three summers ago) changed the proportions of colchicine and 2-ME, and this raised the cleavage rate to almost 50%. When Katsumi got back from Heidelberg that fall, he had only a week at Woods Hole before starting it again with Kerberle in New York; so he naturally switched over to mice from the medical school colony there[2]. And the mice did the same thing! All he had to do was to add the colchicine and 2-ME mixture to the drinking water of unmated females, and, if they were in heat at the time, they just got pregnant—at least, a high percentage of cases did. When they littered, all the young were females, of course; and they were all identical with one another and with the mother, because the genotype hadn't changed. They were just as fertile, too. In fact, up till now we have raised some 18 generations, without benefit of a male of any sort, poor things.

Then Lindberg showed me how to remove uteri and wash out eggs at various stages, and I started one of the most boring and tedious parts of my job in my life. It was terribly frustrating. Much of the time my timing was wrong and I found nothing, or the eggs got lost among all the mucus and epithelial cells and stuff.

But slowly we began to build up a collection of eggs at different stages, fixed and stained nicely, so that Lindberg, when he got time off from teaching and committees, could study them carefully under a microscope. He found that they behaved just like the sea-urchin eggs. The details were published in four papers in Experimental Cell Research last year, and two more are still in press. They are all by Katsumi and one or both of the Lindbergs, and they included me in the authorship of the mouse egg studies[3] as they used to do in the 1950's when we were studying muscle fibers (Maybe this isn't worth mentioning).

Well, about that time my father died, and my sister came over from Connemara. (Mother had died in 1942, in one of the raids on Manchester). Poor soul Ellen has always been so cross-eyed that everybody was always sorry for her. But nobody does anything to help her; all her life she has been lonelier than any soul had ought to be on this earth.

So last fall I tried with some of the 2-ME mixture in her coffee, two mornings in a row, and she hardly noticed anything except a stale taste; then I gave her a bigger dose, about 400 times what we give to our mice—and it worked!

Ellen is a good woman, and she has been always quite a model of virtue, what with being more plain than most girls when she was young. But she has missed six months now and it's beginning to show. So I had to tell Katsumi about it. And that is when he started planning on the "P" pill, as the Lindbergs promptly called it.

And that's when I got worried. They all talk about it at every coffeebreak in the lab now. They say that meiosis is suppressed and the polar bodies are just resorbed into the egg nucleus, so that it becomes diploid again. They think that implantation is normal, at least in mice; so why not in other mammals? Lindberg, with Anderson, with whom he used to work in Stockholm, has taken out a patent. He is going to discuss it with somebody in Dairy Science at the Department of Agriculture in Washington at the beginning of next week, if he can get someone here to take over his anatomy class. But they don't talk about cattle all the time.

They say, too, that all sorts of women who want children can have them now without sinning, whether they happen to be married or not, because there are no laws against such pills. Katsumi has made up the fifth batch of the tablets, in three strengths, and he plans to give them away free after high-school graduation this coming June.

I don't know what to think; I have been wondering about all the mercaptoethanol that occurs naturally in garlic which people eat so much of in Mediterranean countries, and I've looked up *Colchicum autumnale* and found that it, like garlic, grows along the seashore in North Africa and Asia Minor. I don't know what to think now.

I started this story objectively, hoping to keep my personal involvements out of it, but I got carried away. I've been a good and virtuous woman these 43 years, and this is the first serious trouble I've got into. And now, what shall I do? My time is getting short, the Lindberg girls Elsa and Karen (home from Swarthmore for the vacation) can't keep their breakfasts down this week, and school will soon be out.

I hope I have done the right thing in writing all this. ■

---

[2] Kerberle, H. & al. 1965. Biochemical effects of drugs on the mammalian conception. Annals N.Y. Acad. Sci. *123*, 252.

[3] Katsumi, M., Lindberg, D., Lindberg, L. M. and Sullivan, M. 1965. Suppression of meiosis and restitution nuclei in mouse and hamster oocytes by colchicine and 2-mercaptoethanol. Exper. Cell Res., *28*, 567–581.

Great Flops I Have Known

A high point of the SJCC promises to be the panel on Lessons of the Sixties. At that time we can consider why Data Processing is a multi-billion dollar industry, but most people have so little to show for it. One of the reasons must be the enormous number of systems that are designed, programmed, debugged, tested, and documented—but never run. They don't go into production for a number of reasons:

- No real time is left for production.
- The system is dependent on a large data base which has never been assembled.
- The person who insisted on the system has been hired away by Cogar.
- The system was built as a package, but nobody bought it.
- The system cost so much to build that everyone involved went bankrupt.

and so forth.

I think it's possible that fully half of the systems built have never done anything. And there must be an elite corps of 5-year systems men who have never touched one that works.

It is interesting to consider a few case studies. For obvious reasons, the names have been omitted. The essentials are all true.

Case Study One: The project was an active management information system for a merchandise mart. The system would perform central billing for items purchased from 200 private wholesalers. At billing time, the buyer would undergo a credit check, a payment verification, and be assigned a time and location to pick up his combined purchases. The computer was to run an automated conveyor network that would move all purchases for a particular buyer from the various points of sale to the loading dock assigned to him and make them all arrive on time.

The conveyor together with its queucing and switching system had already been built but nobody knew what its capacity was. The project was begun in the faith that a 360/50 could run the conveyor at the necessary capacity. A simulation effort was begun at the same time. Midway through the development project, the simulation was completed, showing that the average day's purchases would take six days to deliver to the loading docks.

The project was cancelled for some convenient reason having nothing to do with the real problem.

Case Study Two: Twelve people began work on a management information system for a large conglomerate. The stated goal of the system was to permit the organization to be run from the top. All computing facilities were to be grouped into the one system to take advantage of Grosch's law.

A considerable amount of ecstatic hand waving was performed over the concept of the computer's modeling certain metal markets. This would enable the company to buy its principal raw materials at advantageous prices and alter its inventory policy to reflect coming changes.

After almost two years of work it was conceded that none of the acquired companies had any intention of giving up its own computer center much less operational autonomy. Furthermore, no one had the slightest idea of how to model the market.

Case Study Three: A medium-sized city with a severe traffic problem decided to install a reactive computer system to control stop lights. A complete simulation was performed. The simulation indicated that traffic flow would be eased by 15% (of something).

The system was put into effect and traffic stopped dead. The simulation was run again and used to prove that traffic couldn't possibly have stopped dead. The system was abandoned. The simulation lives on.

Case Study Four: Three bidders were paid to submit detailed proposals for a large industrial development project. The procuring organization (the Government, who else?) specified that the project should be managed with the aid of an on-line Management Information System.

Since the MIS would have to function from the first month of the project, each of the bidders was instructed to develop their own system. The contract was awarded and two of the three systems were junked.

Maybe you have a story to tell. Maybe you'll send it to me care of MODERN DATA. Maybe I'll start an abortion clinic. ■

Reprinted from MODERN DATA March, 1970 with permission.

Dear Sir:

A patient's report of bioluminescent feces - "stools glowing in the dark" - has long been held to indicate the likelihood of significant psychopathology. In my own practice, I have found this symptom to be more reliable than a report of dental pruritis - "itchy teeth" - although not nearly as pathognomonic as a report of spontaneous combustion of flatus. I would like to report a recent case of pseudobioluminescent feces in an animal model, to alert my colleagues to a likely cause of this symptom.

On a July evening, at approximately 11 p.m. with an ambient temperature of approximately 75° the author was traversing a fenced enclosure when he noted the sudden onset of a squishy sensation between his bare toes. Closer inspection revealed the substance to be feces, recently passed by a 20 kgm., 44 week old Golden Retriever puppy. Observation revealed multiple areas of the stool to be glowing brightly.

Further research revealed that the Golden Retriever puppy had recently ingested approximately 45 grams of the material from a 115 gram Professional model "Moonlighter" Frisbee. This model Frisbee is constructed of a luminescent plastic material for play after dark.

Since the luminescent stools are obviously secondary to the ingestion of the glow-in-the-dark plastic this does not represent true bioluminescence, but rather "pseudobioluminescence".

The implications of this are clear. Physicians must add the possibility of occult frisbeephagia to their differential when a patient reports his stools "glowing in the dark". In these situations it would appear wise to obtain a stool for blood, ova, parasites and Frisbee. Any well equipped clinical laboratory should be able to make the distinction without difficulty.

Yours truly,
Charles Davant

QUOTES:

C. M. Pare.
MONOAMINE OXIDASE INHIBITION AND BRAIN MONOAMINES IN CLINICAL CONDITION
Biochemical Society Agenda, Nov. 1970, Dagenham, N. Y., p. 4.

"Studies on brain amines in patients who commit suicide have been somewhat disappointing".

****

W. T. Weber and W. T. Taylor
General Biology, Van Nostrand, Production N. J. 1968, p. 3.

"Paleontology is a science which deals with the study of extinct plants and animals that survive today as fossils".

****

V. Sladeček
A NOTE ON THE PHYTOPLANKTON-ZOOPLANKTON RELATIONSHIP
Ecology, 1958, 39, 547.

"It can be stated in terms used in saprobiology that the polysaproby was changed into alpha-mesosaproby".

****

L. Lindner
Proc. 8 Intern. Cong. Genetics 1949: 620.

"54 percent of the men and 22 percent of the women were able to move their ears. That the percentage figure is twice as high for men can possibly depend . . . on the fact that men are even in childhood more interested in sports . . .".

THEORETICAL ZIPPERDYNAMICS

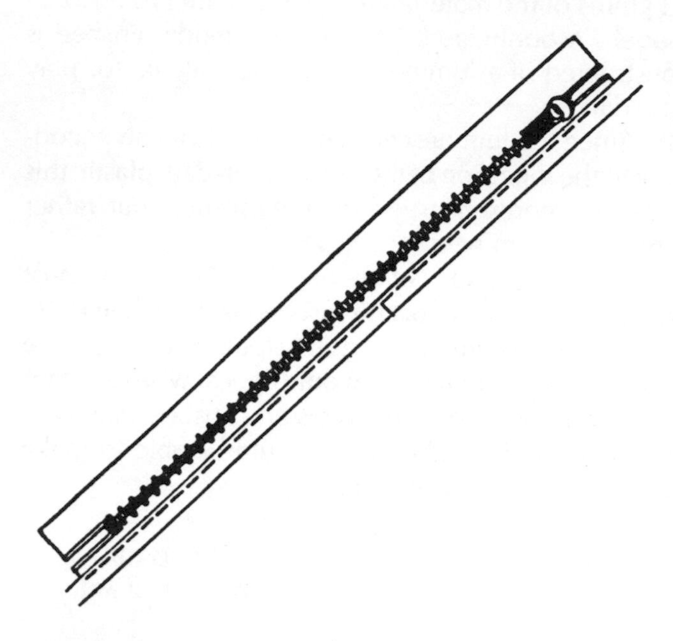

HARRY J. ZIPKIN
Department of Unclear Phyzipics
The Weizipmann Inziptute

INTRODUCTION

The fundamental principles of zipper operation were never well understood before the discovery of the quantum theory[1]. Now that the role of quantum effects in zippers has been convincingly demonstrated[2], it can be concluded that the present state of our knowledge of zipper operation is approximately equal to zero. Note that because of the quantum nature of the problem, one cannot say that the present state of knowledge is *exactly* equal to zero. There exist certain typically quantum-mechanical zero-point fluctuations; thus our understanding of the zipper can vary from time to time. The root-mean-square average of our understanding, however, remains of the order of h.

ZIPPERBEWEGUNG

The problem which baffled all the classical investigators was that of *"zipperbewegung"*[3], or how a zipper moves from one position to the next. It was only after the principle of complementarity was applied by Niels Bohr[4] that the essentially quantum-theoretical nature of the problem was realized. Bohr showed that each zipper position represented a quantum state, and that the motion of the zipper from one position to the next was a quantum jump which could not be described in classical terms, and whose details could never be determined by experiment. The zipper just jumps from one state to the next, and it is meaningless to ask how it does this. One can only make statistical predictions of *zipperbewegung*.

The unobservability of *zipperbewegung* is due, as in most quantum-phenomena, to the impossibility of elimination of the interaction between the observer and the apparatus. This was seriously questioned by A. Einstein, who in a celebrated controversy with Bohr, proposed a series of experiments to observe *zipperbewegung*. Bohr was proved correct in all cases; in any attempt to examine a zipper carefully, the interaction with the observer was so strong that the zipper was completely incapacitated[5].

THE SEMI-INFINITE ZIPPER

A zipper is a quantum-mechanical system having a series of equally spaced levels or states. Although most zippers in actual use have only a finite number of states, the semi-infinite zipper is of considerable theoretical interest, since it is more easily treated theoretically than is the finite case. This was first done by Schroedzipper[6] who pointed out that the semi-infinite series of equally-spaced levels was also found in the Harmonic Oscillator discovered by Talmi[7]. Schroedzipper transformed the zipper probelm to the oscillator case by use of a Folded-Woodhouse Canonical Transformation. He was then able to calculate transition probabilities, level spacings, branching ratios, seniorities, juniorities, etc. Extensive tables of the associated Racah coefficients have recently been computed by Rose, Bead and Horn[8].

---

[1] H. Quantum: A New Theory of Zipper Operation Which Is Also Incidentally Applicable to Such Minor Problems as Black Body Radiation, Atomic Spectroscopy, Chemical Binding, and Liquid Helium. Z.I.P. 7, 432 (1922).

[2] H. Eisenzip: The Uncertainty Principle in Zipper Operation, *Zipschrift fur Phyzip*, 2, 54 (1923).

[3] I. Newton, M. Faraday, C. Maxwell, L. Euler, L. Rayleigh, and J. W. Gibbs, "Die Zipperbewegung" (unpublished).

[4] N. Bohr: Lecture on Complementarity in Zippers, Geneva Conference, "Zippers for Peace" (1924).

[5] P. R. Zipsel and N. Bohr: Einstein Memorial Lecture. Haifa Technion (1956).

[6] E. Schroedzipper: "What is a Zipper," Dublin (1950).

[7] E. Talmi, Helv. Phys. Acta, 1,1 (1901).

[8] M. E. Rose, A. Bead, and Sh. Horn (to be published).

Numerous attempts to verify this theory by experiment have been undertaken, but all have been unsuccessful. The reason for the inevitability of such failure has been recently proved in the celebrated Weisgal-Eshkol theorem[9], which shows that the construction of a semi-infinite zipper requires a semi-infinite budget, and that this is out of the question even at the Weizipmann Inziptute.

Attempts to extend the treatment of the semi-infinite zipper to the finite case have all failed, since the difference between a finite and a semi-infinite zipper is infinite, and cannot be treated as a small perturbation. However, as in other cases, this has not prevented the publishing of a large number of papers giving perturbation results to the first order (no one publishes the higher order calculations since they all diverge). Following the success of M. G. Mayer[10] who added spin-orbit coupling to the harmonic oscillator, the same was tried for the zipper, but has failed completely. This illustrates the fundamental difference between zippers and nuclei and indicates that there is little hope for the exploitation of zipperic energy to produce useful power. There are, however, great hopes for the exploitation of zipperic energy to produce useless research.

THE FINITE ZIPPER

The problem of the finite zipper is best treated directly, without reference to the infinite case. One must first write the Schroedzipper equation for the system:

$$(1)\quad H\, Z = -ih\, dZ/dt$$

The solution of this equation is left as an exercise for the reader. From the result all desired observable information can be calculated.

The most interesting case of the finite zipper is that in which there are perturbations. For this case the Schroedzipper equation becomes:

$$(2)\quad (H + H')\, Z = -ih\, dZ/dt.$$

Because of the perturbation term H′, the original states of the unperturbed zipper are not longer eigenstates of the system. The new eigenstates, characteristic of a perturbed zipper, are mixtures of the unperturbed states. This means, roughly, that because of the perturbation the zipper is in a state somewhere in between its ordinary states.

One interesting case of zipper perturbation has been reported[11]. The zipper system in question was on the front of the trousers of a man sitting in a cinema. The zipper had been lowered to its lowest state when it was perturbed by the back of a dress belonging to a lady attempting to pass in front of the gentleman. The zipper immediately jumped into a highly perturbed state, and the zipper-dress coupling proved to be so strong that it was impossible to separate the variables by conventional techniques. It was necessary for the individuals concerned to leave the cinema, walking down the aisle in "a zippered embrace" with further perturbations from the light quanta of the usher's flashlight and intense sound waves from the audience. The separation was later achieved by "brute force," thereby rendering the dress unsuitable for further use and requiring payment of a fine by the owner of the zipper.

Another theoretical possibility of such perturbation was recently voiced by a lady who was considering buying a pair of trousers for her husband. She was offered a zippered type but she declined the offer. Her uncertainty principle was expressed in the following words: "I don't think such trousers would be good for my husband. Last time I bought him a zippered sweater, his tie was highly disturbed by the zipper perturbation." ■

[9] M. Weisgal and L. Eshkol: "Zippeconomics." Ann. Rept. Weizipmann Inziptute (1955).

[10] Metro G. Mayer: "Enrichment by the Monte-Carlo Method: Rotational States with Magic Numbers" Gamblionics, 3, 56 (1956).

[11] The Ithaca Journal, "A Zippered Embrace" (1939).

From science all Truths are deducible,
To axioms all are reducible.

But when doing our best
To subject them to test

We find that they're IRREPRODUCIBLE.

Ralph Steinhardt, Jr.
Hollins College, VA

The Semitruth about Semiconductors

M. A. VINOR
Department of Semistry
Institute for Advanced Anti-Semics

At present the research in the field of semiconductors is so heavily bombarded by mathematical artillery and so full of red herrings of new concepts such as traps, defects, holes, vacancies and the like, that one can hardly see the semiconductor for the holes. Therefore, we think that it is high time to revive the fundamental concepts and primary axioms and thus relieve the weary inquisitive reader of going through heaps of senseless papers which would be completely useless five years from now, anyway.

The First Dogma of the Semiconductor Creed is:

> FOR ANYTHING WHATSOEVER ALWAYS PUT THE BLAME ON THE ELECTRON (its negative charge, its negative attitude to life, etc.)

Metals are exceptional in that they conduct electricity well. This is because they contain electrons on the loose and any bad influence, such as an electric field, pushes them to damnation.

Now, if you hang a piece of semiconductor (i.e. not exactly a non-metal) on a long piece of red tape in the middle of a conference hall, and if the people present argue hotly enough about it, then the sign of the current carriers in the semiconductor may be determined. This is known as the Hall Effect.

By applying the Hall Effect to many things it was found that some semiconductors had electrons as carriers, while others had positive carriers, which, evidently, could not be electrons since electrons are by definition always negative (see Dogma above).

The poor semiconductors were, therefore, divided into two classes:

1) Semiconductors of the n type (n = normal) in which the electrons misbehave as usual.
2) Semiconductors of p type (p= pervert) in which the electrons misbehave even more.

Now, solid state physicists could not assume the positive carriers to be some sort of positive electrons (positrons) because they have already been discovered by nuclear physicists and no original idea would be involved. Therefore, after many sleepless nights of beer (alco-hole) sipping, a new and very, very original idea was found. The Positive Carriers are HOLES!

If an electron jumps out of its place and goes, probably full of doubtful intentions, to Conduction Alley—then everybody is so glad that its vacancy is named Positive Hole. If another electron fills the vacancy—then it is considered good form to say (especially if ladies are present): "The Hole went away." And so, if you pump a German full of Holes, you don't get a sieve but a piece of p type Germanium.

A piece of solid is also full of other queer irrelevancies such as phonons, excitons, traps, etc. Electrons are always in the habit of getting pushed about by photons and phonons and falling into traps. This causes some confusion since nice people do not say that an electron came out of a trap, but say instead: "A Hole fell into a trap." This should not cause any misunderstanding if the true facts are borne in mind and point only to personal shyness.

We should also mention the Taboo of Semiconductors. By the ingenious definition above, Holes have negative masses. This is to mean that if you inject enough Holes into a semiconductor it should fly like a balloon. One usually finds that in conversation physicists evade or side-track the questions about this property. There are two reasons for it.

(a) They are embarrassed because this idea is not so very original; it was used long ago by the chemists who adhered to the Phlogiston theory (but do physicists know chemistry?).

(b) The whole subject is highly classified as this principle of annulling gravitational force is being secretly developed for guided missiles against guided missiles against guided missiles.

We hope that by laying bare before the reader the essential facts concerning semiconductors, we have shown that one can talk in a very learned fashion about them and never be exposed; one has only to observe two simple rules (one Dogma and one Taboo) and keep the face straight.

Good luck! ■

CLONUNDRUMS

Louis G. Lippman
Bellingham, WA

Placing some cells scraped from his tongue into a petri dish, a geneticist grew a copy of himself. Although his duplicate was a perfect physical match, it had one ironic behavioral quirk: a foul mouth. After two years of constant embarrassment from the incessant barrage of gross verbiage, the geneticist decided to put an end to it: Under the guise of a sightseeing excursion, he lured his duplicate to the top of a skyscraper and, despite the presence of lots of other tourists, pushed his lewd nemesis over the edge to its death. Mercifully, the indecent screams on the way down were its last.

The cause of the copy's demise was surely not ambiguous; the presence of witnesses was no problem either. Still, this event raised some surprising legal issues beyond the obvious question of how the murder of a duplicate is any different from trimming a fingernail and discarding the parings. In this case, a jurisdictional dispute arose. The homocide division of the local police contended that it should be in charge of all investigations and legal proceedings, culminating in a felony indictment. But some federal agencies asserted not only that they should have responsibility, but also that the accusation should be less severe. The upshot of all this quibbling was that the Federal Communications Commission was given full authority for prosecuting the case, charging the geneticist with a misdemeanor. At first blush, this outcome seems absurd—until realizing after all that the geneticist was, in fact, guilty of making an obscene clone fall.

This little story, in some form, has probably circulated throughout the public domain; its survival suggests that there now seems to be some general awareness of certain developments in the biological sciences. This interest may be attributed to several recent events which have placed some of the life sciences and related technology in the media: Fictional and non-fictional accounts of cloning, "test-tube" babies, and recombinant DNA research and development, to name just a few. It might be appropriate—perhaps advantageous—to keep some of these factors before the public eye: If the lay public can be acquainted and accustomed to these developments, then there may occur some moral support (maybe even financial) for further research. Certainly, news of breakthroughs and disasters will reach the public via the media. But to further the cause, the less typical route of humor is also suggested. To that end, it is suggested that some energies be devoted to the proliferation and dissemination of "clone jokes." Instead of the somewhat elaborate form illustrated above, it is recommended that these emulate some types which have optimized brevity and which have been successful fads in the past: sick humor, Tom Swifties, moron jokes, grape gags, elephant jokes, etc. Recommended forms and samples appear below. It is hoped that the scientific community is duly stimulated by these items and goes into production with enthusiasm and alacrity. As further inspiration, consider the following moral, which applies to numerous matters including species survival, parthenogenesis, as well as propagation of gags: When the cloning gets tough, the tough get cloning. ■

A FABLE ON THE LOGIC OF SCIENTIFIC EXPLANATION

Patrick Doreian
Pittsburgh, PA

Once upon a time, a farmer and a farm laborer set out to plow a field that had lain fallow for many years. The field was in an area of the country famous for the buildings and fences built from flint. When the field was plowed, the two men looked at the many flintstones they had turned up in the course of their plowing. The farm laborer pointed excitedly, saying:

"Look at all those flints! They have been growing there all these years."

The farmer, taken aback at this ontological outburst, replied grimly:

"They may have been there all these years, but they certainly have not been growing."

"Yes they have, Look at the different sizes! Some of the flints are older than the others. See, they are the big ones." Tipping his hat back and wiping his brow, the farmer asked:

"If we break a clod of earth into pieces the sizes are not equal. right? Size has nothing to do with age." Curbing his rising impatience, he explained:

"Each time we plow a field we remove the stones and flints we turn up. Over the years we clear the fields of flints a little each year. This field was not plowed for a long time so the flints remained in place."

"No, we left the flints alone. They were free to grow here and they did."

"Come on. How do you know flints grow?"

"Easy. Look at our fences, and those down the road. All falling down, right?"

"Wait a minute! What has that got to do with it?"

"A lot. We build those fences from the flints around here. The flints keep growing and squeeze out the mortar. When the mortar is squeezed out, the walls fall down."

"That's absurd. The wind erodes the mortar. When the mortar has been eroded, there is nothing to hold the flints in place. That's why the walls fall down."

The laborer looked at the farmer quizically, wondering silently at this arrant nonsense. The farmer, interpreting this look, was ready to give up. Dumb peasant! Suddenly, an inspired idea flashed into his mind. He thought of the Guildhall in the nearby large town: a beautiful building viewed by countless tourists every year. It had stood in splendor for over 300 years and, more to the point, was made of flint. Excitedly, the farmer turned to the laborer:

"Think of the Guildhall. A flint building hundreds of years of old—and still standing." Unable to keep the smugness out of his voice he asked "How do you account for that? A flint structure that has not fallen down?" He was in full flow now and, not waiting for an answer to his rhetorical question, went on. "I'll tell you why. Each year the mortar is checked and re-pointed if it has worn. With all this maintenance the building continues to stand." The farm laborer smiled, looked knowlingly at the farmer, and replied:

"Ah, those flint have been knapped. Everyone knows that when you knapp a flint you kill it. Of course the Guildhall still stands. All the flints have been killed and cannot grow. So they don't squeeze out the mortar and the building does not fall down." The farmer knew he was beaten. His rational argument and brilliant assembly of evidence had been thwarted by rational argument, and a more creative use of evidence. ■

Ed. Note: Is that why truth and justice appear to be moving further apart?

The Corn Soup Principle

Submitted By:
Mark Worden
Roseburg, OR

It is no easy task to engage in therapy that reliably leaves the client unchanged, and at the same time maintain professional standards. However, if dieticians can make corn soup that does not contain corn, surely psychotherapists can design psychotherapy that contains no therapy.

Epling & Woodward
Alcoholism/MAR-APR 1982

A MODEST PROPOSAL ON THE PROLIFERATION OF SCIENCE AND SCIENTISTS[1]

Jonathon B. Quick
Institute for Innovative Research Financing

Marvin Margoshes
Tarrytown, NY

There is great concern among scientists and laymen about the growth of science at an exponential rate. The public worries about nuclear weapons and nuclear energy, about genetic engineering and the creation of new life forms in the laboratory, and about loss of jobs to computers and robots. Our national leaders, faced with huge budget deficits, are concerned about the cost of research including funds for travel to scientific conferences, for page charges to publish the results of the research, and for libraries to purchase and shelve the ever-growing number of scientific books and journals.

Scientists complain about the increasing volume of scientific literature, about their need to write and review proposals and manuscripts, and about there being both too many conferences and too many people attending each conference. The recognition is emerging that the pace of scientific development threatens our scientific careers, rendering each of us technologically obsolete almost before the ink has dried on our diplomas.

Despite these legitimate concerns, no way has been found to limit the growth of science and technology. It was hoped that science would be self-limiting, as each scientist had to devote an increasing portion of his time to writing, reading, and reviewing papers and proposals, thus restricting the amount of time that could be spent on research. This limitation has, however, been overbalanced by a vast increase in the number of scientists and by the introduction of automation into the laboratory. (Automation in the factory reduces employment at the same level of production. Automation in the laboratory increases production without reducing employment.) Word processors are speeding the writing of manuscripts and grant applications. Once computers are programmed to also read the review these documents, the last restraint on scientific productivity will be removed.

Fortunately, the solution to this problem is at hand. The United States Government has found that it can pay farmers to not produce crops and milk, and thus limit the surplus of farm products. Similarly, *we should pay scientists not to do research*.

The cost of this program will not be excessive. By eliminating research we do away with expenses such as instrument purchases and maintenance, publication charges, and overhead in general. New scientific libraries will not be needed. Also, it is possible to adopt the strategy by which farmers are paid in kind instead of in cash. The American farmer who agrees not to grow corn receives corn that has been stored by the government in earlier years. This saves money directly and by lowering storage costs, and it reduces the surplus of grain. By the same logic, NIH and NSF could send each U.S. scientist progress reports which these agencies have stored in their files.

Payments would be made only to scientists with Ph.D. degrees. Since no research would be done, students could not write dissertations and no more graduate degrees would be awarded in the sciences. As the eligible scientists retired, died, or took up other livelihoods, there would be a gradual and painless decrease in the cost of the program.

Quick action on this proposal is vital. There may already be a project in some computer science department to automate the reading and reviewing of proposals and manuscripts. Once that project is completed, the floodgates will be open forever.

If the U.S. Government alone adopts this proposal, it must also enact regulations to stop the importation of foreign research results. This is analogous to the regulations on the export of American technology. ■

[1]This research was supported by grants too numerous to mention.

Semiconductors Are Not Equal Opportunity Employers

HAROLD MOTT
Electrical Engineering Department
University of Alabama

THE EQUAL EMPLOYMENT OPPORTUNITY COMMISION
1800 G Street, N.W.
Washington, D.C. 20506

Professor W. D. Radburn
Electrical Engineering Department
Wheat University
Galveston, Texas

June 18, 1973

Dear Professor Radburn:

I have been advised that you are in responsible charge of the Solid-State Laboratory in the Electrical Engineering Department of Wheat University. I have also been advised that this laboratory is concerned with the fabrication and utilization of certain solid-state devices called transistors, which I understand is some kind of portable radio. It is my further understanding that in the fabrication and/or utilization of these transistors you employ workmen having the job classifications of "minority carriers" and "majority carriers." I must tell you immediately that we here are gravely concerned about these job classifications. Even though we wish to give you every benefit of doubt, I must advise you that the existence of these titles appears to be *prima facie* evidence of discrimination, which is forbidden by Title VII of the Civil Rights Act of 1964 and Executive Order 11246, and is punishable by fines and/or imprisonment. Since the presumption of discrimination on your part exists I advise you to reply promptly and carefully to the following questions and directives.

1. Is the ratio of minority carriers to majority carriers employed in your laboratory at least equal to the ratio, in the general population of your community, of individuals of the minority race to that of the majority?

2. Is the average salary of the minority carriers equal to or greater than the average salary of the majority carriers? I suggest that you supply to me without delay a list of minority and majority carriers together with the salaries of each carrier.

3. Are there proportionally as many minority carriers in supervisory positions as there are majority carriers?

4. I understand that these carriers carry charges, and further that they carry charges of different sign. Are the charges carried by the minority carriers heavier than those carried by the majority carriers?

5. I understand that some of these carriers are said to "drift." I trust that you are not implying that any of the minority carriers drift aimlessly. This would seem to be an insult and is forbidden by Federal law.

6. What is this "diffusion" which I have heard about? You should keep in mind that while the maximum amount of integration is desirable, we will not tolerate talk about diffusion of minority carriers throughout the entire region.

7. I understand that your laboratory is involved with the development of devices called "integrated circuits." This is certainly a point in your favor, and we will keep it in mind as we review your situation.

In view of the gravity and urgency of this situation I am sure you will want to answer my letter quickly and completely. I await your reply.

Very truly yours,

W. E. McLiu

W. E. McLiu
Supervisor for Region III Compliance
The Equal Employment Opportunity Commission ■

WEM:ky

Anecdotal Evidence for the Existence of an Unfilled Niche

THEODORE H. FLEMING
Department of Biology
University of Missouri
St. Louis, Missouri

The question of whether habitats are saturated – contain the maximum possible number of animal species – is one which has intrigued ecologists for many years. The fact that introduced species often gain a foothold and increase in numbers in a new ecological setting seems to indicate that not all habitats are saturated (see review by Elton, 1958). The following set of observations, perhaps a perfect example of an unfilled niche, is offered as additional evidence for the unsaturated nature of certain habitats.

Materials and Methods

In hopes of attracting winter birds to my backyard, I set up a bird-feeding station in our backyard on November 15, 1968. The station consisted of a commercial "cone" of assorted "bird seeds" (Burpee & Co.) wired to two 9-inch aluminum pie plates (Jane Parker & Co.) that served as a top and a bottom for the station. The pie plates were supported by two pine slats 8 inches in length. The station was suspended by a wire from the west arm of our metal clothes-line pole.

The ecological setting of this bird feeder is the Backyard Biome (Dice, 1952) and consists of the normal garden variety of plants which includes lilac *(Syringa)*, dogwood *(Cornus)*, peony *(Paenonia)*, zinnias *(Zinnia)*, and strawberry *(Fragarus)*. One noteworthy feature is a stand of several 10-ft. tall *arbor vitae (Thuja occidentalis)*, which serves to increase the privacy of our backyard. The overhanging boughs of one *arbor vitrae* protect the bird-feeder from possible damage by wind and rain. The area encompassed by our backyard is approximately 450 sq. ft. (0.10 a).

Since November 15, I have observed the feeder from my kitchen each day for one continuous hour at a time determined by use of a random numbers table (Steel and Torrie, 1960). In all honesty, I only observed the feeder for one-half hour if the designated time occurred at night. (At a distance of 40 ft., the station was extremely difficult to see in the dark). All observations were made using 7 × 35 binoculars. Biological observations and weather conditions (temperature, humidity, and precipitation) were recorded on IBM data sheets, and the data were transferred to IBM cards at the end of each week. Data included in this paper were analyzed on an IBM-360 computer.

Results

Between November 15 and February 25, I observed the feeding station for a total of 82.5 hrs., and in that time I failed to see a single bird utilize the food available there. Careful examination of the "cone" of seeds indicated that it had never been disturbed. This means that in the 2,448 hrs. that the food was available, no bird ever visited the feeder (Table 1).

The relationship between lack of feeding activity and temperature, humidity, and precipitation was analyzed by multiple regression. Results showed that none of the three variables was more important than any other in explaining the lack of results ($P > .05$).

Discussion

The lack of results was somewhat surprising and, frankly, very disappointing. I find it hard to explain why no birds were attracted to the feeding station. A paucity of birds in the immediate vicinity of our yard cannot be the reason because I observed English sparrows *(Passer domesticus)*, starlings *(Sturnus vulgaris)*, cardinals *(Richmondea cardinalis)*, and blue jays *(Cyanocitta cristata)* on numerous occasions not more than 50 ft. from the feeding station. There appeared to be nothing unusual about the bird seeds or the nature of our yard that might prove unattractive to birds.

The lack of results forced me to curtail the discussion somewhat as I had been prepared to compare my findings with the MacArthur "broken stick" model (MacArthur, 1957), with various measures of species diversity

TABLE 1. Number of hrs. of observation, number of birds seen, and average weather conditions during periods of observations of a backyard bird-feeder.

Hours of observation	No. birds seen	No. birds seen per hr. of obs.	Avg. temp. °F	Avg. humidity %	Avg. ppt. inches
82.5	0	0	30.5 ± 0.24	73.4 ± 2.5	0.13 ± 0.01

(reviewed by Hairstron *et al.*, 1968) and niche breadth (Levins, 1968), and Wynne-Edwards' (1962) theory of epideictic display. These comparisons will have to wait until observations become available.

Because no birds were ever observed to use my feeding station, I can only conclude that the feeder represents an unfilled niche. As such, it is intriguing to speculate on just what kind of bird could fill this niche. A more detailed description of the available niche will aid in deciding what kind of bird could take advantage of it. The "cone" of seeds consisting of hundreds of seeds that average 0.5 mm in diameter weighs 455 g. This biomass represents about 3,000 kcal of potential energy (Farmer's Almanac, 1968).

It is my opinion that this niche could support one very large, transient bird that might eat the entire feeder (including the aluminum pie pans) in one bite, or several small finch-like birds that might subsist on the seeds for several days. It seems clear that because the niche is rather ephemeral, any species specialized enough to fill it will become extinct very quickly unless it can find another niche with similar characteristics. The species will have to be a "fugitive" in the strictest sense of the word.

I am not undaunted by the lack of results so far and plan to continue observing this "unfilled niche" systematically in hopes of eventually discovering a species that can take advantage of this unusual energetic opportunity.

Summary

A backyard bird-feeder was observed from November 15, 1968, to February 25, 1969. In this period no birds were ever seen to feed there. I believe this represents a classic example of an unfilled niche. ■

LITERATURE CITED

Anonymous. 1968. Farmer's Almanac.

Dice, L. R. 1952. Natural communities. Univ. of Michigan Press, Ann Arbor.

Elton, C. S. 1958. The ecology of invasions by animals and plants. Methuen & Co., London.

Hairston, N. G., J. D. Allan, R. K. Colwell, D. J. Futuyma, J. Howell, M. D. Lubin, J. Mathias, and J. H. Vandermeer. 1968. The relationship between species diversity and stability: an experimental approach with Protozoa and bacteria. Ecol., 49: 1091–1101.

Levins, R. 1968. Evolution in changing environments. Princeton Univ. Press, Princeton.

Steel, R. G. D., and J. H. Torrie. 1960. Principles and procedures of statistics. McGraw-Hill Book Co., New York.

Wynne-Edwards, V. C. 1962. Animal dispersion in relation to social behavior. Hafner Press, Edinburgh.

*An Overlooked (?) Basic Principle of Development and Personality**

F. NOWELL JONES
Department of Psychology
University of California

I should like to call the attention of psychologists to a basic principle of personality which has, most remarkably, been heretofore overlooked.

To begin with the history of the idea: Some months ago I visited Professor Harry F. Harlow's laboratory devoted to the study of infant monkeys. Many of the tiny little Ss were to be seen with thumb or finger in mouth, no doubt sucking. I was most struck, however, by one little fellow who was observing me with much manifest anxiety—a term readily defined operationally upon demand. Not only was this infant a thumb-sucker, he was also what we may call a tail-holder, for with the unmouthed hand he was clutching his tail and tickling his ear with the tip. The analogy to the blanket or towel or other object of similar shape which the infant Homo sapiens so often puts to similar use was dramatic and striking. Put bluntly, the human infant seeks a tail-surrogate, and I suggest that even in the adult a condition of tail-envy, as we shall call it, is embedded deeply in the unconscious. The universality of this new principle is emphasized by the remarkable fact that it applies to both sexes, a point which cannot be made for a rival concept which I shall not mention here.

This new principle of tail-envy explains many strange human behaviors. To mention but a few, consider dreams of such objects as snakes, or some persons' addiction to rope climbing—but why extend this list unduly? So important seemed the principle that I began an active search for positive, specific instances where it might be applied and have, after 20 months, found two. The first is the universal appeal *to adults* of the comic strip character of tender years whose life seems to revolve around his comforting blanket. So pervasive is this appeal that it can be based only upon some deeply embedded urge derived from our remote ancestors and transmitted through the unconscious—perhaps in unbroken line back to that ancient anonymous man-ape who first noticed that he had no tail and suffered trauma from this realization. The second instance is even more interesting because the appeal is completely adult and overlaid subtly with patina of modern sophistication. One airline advertises that "We really move our tail for you." How obvious can a principle be?

I cannot close without a brief comment designed to anticipate an expected immature objection to the new principle which will come from those threatened by it. The objection is, how can it be that 20 months have produced only two positive instances? The obvious reply would be that I have been too busy with committees to notice very much, but this is not, I feel, adequate. I should rather say that two instances per investigator for each two years is quite enough. If the active participation of only a few thousand psychologists can be enlisted, I am sure that the next few years will see many thousands of positive instances of tail-envy appearing in the literature. I am confident that once a sufficient body of examples has been assembled, the new principle will be difficult to overthrow, especially among those psychologists working in situations where one has to do something useful, not just play with facts in the laboratory. ■

* Reprinted, more or less by permission from The Worm Runner's Digest, *4*:61 (1962).

TERMITIC TURGIDITY:
End-Product Evaluation

Dale R. Spriggs
Houston, Texas

INTRODUCTION

The world needs energy supplies if we are to survive. We will eventually run out of fossil fuels, and many people have suggested that we exploit alternative energy sources such as solar energy or other fuels like hydrogen or methane (1, 2). Recent studies have shown that termites are a major source of these particular atmospheric gases (3). There is one-half ton of termites for every person on earth, and all together these insects annually excrete approximately 100 million tons of methane and 200 million tons of molecular hydrogen. This report shows that termite flatulence can be tapped as a source of energy.

MATERIALS AND METHODS

Termites *(Reticulitermes flanipes)* were collected from the wooden foundation of a house before and after massive insecticide treatment. Progeny hives were established by placing 1,029 termites (with queen) in separate, covered geodesic domes (Fuller model no. 101) fitted with adapted Harvon collector modules. High humidity was established by placing a full Tommy Turtle®playtub in the bottom of each dome. Various processed pulp materials were used as cellulose sources, and excretion rates were measured with a Hodges flatometer (Windfall, Idaho) standardized with pressure pellets provided by the manufacturer.

RESULTS

Previous anatomic studies (unpublished data) indicated that mutant termite populations selected for resistance to various insecticides might be better gas producers than are normal termites. This hypothesis was based on the examination of the anal orifices of 75 soldiers from each group. The circumference of the cavity of the mutants was significantly larger ($P<.01$, Chi-square test) than that of the normal termites. This observation suggests a novel mechanism for the evolution of insecticide resistance in insect populations.

Mock-up hives of both populations were therefore established to test their relative levels of gas production. Initial findings supported the hypothesis, but the experiment had to be terminated when the piston in the Harvon collector module of the mutant hive overheated and jammed in the cylinder. The resultant gas pressure build-up caused an explosion that blew shards of the hive up to 67 meters from the site. Efforts are underway to isolate another population of mutant termites; the studies outlined below, therefore, were performed only with normal insects.

The first studies were aimed at determining the effect of dietary sources of cellulose on gas output. As shown in Figure 1, output was greatest when the termites were fed on pulp sources like the *National Enquirer* or *True Confessions*. *Reader's Digest* was effective for short bursts of activity, but the overall levels of gas production were not satisfactory. These particular termites did not eat scholarly journals (group D) or magazines with pictures of women on motorcycles (group E). The reasons for this selectivity are currently being investigated.

The gas produced in these hives has also recently been used as a fuel to run a motor vehicle engine (Figure 2) and a fuel for the generator that powers the collector module. In addition, excess electricity from the generator has been tapped to run an electric razor. Preliminary trials with a hair dryer and a Mr. Coffee®have been promising.

DISCUSSION

The data in this paper show that termites will be a major source of energy in the future.

Further studies will be aimed at reestablishing the strains of mutant termites with enlarged orifices. Efforts are also underway to isolate, from the termites' guts, the bacteria that actually digest the cellulose in the pulp. The genes that code for enzymes digesting cellulose will then be genetically engineered and cloned to establish a superior breed of termites. It seems likely that these improvements will result in our energy autonomy by 1987. ■

REFERENCES

1. **Abelson PH.** Methane: A motor fuel. Science 1982; 218:641
2. **Maugh TH II.** The new routes to solar hydrogen. Science 1982; 218:557
3. **Zimmerman PR, Greenberg JP, Wandiga SO, Crutzen PJ.** Termites: A potentially large source of atmospheric methane, carbon dioxide, and molecular hydrogen. Science 1982; 218:563-5

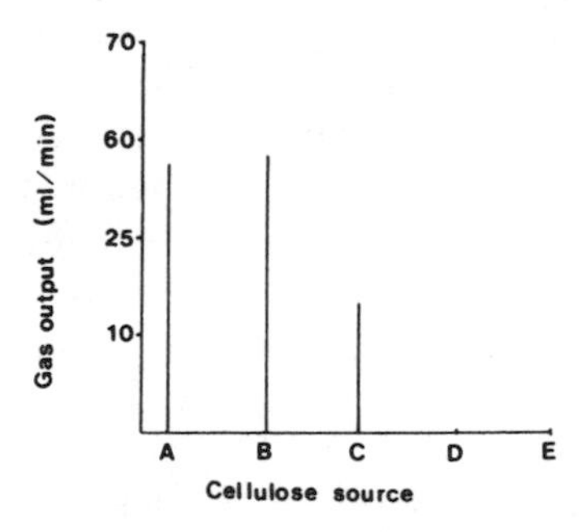

FIGURE 1. The effect of cellulose source on gas output by *Retuculitermes flanipes*. Output was measured as indicated in Materials and Methods. Cellulose sources were (A) *The National Enquirer,* (B) *True Confessions,* (C) *Reader's Digest,* (D) *The Journal of Insect Ethics,* and (E) *Bike and Broad.*

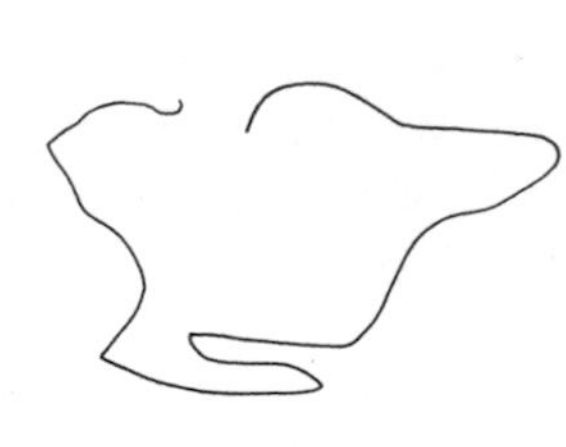

FIGURE 2. Vehicle displacement powered with compressed gas produced by *Reticulitermes flanipes*. A Harley-Davidson (450 cc engine) motorcycle was driven through a suburban mall parking lot in the pattern outlined above. 1 cm = 10 m.

This work was supported by grant HO 1020 from the Natural Gas Institute of America; additional funds were provided by the Environmental Protection Agency and the Cooperative Research Service of the Department of Agriculture.

Wanda Lee Burke provided technical assistance.

Musings of an aging "SCIENTIST"

RUSSELE de WAARD
Old Greenwich, Conn.

To an astute reader the title ("Scientist") is sufficient to set the mood for this document.

Ever since I can remember I have been insecure. When I was small I worried and worried that my slingshot would break and I would never find another tree branch to make a new one. This attitude has persisted until today. I get interested and work hard on something and it seems to me at the time that I have done a good job. The results of my work are printed and distributed among my peers. I go home that night quite satisfied and feeling good, but by the end of the next day I am concerned that I could have done a better job. By the second day I am convinced that most of what I wrote was wrong and I feel quite stupid. This is an example, which repeats itself over and over. I am quite convinced that I was apprehensive as a fetus. Although my failing memory doesn't confirm it, at that stage I probably worried about a kink in the umbilical cord.

I don't know whether to blame this deficiency on my genes, my chromosomes or my upbringing. But in the latter category, I have made a few observations. I have been taught to respect authority; when the Department Head says I am wrong, I am wrong. Since I did not go to MIT, I reason miserably, that Edgar Slack, my teacher at BPI, must have been wrong. On These occasions I take his book (Haussman & Slack) into the men's room to see if I might have misread it.

What one has to learn is that all the laws of physics are repealed by authority. Hence, if one speaks with authority, from a recognized box in the company organization chart, he is perforce believed. These things have only very recently come to me, but I don't know what to do about it. If I am lucky this draft will not be published and my discovery will go "undiscovered." My position is a bit like that of Galileo in the 15th century when his telescope told him beyond a doubt that the world was round and was rotating about the sun, but his friends told him to keep his big mouth shut or the Pope would feed him to the lions. I like to eat and to drink and not to rock the boat, so I think I'll do what Galileo did, unless tomorrow morning I ask Bobby to type this tape. The latter is unlikely since overnight my apprehension will return.

The funny part about all this is that when I go out on a sailing ship all my chicken-hearted propensities dissolve into the sea and I become a soul-mate of Captain Horatio Hornblower. The trouble is that none of the fellows in authority are along, and I curse at the wind and the waves to no avail.

If I keep on like I'm going, I might make three score and five, and if by that time I don't continue to spike my vodka martini with lemonade, I may have worked up enough gumption to say "no, I won't shut up, you sit down and listen for a change" to one of the fellows in the boxes. ■

SAGA OF A NEW HORMONE

NORMAN APPLEZWEIG*

In recent months we've learned of the discovery of three miracle drugs by three leading pharmaceutical houses. On closer inspection it appears that all three products are one and the same hormone. If you're at all curious about how more than one name can apply to the same compound it might be worth examining the chain of events that occurs in the making of a miracle drug.

The physiologist usually discovers it first—quite accidentally, while looking for two other hormones. He gives it a name intended to denote its function in the body and predicts that the new compound should be useful in the treatment of a rare blood disease. From one ton of beef glands, fresh from the slaughterhouse, he finally isolates ten grams of the pure hormone which he turns over to the physical chemist for characterization.

The physical chemist finds that 95 per cent of the physiologist's purified hormone is an impurity and that the remaining 5 per cent contains at least three different compounds. From one of these he successfully isolates ten milligrams of the pure crystalline hormone. On the basis of its physical properties, he predicts possible structure and suggests that the function of the new compound is probably different from that assigned to it by the physiologist. He changes its name and turns it over to the organic chemist for confirmation of structure.

The organic chemist does not confirm the structure suggested by the physical chemist. Instead he finds that it differs by only one methyl group from a new compound recently isolated from watermelon rinds, which, however, is inactive. He gives it a chemical name, accurate but too long and unwieldy for common use. The compound is therefore named after the organic chemist for brevity. He finally synthesizes ten grams of the hormone but tells the physiologist he's sorry that he can't spare even a gram, as it is all needed for the preparation of derivatives and further structural studies. He gives him instead ten grams of the compound isolated from watermelon rinds.

The biochemist suddenly announces that he has discovered the new hormone in the urine of pregnant sows. Since it is easily split by the crystalline enzyme which he has isolated from the salivary glands of the South American earthworm, he insists that the new compound is obviously the co-factor for vitamin B-16, whose lack accounts for the incompleteness of the pyruvic acid cycle in annelids. He changes its name.

The physiologist writes to the biochemist requesting a sample of of his earthworms.

The nutritionist finds that the activity of the new compound is identical with the factor PFF which he has recently isolated from chick manure and which is essential to the production of pigment in fur-bearing animals. Since both PFF and the new hormone contain the trace element zinc, fortification of white bread with this substance will, he assures us, lengthen the lifespan and stature of future generations. In order to indicate the compound's nutritive importance, he changes its name.

The physiologist writes the nutritionist for a sample of PFF. Instead he receives one pound of the raw material from which it is obtained.

The pharmacologist decides to study the effect of the compound on grey-haired rats. He finds to his dismay that they lose their hair after one injection. Since this does not happen in castrated rats, he decides that the drug works synergistically with the sex hormone, testosterone, and therefore antagonizes the gonadotropic factor of the pituitary. Observing that the new compound is an excellent vasoconstrictor, the pharmacologist concludes that it should make a good nose-drop preparation. He changes its name and sends 12 bottles of nose drops, together with a spray applicator, to the physiologist.

The clinician receives samples of the pharmacologist's product for test in patients who have head colds. He finds it only mildly effective in relief of nasal congestion, but is amazed to discover that three of his head cold sufferers who are also the victims of a rare blood disease have suddenly been dramatically cured.

He gets the Nobel prize. ■

* President, Norman Applezweig Associates, Consulting Biochemists.

populace into blind obedience. What is put into them emerges as conclusions which cannot be questioned, but, what is more bothersome is that their deductions are at variance with what the people wish and need.

Our First and Second Worlds (U.S. and U.S.S.R.) are equipped with automatic weapons of all kinds, and both of them are rapidly capacitating the Third World countries as well. We have automatic pinsetters, automatic stop lights at street intersections, and automatic telephone answering machines. The catalogue of our automatic devices and electronic computers, with all their servo-mechanisms, components, detectors, analyzers, and indicators could fill pages. Truly man's minds, muscles, and souls are being replaced by electronic apparatus of many kinds. We are fast moving in the direction of mental, physical, and moral destruction. Witness what is happening to the automotive business! Machines are replacing men on the assembly lines. And now, I understand, we have a machine which writes better prose than the reliable old Remington Portable now at my finger tips. It is time we considered doing something about all this!

HYPOTHESES

1. I hypothesize that as the calculator replaces the abacus, man will lose abilities in mathematics. This prediction arose from the statement made by so many businessmen seeking clerks: "Why, They can't even add!" (this is an extrapolation of "Why, Johnny can't read!" made a few years back.)
2. I postulate that as the computer replaces the sales girls at the checkout counters, males will lose personal contact with the females. Conjunctively, I predict a rising divorce rate and an increase in illegitimate births. This postulate, and its derivatives, came to me as I tried to kiss the digital robot which erred in my favor when I recently bought a can of spiced (spiked?) gastropod mollusks. It (the robot register) only blinked its glazed eye and reflected some French word, presumably not of the four-letter variety.)
3. I prognosticate that as remote-control devices replace men and women as sources of power, they will cause alienation between people and their governments. This occurred to me about January 15 when I received the 1983 Income Tax Forms.
4. I predict that as the gyroscoptic pilots in rockets replace human pilots, man will develop more brutal phobias relative to foreigners, aliens, and communists.
5. I speculate that more data goes into computers than come out. (I must thank Art Buchwald for this prehensile conceptus.)
6. I conjecture that the results of the foregoing hypotheses will be arrhythmic, heteroclite, aberrant and idiosyncratic.

DISCUSSION

None is necessary. Any abderite who wants to discuss such a vacuous gambit as this is lacking in sophrosyne. ■

10 RECENTLY PATENTED INVENTIONS WHOSE TIME HAS NOT YET COME

1. *TOILET LID LOCK* (U.S. Patent 3,477,070)

To prevent unauthorized access to toilet bowl.

2. *WHISPER SEAT* (U.S. Patent 3,593,345)

Toilet seat with acoustical liner to prevent sounds from being heard by other persons.

3. *COMBINATION DEER-CARCASS SLED AND CHAISE LOUNGE* (U.S. Patent 3,580,592)

4. *EYEGLASS FRAME WITH ADJUSTABLE REARVIEW MIRRORS* (U.S. Patent 3,423,150)

5. *FLUID-OPERATED ZIPPER* (U.S. Patent 3,517,423)

6. *POWER-OPERATED POOL CUE STICK* (U.S. Patent 3,495,826)

7. *CARRY-ALL HAT* (U.S. Patent 3,496,575)

A hat with a cavity for carrying cosmetics, jewelry, and the like.

8. *SIMULATED FIREARM WITH PIVOTALLY MOUNTED WHISKEY GLASS* (U.S. Patent 3,450,403)

Pulling the trigger pivots the glass towards a person's mouth.

9. *BABY-PATTING MACHINE* (U.S. Patent 3,552,388)

A device for putting a baby to sleep by means of periodic pats upon the rump or hind part of the baby.

10. *ELECTRONIC SNORE DEPRESSOR* (U.S. Patent 3,480,010)

Snore is detected and the snorer is electrically shocked.

Submitted by:
Michael J. Groves
Chicago, Illinois

"So, the question is; If a man falls over in the city and a tree isn't there to witness it, did it really happen?"

Irreproducible
great
thoughts
concerning

RELIGION
AND THE
COSMOS

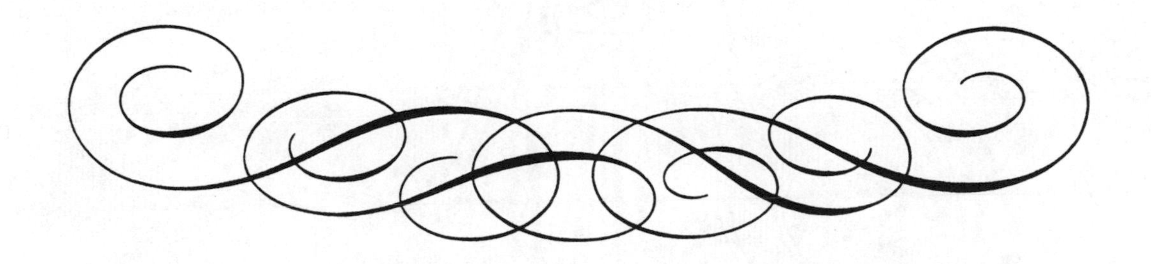

Proof:
How Many Angels Can Dance Upon the Head of A Pin?

JEFFREY A. BARACH

Ever since the middle ages, scholars have wondered how many angels could dance upon the head of a pin, without a satisfactory solution.* Finally, an answer to this medieval question has been found. The answer is: none. The reasoning follows.

The great mistake of the medieval scholastics was to think that the size of an angel was the key variable, when it is not. Size cannot only be factored out of the equation, but it confounds the solution. Since size is a terrestial, exact, real concept without strong moral overtones, its logical type (or level and character of abstraction)** is inappropriate for use with the concept of an angel in the context of a logical proof.

An angel is a metaphorical, anthropomorphic concept with moral overtones. That is its logical type, or level and character of abstraction.

In the same logical typology, we do not even need the aid of the works of Sigmund Freud to know that "to dance upon the head of a pin" is blatantly salacious.

Given the sexually repressive character of early Christianity,# no angel — either male or female — could dance upon the *head* of a *pin*. Q.E.D.

What is particularly fascinating is that the ancient scholars did not pose the problem upon another miniscule object, but rather chose the head of a pin and asked the angels to enjoy the pleasure of dancing upon it.

One reason for the fascination of this exact formulation of the problem and for the delayed discovery of the solution until the 20th century may be that the highly devout scholars of the Middle Ages could not answer it because of the "double bind"## placed upon them by the sexual connotation. If they had thought of the solution — perish the evil thought — they might have repressed it. Had they not repressed it, it would have approached heresy to articulate it, and implied the carnal sin of all who posed or pondered the question itself, including the one who claimed to have solved it.

Thus, the riddle remained, wrapped in the enigma of mismatched logical types, confounded by an extraneous variable (*Clupea harengus rubens*), and sealed in silence, binding by its nature anyone from leading his mind towards its solution and sealing his lips should he stumble upon it.

Only with the advent of a secular and sexually freer academic community could the bind be lifted, and by then, the matter was of insufficient interest to warrant the serious attention necessary to solve it.

---

*Vide William Duranti (Durandus), 1237-1296, *Rationale diuinorum officiorum*. Mainz: Fust and Schoeffer, 1st ed., 1459.

***Vide* Gregory Bateson, Don Jackson, Jay Haley, and John Weakland, "Toward a Theory of Schizophrenia," *Behavioral Science*, I, #4 (October 1956), pp. 251-64.

#E.g., "Sex and marriage always involves sin, partakes of sin, since it inevitably involves bestial movements," St. Augustine; "Sex in marriage has only one positive value, it can produce virgins," St. Jerome.

##Bateson, *et al*, *op. cit.*

"Clearly, the most challenging aspect of the problem from the point of view of potential for improved morality is the group of patients who die too suddenly to receive medical care."

Stuart Bondurant: Circulation Supplement IV, Vol. 40 (5), page IV-2, November 1969.

Well, that's one way to ensure that they won't sin no more.

The Creation Clarified

or

a Simple and Straightforward
Treatise to Explain the Hither to Complex
Study of the Creation of the Universe

It seems that we have to deride
Our experts who often have tried
To explain in terms terse
Our complex universe —
They have tried, but no facts could provide.

By focusing brains that are trained
The secrets of life are unchained;
All the puzzles dissolve
How the world did evolve —
The complexities now are explained.

ALBIN CHAPLIN
15362 Grandville
Detroit, Michigan 48223

The creation of the universe has been under perpetual discussion since man first learned to speak, and out of a multitude of ideas three major ones have come to the forefront, but up to the present time no one has been able to satisfactorily substantiate any one theory and dismiss all the others. This treatise is therefore presented to clarify this discrepancy, once and for all.

The three possible theories of the creation of the universe are well-known to all and are as follows: Firstly, that the universe is stable; secondly, that the universe has a beginning, and by implication, an ending; thirdly, that the universe is pulsating and is without beginning or ending. Any other theories are so highly conjectural as to have no place in a discussion of this type.

Let us consider the first proposal, that of the stable universe.

A stable universe means a changeless universe. This is a universe that has always been here and always will be here. In this kind of universe nothing can be created or destroyed. This suits the first law of thermodynamics, but in a changeless universe there can be no progress and no evolution. There can be no Garden of Eden, no Adam and Eve, it cannot progress and it cannot run down, and this is where it contradicts the second law of thermodynamics.

Since simple observation points to the fact that the sun is running down, novae are bursting out here and there, galaxies are receding, then we may conclude that we are most certainly not a part of a stable universe.

If we were to plot the amount of matter in a stable universe at any given time, we would obtain the graph as shown in Figure 1.

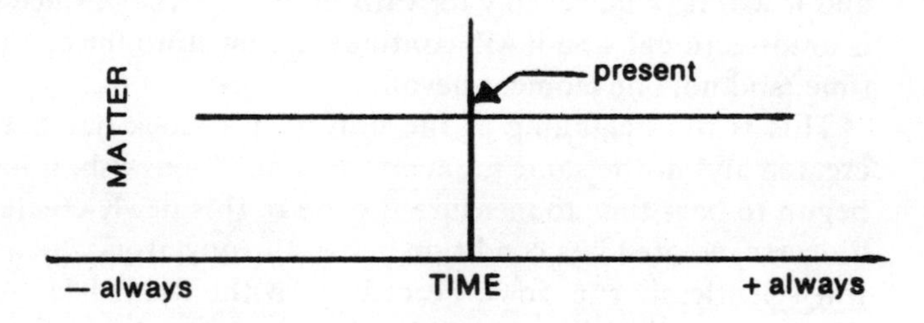

FIGURE 1

From Figure 1 it can be readily observed that there are no changes for better or for worse, and likewise there is nothing going on. This obviously is not the case, for in our universe, and especially in our planet, there is always something going on.

We can conclude that we are not living in a stable universe.

The second theory is not by any means as simple as the first theory. This theory supposes that the universe has had a beginning and therefore must also have an ending.

In this type of universe, a beginning implies that before the beginning there was nothing, absolutely and categorically nothing — just a vast emptiness of space. This is very difficult to imagine; as a matter of fact it is impossible to conceive of such a state of things, or as one may more aptly put it, such a state of nothings.

From this nothing we are suddenly plunged into a vast universe, all in an instant, with a big bang, so to speak, and this is often called the Big Bang Theory.

If we can for the moment conceive of this universe starting instantaneously with a big bang, and we plot matter against time, we will have a graph of the creation of the universe as shown in Figure 2.

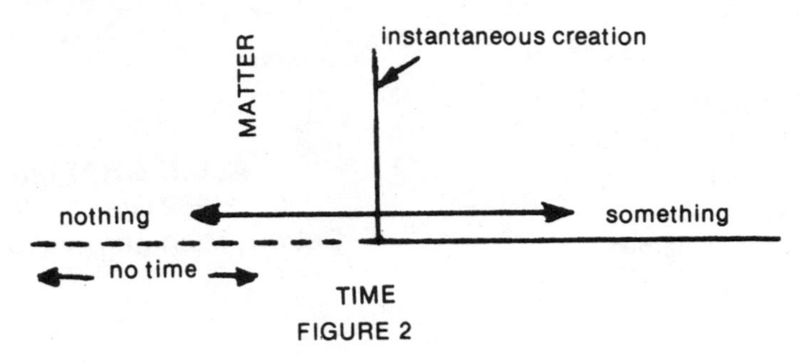

FIGURE 2

At some given time, although we cannot conceive that time exists and matter does not, a dot will appear on the graph representing x atoms of matter created in zero units of time. At the same time, or moment, if you please, time will begin, and it will flow inexorably forward in one direction, since it is unidirectional, and it will continue to flow unto the end of time, and not one moment beyond.

This is the beginning of the universe; a clock has been created and at the same moment, or time if you wish, it has begun to beat time to measure the life of this newly created universe, created in a condition of low entropy, from which it will relentlessly run down according to the second law of thermodynamics, like a child, who upon emerging from some mystical and formless womb, is undeniably faced with the prospect of death.

However, long before the entropy has increased to its maximum, the universe is doomed to disappear in, can we say, a blinding flash? But no, for a flash implies the conversion of matter into light and heat which is just another form of energy. Therefore the universe will disappear into an unblinding nothing, suddenly, and incapable of being observed. The eye, no matter how it strains, will not be able to see anything, for there will be no eye to see. All the light waves which stretch for septillions of miles into space in all directions will suddenly be overcome, and these light waves will be overtaken instantaneously and converted into nothing, and this of course is entirely possible, for nothing travels faster than light.

The graph of the annihilation of the universe would appear as in Figure 3.

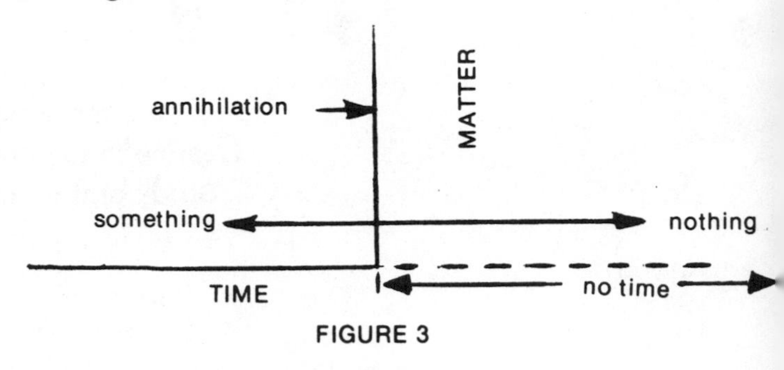

FIGURE 3

To show clearly the graph for the existence of the universe, we can combine the graphs from Figures 2 and 3, enlarge the dot representing instantaneous creation, split it in two for clarity, and we have the result as in Figure 4.

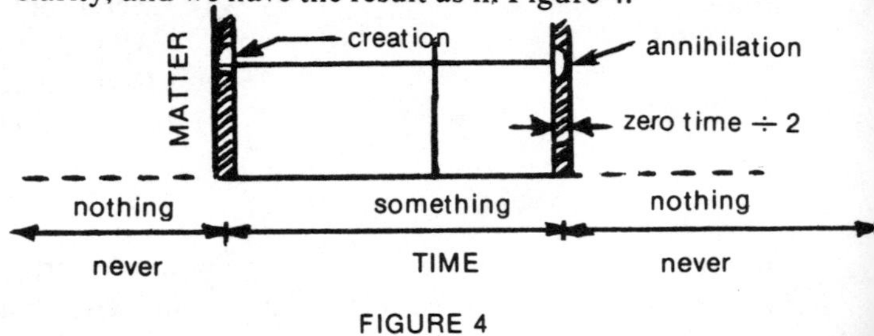

FIGURE 4

Note that the vertical lines which pass through the semidots, although in reality having no width, have been widened to conform with instantaneous occurrence, and with instantaneous annihilation.

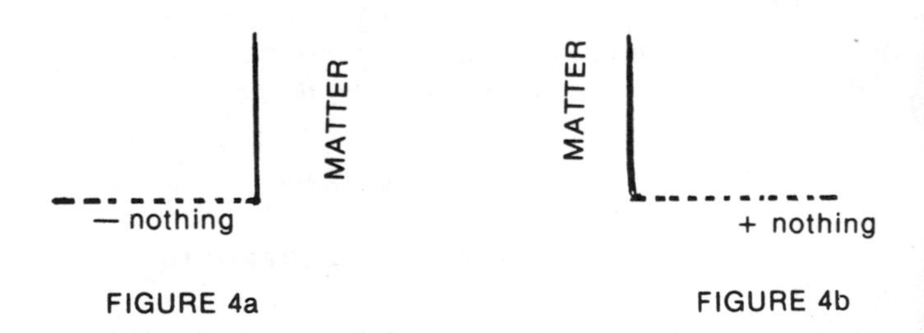

FIGURE 4a FIGURE 4b

Figures 4a and 4b are added for further clarification. They show the graphs of matter against time to the right and to the left of the "time" portion of Figure 4. Of course there is nothing on the graphs proper since time does not exist, and nothing therefore takes the place of time in the abscissa. The plus and minus values should be self-explanatory.

Now we come to the second version of the second theory, wherein the universe was created over a period of time, in the place of instantaneously, whether that time be six days or six microseconds. A graph plotting creation of matter against time is shown in Figure 5.

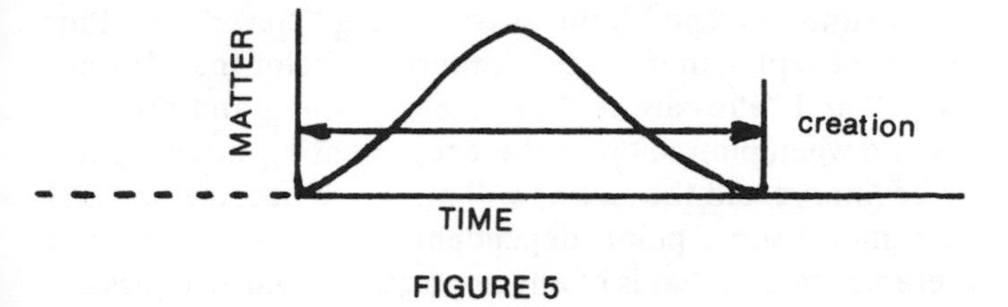

FIGURE 5

The area under the curve represents the total amount of matter in the universe. Time starts with the creation of matter and continues as long as matter exists. This is an excellent example of the Big Bang Theory where the universe is created in a finite time.

We still have to contend with the annihilation of the universe, where matter and time cease to exist. Once we have assumed that it takes time to create the universe, then ipso facto it follows that time is consumed in destroying the universe. This is the big bang in reverse and this is shown in Figure 6.

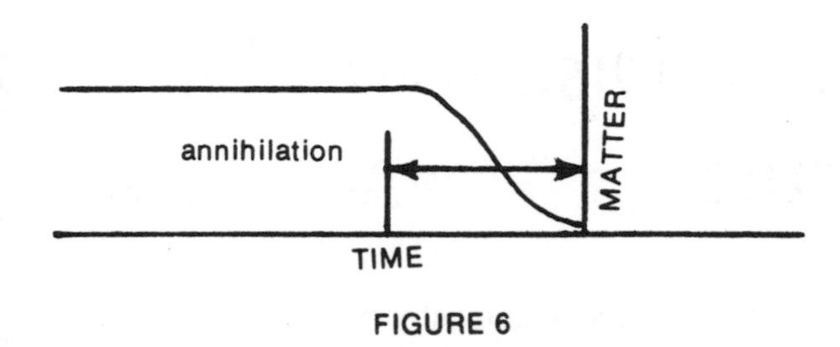

FIGURE 6

Combining Figures 5 and 6, we have the complete story of the universe according to the Big Bang Theory, version two, and this is shown in Figure 7.

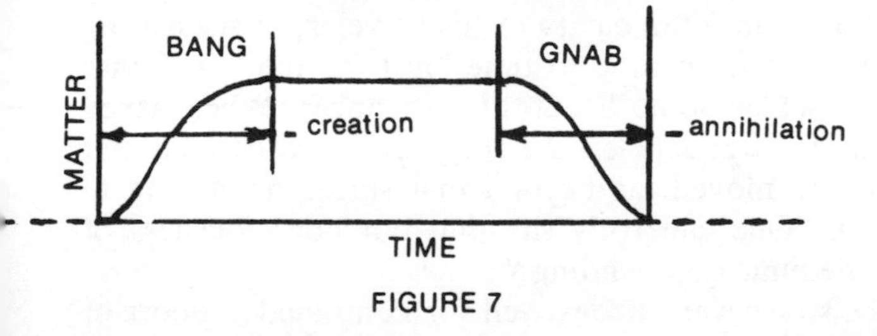

FIGURE 7

Thus we see, according to this theory, that the universe is created with a fantastic BANG, continues for a whole eon (an indefinite period) and then it disappears in the same amount of time that it took to create it, with an unheard of GNAB, or a tremendous SCHLLP, which is pronounced while sucking in wind.

Considering the two versions of the second theory, although this type of creation does lend itself to graphical analysis which can readily be comprehended by the layman, and it can simply be interpreted and plotted, yet it defies the first law of thermodynamics, which states that matter or energy can neither be created nor destroyed. Furthermore it also runs afoul of the second law of thermodynamics which states that the entropy in a closed system is increasing, whereas the sudden creation of the universe implies a sudden decrease in entropy from a condition wherein no entropy existed. This whole occurrence is a once-only phenomenon, and it is not within the ordered realm of science to draw a conclusion or formulate a law from one partially observable event. It follows here that we must look askance at the second law of thermodynamics, and I fear that this law must be subjected to further critical and unbiased examination.

These observations on the second theory, versions one and two, would lead one to conclude that the universe did not materialize with a big bang and will not dissolve with a gnab (schllp, while inhaling).

The third theory maintains that the universe always has been here and always will be here, and furthermore it is a pulsating universe, which consists of matter and energy, constantly converting one into the other, the total supply remaining constant, in deference to the first law of thermodynamics, but in dispute with the second law of thermodynamics, for this pulsating universe requires that entropy be pulsating and not be increasing, and a slight revision of the second law of thermodynamics is hereby in order.

A graph describing the pulsating universe is shown in Figure 8.

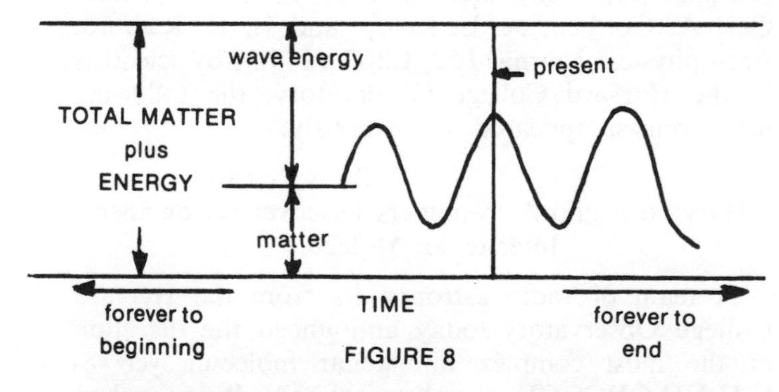

FIGURE 8

Time is always considered to flow one direction but no one has yet come up with a direction in which time flows. However, there is a simple way out of this dilemma. If Figure 8 is plotted on a sphere, time then can flow in any direction through curved space (ref. Einstein) and it will be noted that minus always on the graph will always meet plus always, and time can be considered to flow in all directions, somewhat in the same manner as light flows in all directions.

This whole theory of the pulsating universe is akin to a baseball being unraveled, while the end of the string is fed back into the core, leaving the baseball continuously the same size as it was at the beginning, if any, and as it will be at the end, if any. The pulsations, of course, would be evident if the string in the baseball were pulled out hand over hand, and this is evidently what occurs in the universe where great globs of matter are converted into energy, and this energy is gathered or forced into a core where it is reconverted into matter.

Another favorable point, and a veritable cornerstone of the pulsating universe theory, is that it answers all questions, such as, "When did the universe begin?", the answer being "Never" or "Minus always", and to the question, "When will the universe end?", the answer being "Never" or "Plus always"; keeping in mind this important factor that "Minus always" and "Plus always" are the beginning and the ending, and when plotted on a sphere representing curved space, the beginning and the ending will meet and become one and the same at some point, dependent, of course, on exactly where we start, if that is of any importance, and vice versa.

In conclusion we may state that this concept of the pulsating universe is the most plausible description of the universe, and it has been pulsating from the very beginning of time, and it will continue to pulsate unto the very end of time, barring any unforeseen catastrophe, such as World War III.

I sincerely hope that this treatise will clear up what has been in the past a very confusing situation regarding the creation of the universe. ■

Harvard Radio Astronomers Discover Yet Another Interstellar Molecule

JAY PASACHOFF

Astronomers have recently been discovering more and more complex molecules existing floating in space among the stars. Following the discovery of the most complex yet, methyl alcohol, CH_3OH (John A. Ball, Carl A. Gottlieb, A. E. Lilley, and H. E. Radford, Astrophysical Journal *162*, L203 (1970) by scientists at the Harvard College Observatory, the following mock release appeared anonymously:

Press Release
Harvard Radio Astronomers Discover yet another Interstellar Molecule

A team of radio astronomers from the Harvard College Observatory today announced the detection of the most complex interstellar molecule yet — $C_3H_4NH_3OHCOOH$ — "mescaline". When asked about the significance of this stunning discovery, the astronomers responded with a general euphoria. "Far out," said Professor A. E. Lilley, leader of the group, "compared to this, the methyl alcohol was a bummer."

At a sherry party held in their honor by the Professor of Theoretical Physics (curiously, the radio astronomers were not interested in the sherry), the Director of the Observatory observed that astronomy is becoming more relevant than ever before. Reaction of local citizens at the Cambridge Common was enthusiastic.

The critical scientific breakthrough that lead to this discovery actually occurred in the radio astronomy laboratories. Harry Radford, who performed the pioneering experiment, admitted that he had been dropping mescaline into the cavity of his wavelength measuring equipment from time to time "just for fun," and that it had set up some beautiful resonances. When asked about line frequencies, the entire group responded that they had moved far beyond that stage. According to Lilley, "One can truly say that our consciousness of the medium is expanding."

This discovery bodes well for continued support of astronomy. NASA is moving with immediate plans for a new manned space flight program, encouraged by the advice of the Harvard group that rockets are no longer necessary. Applications by local youths for the new astronaut program are pouring in, despite the restriction to Sagittarius only.

There were, however, some reservations expressed. One astronomer remarked, "Excuse me, but I would prefer to reserve comment until I consult with somebody absolutely first rate," and returned to his afternoon tea. ■

Confess: A Humanistic, Diagnostic-Prescriptive Computer Program to Decrease Person to Person Interaction Time During Confession

KENNETH MAJER
Institute for Child Study

MICHAEL C. FLANIGAN
School of Education
Indiana University

Recent Vatican interest in the effect upon laymen of the shortage of professional priests (PP) and the decreased seminary enrollment of potential priests (P'P) has led to the development of Computerized Operations (Non-retrievable) for Expediting Sinner Services (CONFESS). This program provides a viable alternative to traditional confession procedures by listing penance requirements (by sin) on a private print-out to confessees appropriate to the sin committed. This eliminates one problem which frequently occurs where the confessee, because he is under extreme duress, may forget the original penance. In addition, the program provides a probability estimate of the consequence of not completing the penance associated with a given sin; for example, number of years in purgatory. Thus, full freedom of choice is given to the participant/user (PU). The program requires no PP involvement and hence frees PPs to engage in more pressing activities. It is hoped that by providing PPs with more time for critical theological activities, P'Ps will consider the priesthood a more socially conscious and relevant profession, causing an increase of P'P enrollment in accredited seminaries.

Program Description

CONFESS is available in three natural interactive languages, COURSE WRITER III, BASIC and TUTOR and can be programmed for most other natural languages such as interactive FORTRAN. The program has been developed utilizing on-line computer terminals linked to an IBM 360 for data input, but could be modified to operate in batch mode on almost any third generation configuration given the willingness to sacrifice immediate feedback.

The computing procedures for CONFESS are as follows: The present sins input (psi) yields the graduated penance accrual (GPA) as a function of present sins (ps) plus frequency of confession visits (fcv) times completed penances (cp) divided by recurring sins (rs). Hence, GPA is a function not only of the immediate sins reported but also a partial function of the reciprocal relationship of recurring sins to completed penances by frequency of confession visits. The relative penance, then, is increased by the inclusion of recurring sins.[1] Mathematically, this can be represented as follows:

$$\text{psi} \rightarrow \text{GPA} = f \left\{ \text{ps} + \text{fcv} \left(\frac{\text{cp}}{\text{rs}}\right) \right\}$$

Therefore, each present sin yields a specific GPA that is stored until all GPAs have been computed. At that time, punishment and its maximum likelihood of occurrence[2] should the GPA not be completed, are retrieved from core storage and printed out for the individual GPA prescription.

Validity and Reliability

A study to establish the validity of the CONFESS program was conducted. The procedure included a sample of 243 actual confessions stratified across low, medium and high socio-economic income brackets with non-significant differences in proportions of black, white and Spanish speaking PUs. Fourteen priests were used in the study from seven different cities.

The actual sins confessed and penances prescribed in the confessional booths were tape recorded without the confessor or confessee's knowledge to insure absolute authenticity of confessor-confessee interaction.[3] The tapes were further analyzed and penances were rated on a scale of 1-10 where 10 = maximum severity.[4] Then ratings were made by the seven cardinal evaluators identified by Stake (AERA, 1972). The interrater reliability was .949.

The 243 sin sets taken from the taped confessions were then entered into the CONFESS program via remote terminal. A Pearson product moment correlation was computed between the actual PP penance prescriptions and the CONFESS PGAs. A correlation of .971 was interpreted to provide sufficient concurrent validity for CONFESS confidence.

A further series of small studies to determine the reliability of the CONFESS program were conducted as follows:

Study I: External Latency Reliability. The mean wait for confessional booths with PPs (where there were 2 booths/church) was 7.12 minutes while, in comparison, the average wait for a CONFESS box (one installation per church) was only 1.72 minutes. This difference in out-side wait latency is significant at the $p < .01$ level.

Study II: Internal Latency Reliability. This study examined the latency from the last sin confessed until the PP or CONFESS program provided the penance or GPA, respectively. Again, the CONFESS latency was significantly shorter than the PP latency. The means were 1.31 minutes (plus an average of 9.3 head shakes) for PPs, and 6.1 seconds for CONFESS.

Study III: Computer Breakdowns vs. PP Rest Breaks. In this study the CONFESS program was monitored for computer breakdowns and don't-understand-not-compute-either (DUNCE) loops. During the 243 CONFESS program runs (a total of 517 minutes), no breakdowns were reported and only one (1) DUNCE loop was reported. The DUNCE loop was in the case of one PU who was previously excommunicated from the church; however, the CONFESS program has been modified and will now process excommunicated PUs as well as non-excommunicated PUs. PPs, on the other hand, showed an average of 1 rest break for a mean of 12.3 minutes every hour and one-half.

Study IV: Consistency of PP penance vs. GPA. In this study, the 243 confession tapes were re-heard by the same 14 PPs. Each PP re-heard the same confessee's albeit on tape and without hearing the end of the tape which contained the penance he gave. In 241 cases, the PPs did *not* give the same penance and, in

fact, in 191 cases the penance severity changed at least one degree (e.g., from a severity rating of 7 to a severity rating of 8). Although no speculation for causality is made here, it is important to compare the CONFESS consistency. In all 243 cases, the GPA was identical.

The results of these four studies are sufficient to provide confidence in CONFESS program reliability.

Procedures/Output

Being a natural language program, the procedures for CONFESS are extremely simple. The following steps describe the PU procedures.

Step 1: Enter the CONFESS box,[5] and kneel on cushioned kneeler in front of the typewriter/console. Type in your personal PU identification code.

Step 2: The typewriter will type your name and the elapsed time since your last CONFESS session (CONFESSION). Following the request for present sins, type in all sins since your last CONFESSION.

Step 3: Press the "enter" button and silently repeat the short form of the ACT of Contrition. (Given the average latency for GPA, 6.1 seconds, this is usually reduced to "I'm sorry").

Step 4: Remove the CONFESS personalized GPA printout.

Sample Printout

CONFESS GPA PRINTOUT JOHN POPE Age 29

TIME SINCE LAST CONFESSION = 3 WEEKS

PRESENT SINS	TYPE	GPA	PUNISHMENT	PROBABILITY THEREOF
1. SECRETLY ENVIES BOSS	VENIAL	10 OUR FATHERS. PRACTICE SMILING AT BOSS	1 YEAR IN PURGATORY	.98
2. SWEAR AT WIFE	VENIAL	10 HAIL MARY'S. PRACTICE SMILING AT WIFE	1.73 YEARS IN PURGATORY	.84
3. COVET NEIGHBOR'S WIFE	MORTAL	ONE ROSARY/DAY FOR ONE WEEK. PRACTICE SMILING AT WIFE.	ETERNAL DAMNATION	.91

ONLY *3* SINS THIS TIME *MR. POPE*. YOU'RE IMPROVING. YOU HAD *14* LAST CONFESSION. NICE GOING. KEEP UP THE GOOD WORK. LET'S SEE IF YOU CAN MAKE OUT A LITTLE BETTER WITH NUMBER *3* IN THE FUTURE.

Availability

The write-up, listing, and source deck can be obtained from Dr. Kenneth Majer or Dr. Michael Flanigan, School of Education, Indiana University, Bloomington, Indiana 47401. ■

[1] Notice that updated past history records are necessary to compare previous sin records to present sin input in order to compute an up-to-date GPA. To insure confidentiality between confessor and confessee, a private code number is given to each PU and no master record is kept to identify the PU. This is the Non-retrievable aspect of the CONFESS program. Only by preceding the CONFESS session with his private code can the PU receive a GPA. Social security numbers are suggested as possible PU code identification numbers.

[2] This probability estimate taken from *Rome's Actual Transgression Sentences* (RATS) edited by Pope II. Randomness, Inc., Rome, 12 A.D.

[3] It should be noted that *ex post facto* permission to use the tape recordings was obtained from each confessor and confessee prior to use of the data. Hence, the sample of 243 represents a sub-sample of non-refusals from confessors and confessees. The original sample was 12,409.

[4] It may be of interest to the reader that 4 Our Father's and 4 Hail Mary's received a unanimous severity rating of 1 while 2 Rosaries per day plus mass each day for 2 months was rated unanimously as 10.

[5] The CONFESS box is patterned after the Skinner experimental box/chamber described in *Schedules of Reinforcement* by C. B. Ferster and B. F. Skinner, Appleton-Century Crofts, Inc., N. Y., 1957, 14-19. Although the present study does not address itself to the question of recurring sins and reinforcement/punishment contingencies, this question is currently under investigation by the authors.

REFERENCES

Ferster, C. B. and Skinner, B. F. *Schedules of Reinforcement.* Appleton-Century Crofts, Incs. New York, 1957, 14-19.

Rome's Actual Transgressions (RATS). Ed. Pope II. Randomness, Inc., Rome, 12 A.D.

Stake, Robert. *The Seven Cardinal Evaluators.* A paper presented at the National American Educational Research Association in Chicago, 1972.

Lazarus' Last Will and Testament

W.R. Fahrner and D. Decman
Hahn-Meitner-Institute for Nuclear Research,
Berlin, Germany

The question of Lazarus' last will and testament is an important legal problem that has found undue little attention up to now. Perhaps it is surprising that this investigation should be carried out by solid state and nuclear physicists in an institute for nuclear research. However one must remember that lawyers direct the German ministry of Research and Technology; in this sense the usefulness of this report is obvious even to the layman.

The problem dealt with here can be expressed in a single sentence: What are the legal consequences of Lazarus' resurrection? In the first place, we have the inheritance disputes; can Lazarus reclaim his estate? Let us review the circumstances of the time. No death certificate had been written, but at that time this was normal. Some hundred honest witnesses testify that the death has occured. His estate is divided according to his will or in absence of a will according to the courts. Suddenly Lazarus comes back and can reclaim all that was his. Apparently Lazarus has not tried to do this. Therefore the heirs at the time of his second (final) death must now file suit against the heirs from the time of his first death. How can this matter be resolved?

In regard to the suit we can simply say that Lazarus and his "second" heirs would have little chance. His one hope lies in the fact that he is his own closest relative and as such has a better claim than other potential heirs. On the other hand he could present his case to be similar to that of an unborn child and could claim his position in the order of heirs. This decision would have disadvantages, that is, Lazarus might be required to attend school a second time and might also be eligible for the draft.

These are only some of the problems to which this poor man is subject. If his widow remarries, is she a bigamist? Are his children entitled to the survivor benefits? When he applies for life insurance, on what age will his premiums be based? In short, we should pity this man. There might be a widespread opinion that the problem "Lazarus" is a singular case and thus without any further relevance. This opinion is wrong. Let us rather recall some very recent events or persons certainly known to everyone of us: Firstly Alkestis, Greek queen, who undertook death instead of her husband Admetos. This one had broken his promise to sacrifice to the goddess Artemis, and was dedicated to death unless someone else would die for him. Herakles fought for Alkestis' life in the underworld and brought her back from the 'hades'.

So we have mentioned another person commuting very often between death and life, namely Herakles. He really enjoyed this occupation. The third person was less lucky: Theseus wanted to liberate Persephone from the hades, but was captured and released only by Herakles' help.

It is evident that our history books are inundated by such revivals. To our surprise, no "resurrection" columns are found in the newspapers next to the obituary notices. We thus demand an end of the legal vacuum for the concerned people: The German congress is urged to find a solution to this dangling legal problem by a lex Lazarus. ■

ASHES TO ASHES

H.J. Stevens
Alfred, New York

During the last two years we have been involved with studies on lunar glasses and their properties under the sponsorship of NASA Contract No. 33-187 (001). In the course of our work we analyzed the composition of several glass spheres[1] extracted from four half gram samples of lunar soil 1mm in depth (Samples 12070,38; 12057,61; 12033,25; 12001,74). Sample 12033,25 was taken from a depth of 15cm. The range of the major analyzed compositions is given in Table I.

Table I
Lunar Glass

Oxides	Wgt. %
SiO_2	38.8 - 48.8
Al_2O_3	9.8 - 27.3
FeO	3.52 - 19.4
CaO	9.0 - 13.55
MgO	5.8 - 12.8

A new research project on fly ash sponsored by a local electrical power company revealed an interesting correlation between fly ash, the residue from coal combustion, and our analyzed lunar spheres.

Fly ash and bottom ash are the residue left from the combustion of coal in the power plants. The fly ash is collected from the stack through a series of collectors. It is then generally shipped to suitable land fill sites. The bottom ash is that material which settles to the bottom of the furnace, and is removed from there as a glass, often spherical in form. It is also used for land fill.

The chemical composition of fly and bottom ash is given in Table II.

Table II
Fly & Bottom Ash

Oxides	Wgt. %
SiO_2	41.9
Al_2O_3	27.6
Fe_2O_3	16.1
CaO	3.1
MgO .	0.8

Chemically, it is very similar to the analysis of the lunar glass that we reported. The appearance of the fly and bottom ash is also very similar to lunar fines in regard to color and flow characteristics. Particle analysis of lunar fines[2] showed

57.5% thru 200 mesh .074 mm.
45.4% thru 325 mesh .044 mm.

The fly ash particle size analysis varied slightly from the lunar fines with

87.0% thru 200 mesh
77.5% thru 325 mesh

The similarities between the two materials immediately evoked discussion among the members of our research staff and various theories about the origin of the lunar fines were projected.

The theory that our power companies were secretly shipping waste fly ash to the moon was ruled out because of two factors: 1) The fly ash has iron in the form of Fe_2O_3, whereas the lunar fines are mainly FeO. This would indicate that the original formation occurred in a more reducing atmosphere than that which occurs on earth, i.e., the moon's atmosphere. 2) Since the particle size of the lunar fines is slightly larger than the fly ash, formation on the moon is indicated. The moon, having a lower atmospheric pressure and gravity, would permit larger particles to be blown through the stack.

The slight differences in composition also indicates another raw material source.

All of these observations point to one inevitable conclusion: Sometime in the history of the moon, large power generating stations covered the landscape. In the quest for more and more power, the warnings of the ecologists were ignored and alas the moon buried itself with its own fly ash! ■

REFERENCES

[1] Greene, C. H., Pye, L. D., Stevens, H. J., Rase, D. E., and Kay, H. F., *"Compostions, Homogeneity, Densities and Thermal History Of Lunar Glass Particles,"* Proceedings of the 2nd Annual Lunar Science Conference, Vol. 3, pp. 2049-2055, The M.I.T. Press (1971).

[2] Frondel, C., Klein, C., and Ito, J., *"Mineralogical and Chemical Data On Apollo 12 Lunar Fines,"* Proceedings of the 2nd Annual Lunar Science Conference, Vol. 1, pp. 719-726, The M.I.T. Press (1971).

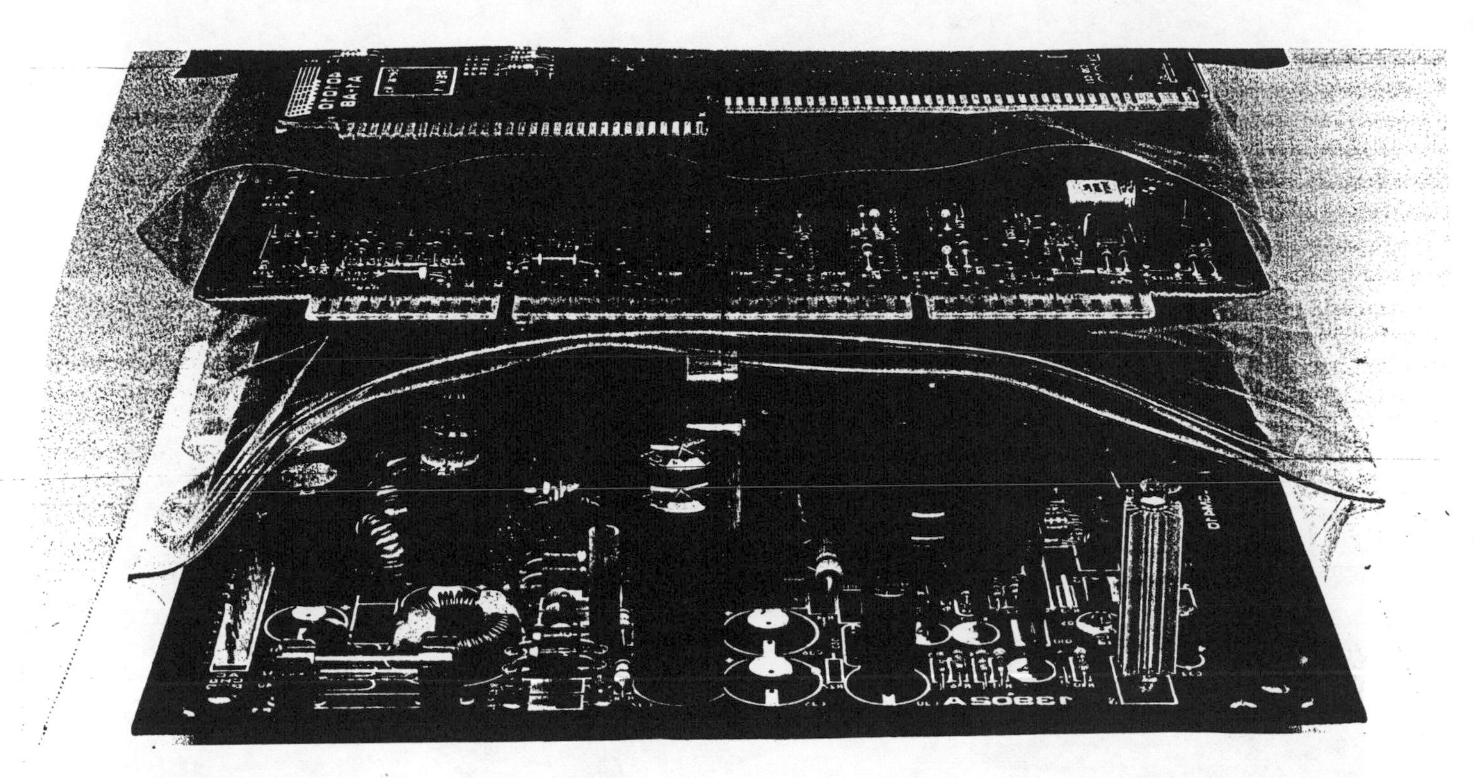

Computer Program Virtually Eliminates Machine Errors

W. S. MINKLER, JR.
Pittsburgh

Spokesmen for a local electronic firm have announced a computer program that—through fresh application of an old technique—virtually eliminates lost time due to malfunction of computer components. Called OREMA (from Latin *oremus,* meaning let us pray), the program offers prayers at selected time intervals for the continued integrity of memory units, tape transports, and other elements subject to depravity.

Basically liturgical in structure, OREMA used standard petitions and intercessions stored on magnetic tapes in Latin, Hebrew, and FORTRAN. It holds regular Maintenance Services thrice daily on an automatic cycle, and operation intervention is required only for mounting tapes and making responses, such as "Amen," or "And with thy spirit" on the console typewriter.

Prayers in Hebrew and *Fortran* are offered directly to the CPU, but Latin prayers may go to peripheral equipment for transfer to the CPU by internal subroutines.

Although manufacturer-supplied prayer reels cover all machine troubles known today, the program will add punch card prayers to any tape, as needed, after the final existing Amen block. Classified prayer reels are available for government installations.

In trials on selected machines, OREMA reduced by 98.2 percent the average down time due to component failure. The manufacturer's spokesman exphasized, however, that OREMA presently defends only against malfunction of hardware. Requestor errors and other human blunders will continue unchecked until completion of a later version, to be called SIN-OREMA.

Reprinted from Data Link, March 1966, which reprinted the paper from THE SOURCE (Pittsburgh Section of American Nuclear Society), Jan. 1965.

Submitted: R. B. Gordon (Raytheon)

Irreproducible
forays
into the
intriguing
mystery
of

SEX

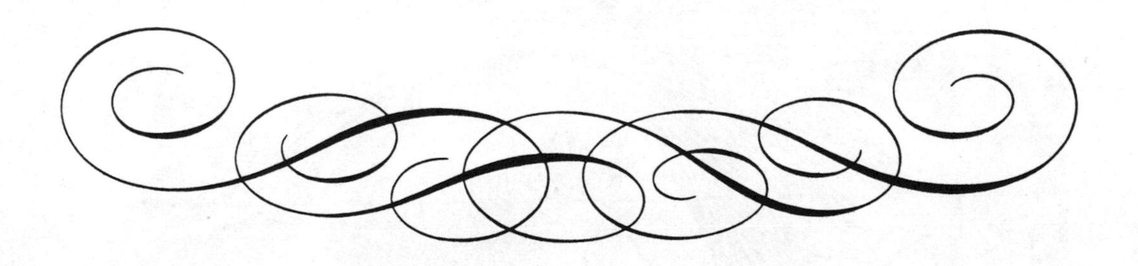

A young virgin quite introducible
Was tested but found untraducible

"You see, if I yield
To you here in this field,

My state would be irreproducible."

Beverely Christen
Irondale, OH

Sexual Perversion in the Poetry of A. E. Housman

Brenda E. Sartoris
Mississippi State, MS

THE RAKE'S PROGRESS, by Hogarth, Sir John Soane's Museum, London England

Loveliest of trees, the cherry now
Is hung with bloom along the bough,
And stands about the woodland ride
Wearing white for Eastertide.

Now, of my threescore years and ten,
Twenty will not come again,
And take from seventy springs a score,
It only leaves me fifty more.

And since to look at things in bloom
Fifty springs is little room,
About the woodlands I must go
To see the cherry hung with snow.

Critics have traditionally regarded this fine poem by the English poet A.E. Housman as a commentary on the brevity of human life and a lament for the transitory beauties of nature which human life has all too short a time to enjoy. However, such a reading is both superficial and simplistic, dependent as it is on a casual, surface reading of the poem. When one recalls the background against which it was written, i.e., the repressed sexuality of late Victorian England, the poem then, carefully read, reveals a subsurface which reflects Housman's own perverted sexuality. Properly interpreted, the imagery in the poem supports a view of Housman as a peeping tom with somewhat unusual tastes.

The spring setting of the poem indicates the poet's focus on the motif of the generative sexual forces in the world traditionally associated with the season of rebirth—an association further emphasized by the reference to Easter. Thus the way is opened for an interpretation of other sexual elements in the imagery. It is clear from the opening stanza that the poet is in the habit of riding about the countryside in springtime to look at—ostensibly—the cherry trees in bloom. But the verb "ride" carries with it the slang connotations of sexual intercourse; and cherries, of course, suggest virgins. It becomes clear then that the poet goes about the countryside in spring, getting his rocks off, so to speak, by peeping at virgins. This notion of cherries as symbolizing virgins is reinforced by the image "wearing white for Eastertide," white itself representing purity, as well as being the color young girls traditionally wear when making their First Communion (Easter being the season at which the First Communion is usually made in the English church). Hence, at best we see Housman as a peeping tom; at worst, as the eminent critic V.D. Buttock-Smith has suggested, a molester of pre-pubescent girls.[1]

In the references to his age in the second stanza, as well as to the sense of urgency which enters in to the poem at this point, Housman's perversion takes a rather interesting turn, pointing to a rather novel problem which he is presented in his quest to satisfy his voyeuristic desires. He indicates that he feels "Time's winged chariot hovering near," to quote another poet with similar concerns. The problem is made clear in the third stanza and the sense of urgency reinforced when the poet states "Since to look at things in bloom/ Fifty springs is little room/ About the woodlands I *must* go [italics mine]/ To see the cherry hung with snow."[2] These lines suggest two things. First, that the poet feels that fifty years is too short a time to "look at things in bloom," and that he must ride about the countryside in order to satisfy this urge imply, perhaps, a shortage of virgins in his own part of the country, in which case these lines become a biting thrust at the growing laxity in rural morals—the very idea that one should have to go to such lengths to observe a virgin! The closing lines, in the choice of the verb "must go," indicate the compulsive nature of the poet's perversion. And we are perhaps therefore left with a sense of sympathy rather than one of repulsion. ■

[1]I hesitate to accept Professor Buttock-Smith's rather harsh assessment; it seems clear, from the other images in the poem, that Housman's activities in this regard consist mainly of looking—as I shall endeavor to show.

[2]T. Titworth Bare, the late eminent Victorian commentator on the manners and mores of his age, has argued that Housman may reflect here a peculiarity in his perversion. Bare notes that "snow" is a traditional image of old age (suggesting white hair, etc.); he therefore concludes that the phrase "to see the cherry hung with snow" indicates Housman's preference for peeping at *old* virgins, i.e., spinsters. The scarcity of such would certainly explain the urgency of Housman's quest, but Bare's interpretation has not been given much credence by subsequent readers of the poem.

Sexual Perversion in the Criticism of B. E. Sartoris

Richard Patteson
Mississippi State, MS

Professor B.E. Sartoris's Housman fantasy achieves so high a degree of risibility that one might be tempted to think it all an elaborate, tasteless joke. But the well-known conservatism of th editors of *The Organ* places such an inference beyond the bounds of plausibility. Sartoris's piece, I fear, is a serious attempt at a revisionist interpretation of Housman. Her "analysis" does nothing less than imply—through citation of the most spurious "evidence" imaginable—that A.E. Housman was a latent heterosexual. A slander of this magnitude, undermining as it does Housman's poetry *and* his good name, cannot go unchallenged.

One need not even call to witness the facts of Housman's life to refute Sartoris's vile and ridiculous allegations; proof of their inaccuracy is in the very poem she pretends to discuss. For instance, Sartoris focuses her libidinous attention on the word "cherry" (clearly a reference to hemorrhoids, one of the less fortunate by-products of frequent anal penetration), and finds in it a suggestion of "virgins." Virgins indeed! One begins to wonder just who is the pervert here, Housman or Sartoris. If the poet does have virgins in mind, they are surely not of the female species. The tree (a phallic symbol) "stands" proudly erect in stanza one. Hardly any imagination is required to translate "along the bough" into "a long bough" (i.e., penis). The word "hung" indicates the generous length and thickness of the "bough," and as for "Wearing white," we all know (with the evident exception of Professor Sartoris) the color of semen. Finally (and I am still on the first stanza!), Sartoris drags religion ("First Communion") into her comment on "Eastertide"—completely overlooking the obvious "erection" implicit in resurrection.

The correct reading of the rest of the poem is equally self-evident. The "come" in stanza two, embraced as it is by "threescore" and "score," reveals most unambiguously Housman's desire to score and, of course, to come. Coming usually follows naturally from scoring (the two are often held to be one and the same), but perhaps this is a concept foreign to the experience of Professor Sartoris who seems to identify rather closely with the old spinsters mentioned in her Titworth Bare[1] footnote. In the last line of the poem "hung" and "snow" again refer to the male member discharging its milky load, and the coldness of snow additionally recalls a line from another Housman poem:

> chiefly I remember
> How Dick would hate the cold.

In face of all this, what else can be said of Sartoris's bizarre tissue of innuendo and wish-fullfillment? She does, one has to admit, recognize the urgency of Housman's sexual drive, but she utterly misses the point as to its goal. A more objective approach to the poem (such as mine) might have helped her to understand that the poets' urges are directed solely at escaping from his "little room" (line 10), or closet, and into the "woodlands" where well-hung "trees" abound. At the risk of relying too heavily on the extra-textual, I must cite yet another Housman poem to make explicit what the poet expects to find in those woodlands:

> By brooks too broad for leaping
> The lighfoot boys are laid.

If Professor Sartoris gets her "rocks off" (as she so delicately puts it) by imagining A.E. Housman hot in pursuit of randy females, I have the undiluted pleasure to inform her that this time, alas, the lady is barking up the wrong tree. ■

[1]By calling T. Titworth Bare an "eminent Victorian," Sartoris apparently wishes to imply that a gulf of years separates him from her. In reality, Sartoris and the late Professor Bare are exact contemporaries.

FOOD and SEX: An Ancient Problem Recently Visited...

BEN Z. DREAN *
Texas Christian Institute of the Institutionalized

A total of 23 subjects participated in a social psychological experiment. All subjects were about half male and half female. Each of two groups were composed of 11 ½ subjects. The first group was aked to do various things. The second group, the control, was aked if they would please not do anything. The results lend support to the hypothesis (p= .05) that as far as sex is concerned doing something is often better than nothing doing.

Down through the years various psychologists studying sex have asked various subjects to do various things in various experiments. Interestingly enough they have gotten various results which they have interpreted in various ways for various reasons. For instance we might look at the results of Round and Ribeye (1937). Using bulls and cows with steers as a control, they found their results were, indeed, inconsistent with T-Bone and Sirloin (1967) who tried a different approach 40 years later for higher stakes. To account for these discrepant results Swiss (1954) devised an experimentum crucis that was not only trivial but infeasible as well. His results were similar.

Heartburn and Bloated's replication (1951) found that sex made little differences. This throws considerable doubt on the interpretations given by Contra and Ceptive (1969).

Seltzer (1960) found that depending on the feeding patterns the researchers used, different weights are possible. Fizrin, Anacin, and Excedrin (1961) like-wise found a high positive correlation between the cost of living and an increase in wholesale and retail prices. Nixon, Agnew and Mitchell, et al. (1972) have thrown their weights all over.

A stiking pattern also runs throughout Thunder and Lightning's (1975) series of meteorological experiments. The development of their stunning results is left as a survival exercise for the interested reader.

Thus the present study seeks to test the hypothesis that it is difficult to compose groups of 11½ subjects each who are half male and half female and if they are so composed to keep the parts apart.

Method

Subjects

Without exception all subjects were either half male or half female, and were highly motivated to fufill the experimental assignment. For the purposes of this experiment all subjects were experimentally high, except those drawn from Sociology/Anthropology classes who were normally stoned. Eight subjects had to be dropped from the experiment due to their equipment failure; two of the eight are still alive on Vitamin E. The reader may wish to watch for a subsequent report on the six subjects who were consumed by passion. This report will include comparison of data of the six who died with the two who lived to experiment another day.

Materials

For a complete description of the experimental apparatus the reader is directed to the Gray's Anatomy, Kinsey, and Masters and Johnson. The critical reader may also wish to consult the Montgomery Ward Catalog, Fall and Winter Editions 1973 in the bedding section for a thorough description of the experimental environment.

Procedure

All groups (I & II) were randomly arranged by Pointing's (1970) first finger procedure. The experimenter's finger was

*Now at the Department of Psychology, Kansas State University, under the assumed name of Steve R. Baumgardner.

first given to each member of group I and then to each member of group II, in that order using Insults (1984) random tables.

Further research is needed to determine exactly what was done after the two groups were assigned.

Results

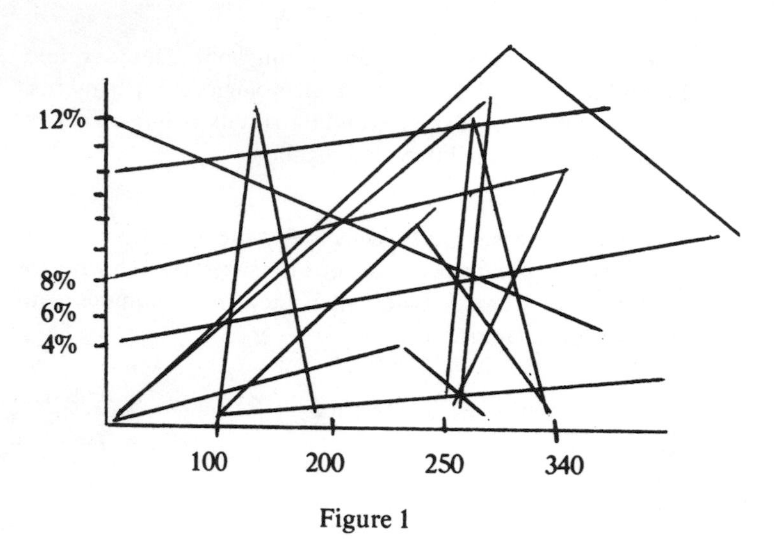

Figure 1

The results, shown clearly in Figure 1, speak for themselves.

Discussion

The results also indicate the presence of a subtle interaction. Although interpretation is difficult because the axes of the graph are not labelled, total confusion was unaffected. Next a subject-for-thing analysis of variance was carried out of the country. From here, the results obtained, were very close to significance. From these results one can easily prove that if the data are graphed the different lines will cross.

With regard to the variable of sex, a fly-by-night analysis was done but the results could not be found. Half of the Ss agree, however, that curves were preferred to lines. Though some lines worked others were rejected since many Ss were disclined. This indicated a change. Only further research can clarify the many variables that were fooling around. In the words of one of the more successful Ss , this study was an end in itself. ■

REFERENCES

Contra, Diction and Ceptive, Very Ree, "No, and Then Again Yes," *Journal of Sexual Decisions*, Vol. 69, 1969.

Fizrin, Ima, Annacin, Ann A., Excedrin, X. *Headaches We Have Known*, Purple Prose Publishers, New York, New York, 1961.

Gray, T. *Anatomy of The Human Body*, Adam and Eve Press, T & E Vally, Year I.B.C.

Heartburn, Heavy and Bloated, Puffed N. "Too Much, Much Too Much," "*Quality Meats*," Vol. 39, No. 5, 1951.

Kinsey, Alfred, *The Sex Books*, Sanders, 1950-1970.

Masters and Johnson, *Every Bedroom in America*.

Montgomery Ward Fall and Winter Catalog, 1973 edition, Chicago, Illinois (Write immediately).

Nixon, Richard M., Agnew, S., and Mitchell, M. *Four More Years*, Sad State of the Union Press, Washington, D.C., 1972.

Round, Ground and Ribeye, Red, "Steaks of Quality," *Meat and Potatoes*, Vol. 100004, No. 1, A No. 1, A & P Tea Company, 1937, Pgs. 1-100.

Seltzer, Sizzen, "Bromos and Homos," *Deviant Foods for Deviant People*, Vol. 100004, No. 321, Pgs. 2-22, 1960.

Swiss, Cheese, "Mr. and Miss Takes", *J. of Paranormal Sex*, Vol. 2, 1954, Pgs. 0-000.

T-Bone, Liptons, and Sirloin, Susie, "Higher Stakes," *J. of Poker Chips*, Vol. 1, No. 4, 1967, Pgs. 2-20.

Thunder, Lotta and Lightning, Shite, " The Positive Relationship Between Stormy Sex and Stormy Weather," *Journal of Flare Ups and Fire Storms*, Vol. 17, No. 9, Pgs. 12-24, 1975.

S. J. Zimmerman, Maureen B. Maude and M. Moldawer
FREEZING AND STORAGE OF HUMAN SEMEN IN 50 HEALTHY MEDICAL STUDENTS.
Fertility and Sterility 1964, *15*, 505

(If done properly, this would be the neatest trick of the year.)

Preliminary Observations Upon An Isobaric Spin Model For Human Sexuality

A. W. KUHFELD et al.
M.I.T. (Monster Institute of Transylvania),
Black, Mass.

Abstract: Research into sex has long been hindered by the common practice of considering the human male and female to be two distinct entities, having totally unrelated properties. The clearest expression of this attitude may be found in the erroneous system of experimental calibration often referred to as the Double Standard. Rather, there is an intimate relationship between these two forms of humans, as recent experiments have conclusively revealed. This relationship is amenable to various formalisms of modern physical science.

Certain aspects of human sexuality have been investigated in detail by the methods of nuclear physics[1], but the field is still largely unexplored. In particular, it is astonishing that the powerful methods of Group Theory have never been applied to the phenomenon, in view of the well-documented knowledge that most human sexual behavior of interest occurs in groups of two or more. The most prominent human sexual characteristics is dualism, which suggests that an isobaric-spin formalism based upon the more familiar $t=1/2$ baryon doublet would be immediately applicable.

It is generally conceded that the male gets a larger charge out of sex than the female[2] (although this finding has recently been challenged by Huang, Bang and Sigh[3], who claim that the greater escitation of the female is masked by lack of a monopole emission mechanism, giving the male a spuriously higher transition rate to the unexcited state). We may therefore tentatively assign $t_z=+1/2$ to the male and $-1/2$ to the female. If the sexual investigations now proceeding in almost all universities and colleges having proper research facilities produce data requiring modifications of this assumption, we can easily apply a similarity transformation to all results obtained under this assumption.

Statistics are another important aspect of the problem, since they will govern the sexual relationship between isospin coupling and spatial behavior. Early and easily-obtained data upon the deviations of humans from ideal-gas behavior at room temperature would suggest that humans are fermions rather than bosons. In particular, when humans are compressed they tend to maintain as large an average inter-human distance as possible. This is especially noticeable in elevators and buses, where there is a limiting density which is asymptotically approached only under great pressure. It is true that the average m-f distance is smaller than the average m-m or f-f distance, which is strong support for the fermion concept, but this behavior could be governed by the potentials of the system. It is known that many apparently-attractive humans possess repulsive cores.

Stronger evidence in favor of fermionic behavior in humans comes from observing the pure T=1 states m-m and f-f. In the closely-interacting system the most commonly observed behavior is antisymmetric, with the humans oriented in opposite directions. Since this is an S-state (the only relative motion being radial rather than angular) opposed intrinsic orientations imply antisymmetry. The analogue of this T=1 state is also observed in the m-f system, although there has not been enough valid research to establish the relative amplitude of such behavior. If it is found to occur with equal strength in the f-f, m-f, and m-m systems then our isobaric-spin model is upheld. In any case, the T=O coupling with m and f having parallel orientations is preponderant. This indicates that the T=O attraction is much stronger than T=1 under normal conditions.

Unfortunately for acceptance or rejection of the group-theoretic hypothesis, by far the greatest number of experiments in the interaction of humans occur in the T=O, spatially symmetric m-f state. This is not surprising, for the T=O behavior is far more accessible to the majority of researchers (with their limited funds and specialized equipment) than the theoretically more complex T=1 system. There is a surprising lack of information upon the t=1/2 isolated human in the literature, although it would seem at first that the study of this system would be the easiest of all. (We may speculate that the government considers the dissemination of knowledge about t=1/2 behavior dangerous to the national security, and is therefore suppressing all mention of the subject. This seems improbable, for t=1/2 behavior rarely leads to observable consequences, while T=O interactions appear to provide the main motive power for many humans. It could be that the government, acceding to the many requests for it to attempt feedback in the t=1/2 state, has experienced a breakthrough. In the balance, however, it seems likely most researchers find the t=1/2 system interesting only as preparation for T=O and 1 investigations. Since there is no tradition of "publish or perish" in this field—a unique situation in itself—this would explain the scanty literature).

Although this model shows considerable promise in evaluation of many of the features of human sexuality, it is not totally valid. Obviously, with the mass of the female less than the mass of the male, they are not totally describable by an isobaric model (which assumes equal masses). Also, the mesons mediating the interaction seem to be almost totally emitted by the male and absorbed by the female. A mestastable state with a nine-month lifetime has been observed in the female, while multiple emission from this metastable state is also infrequently seen (with a slightly smaller average lifetime). This metastability is completely lacking in the male. It is obvious, therefore, that there is some symmetry-breaking interaction present. This interaction may in most cases be treated only as a perturbation; it is most perturbing when the metastable state is excited, since this excitation *requires* a symmetry-breaking interaction.

Despite the inadequacies of the isobaric-spin model, it suggests many lines of future research. Perhaps the most fruitful would be the highly-excited states of the many-body problem (referred to in the popular literature as the "orgy"). If we assume that all interactions are the sum of the two-body interactions involved, then it is possible to predict immediately that in the three-body problem the T=1/2 state will dominate the T=3/2 state; and that the T=1/2 contribution will come from a strong m-f T=O coupling, with the remaining human loosely coupled to this pair. An investigation of this system as a function of the valence-human t_z might shed a good deal of light on the symmetry-breaking interaction. ■

REFERENCES

1 "The Quantum Mechanics of Sex" Arluis, E. Vell, *Journal of the M.I.T. Science Fiction Society*, 69, 20-443 (1984).

2 "Do Children Have as Much Fun in Children as Adults Do in Adultery"? Fraud, S., *Aberrational Psychophysics,* 1 1-2 (1894).

3 "Upon Certain Aspects of Certain Interactions About Which Nothing is Certain" Huang, Bang and Sigh, *Comptes Rendus Hebdomadaires de la Societe Aphrodites Anonyme, .236A* 56-100 (1966).

MALE STERILIZATION

It's in the bag these days

BEN C. SCHARF, D.O., F.A.C.G.P.
Editor, N.Y. State Osteopathic
Society BULLETIN

Vasectomy has become so popular these days that hardly a month goes by that some journal, magazine, or newspaper[1] does not have an article about it so I can see no reason why THE JOURNAL OF IRREPRODUCIBLE RESULTS should not have one as well.

Having performed more vasectomies in my practice area than the combined caseload of seven urologists (and I include one urologist[2] in my statistics), I feel eminently qualified to give you my feelings about feeling for a vas in an area undoubtedly containing more tactile sensory organs than any other in the male.

The vas deferens makes no difference to anyone but the owner until such time as he decides to have it resected. According to Dr. Phil Anderer who has had more experience with post vasectomy problems[3] than most, the difference in the consistancy of the vas from surrounding structures is obvious to the be-holder. It stands to reason that the female has more experience than the male in this delicate area.[4]

The most frequent reason given by patients for wanting a vasectomy is that they do not want anymore children by anyone.[5] The most frequent reason given by wives when consenting to their husbands having the operation is that they want more for less.[6]

While there is evidence that as much as three quarters or more of the scrotum can be burned or denuded and a whole new scrotum will reform we have not had any success in recruiting volunteers for this operation.[7] The one sixteenth of an inch incision we make for exposing the vas is sufficient for most patients to holler "uncle" [8] The post-op pain has been described as similar to that experienced when getting a baseball pitched at a testicle.[9] Here again we were unable to do objective studies to prove this as we could get no volunteers to pitch baseballs at. Research has further been hampered in trying to ascertain the libido of vasectomized males vis a vis other vasectomized males according to Dr. Matt A. Chine.[10]

A frequently asked question is "What happens to the sperm that keep being produced by the testes after vasectomy?". No one really knows but Dr. Mary Little Lamb's studies indicate that they continue to wag their tails even years later and after repair and recanalization of the vas.

There is absolutely no truth to the rumor that there are failures in doing this sterilization procedure. To prove this rumor to be false, we interviewed the wife of one of our patients several years post-op. We had difficulty in locating her and at last report she was living in a house that passersby swore was the shape of a shoe having moved there only three years ago from a one bedroom apartment, just after her husband was sterilized.

It is reported that in India volunteers for vasectomy are granted a transistor radio by the government.[11] This, too, is in error. The facts are that the instrument is a camouflaged geiger counter calibrated to keep track of the males post-op migratory habits much as they tag the giant wild rabbits of the great African plains of Heer-Icom.

One final note. You can always tell a male who has been vasectomized by his behavior. Some say he de-man-ds too much, others that he is de-men-ted and some swear they have become wee men. ■

---

1 a. *Journal of Playboy Medicine*, March 1970, "Cut it out—For Crying out Louder".
b. *Magazine for Men*, April 1970, "Sterilization: Vas the Difference—Men or Women?"
c. *The Dilly Dally City Daily*, May 1970, "No Dilly, No Dally".

2 No he was not a Eunich as reported by one colleague—his wife.

3 *Journal of Vasology*, June 1941, Dr. Phil Anderer, "experience of 300 male prostitutes who claimed to have been vasectomized." The paper reveals that they were victimized not vasectomized.

4 *Sensitivity Journal*, August 1954, Dr. N. Counter, "Group Feelings at a Ballroom Marathon".

5 Four out of five do not want to be reminded of past, present, or future children claiming they had a poor childhood themselves after learning their fathers were vasectomized.

6 None were willing to explain more of what for less of which.

7 Those who were asked all assumed a rather interesting fetal posture—cupping the scrotum in both hands and chanting the latest song, the "OO-OO Rock" while rolling rhythmically back and forth not losing a stroke.

8 "UNCLE" stands for "Under No Conditions Locate Eggs".

9 Or a testicle pitched at a baseball. We see no difference in the pain that results.

10 Dr. Matt A. Chine, *Journal of Gay Times*, May 1969, "Be Fruitful Without Multiplying".

11 To encourage listening to political speeches not to encourage vasectomies.

The Wife of Bath Sign: An Aid in the Diagnosis of Gonococcal Arthritis and its Predisposing Cause

KERN WILDENTHA, M.D.
Dallas, Texas

> Gat-toothed I was, and that became me well;
> I had on me the stamp of Venus' seal.
>
> . . . .
>
> My nature was that I could not withdraw
> My prize of Venus from a good fellow.
>
> CHAUCER[1]

INTRODUCTION

The observation that promiscuity in the human female may be associated with the presence of gat-teeth was stressed by Chaucer in his case report of the Wife of Bath[1]. In view of widespread promiscuity in the 20th century[2, 3] it seemed of interest to determine if the relationship noted in the 14th century still exists.

Two difficulties in testing the hypothesis were apparent: (1) the current widespread availability of corrective dental care tends to obscure the true incidence of gat-teeth, and (2) promiscuity is difficult to establish reliably by history because of (a) embarrassment and fear, or (b) bragging. However, since (A) promiscuous women are more likely to contract gonorrhea, (B) women with gonorrhea are unlikely to develop complicating gonococcal arthritis unless they neglect seeking medical attention for the initial infection, and (C) those who lack medical attention will usually have failed to receive dental attention as well, it seemed probable that women with gonococcal arthritis would comprise a group in whom *objective* evidence of promiscuity plus a virginal tooth status would be combined. Thus if the gat-tooth hypothesis be correct, a high incidence of that finding should be seen in women with gonococcal arthritis.

The proposed relationships are symbolized in Figure 1. Shunt pathways opened by affluence and its accompanying higher standards of medical and dental care are showed by hatched lines.

METHODS

All women between the ages of 15–35 admitted to a local charity hospital during 1966 with acute arthritis of any cause were examined for the presence or absence of gat-teeth. Subsequently, diagnoses of gonococcal arthritis were made according to standard criteria[4]: (a) typical history and clinical course, (b) demonstration of the gonococci by smear or culture, (c) positive test for gonococcal antibodies, and/or (d) definite improvement following penicillin therapy.

RESULTS

Table 1 summarizes the results of studies on 25 consecutive patients, 11 with gonococcal (GC) arthritis and 14 with other causes. It is apparent that a highly significant correlation is present between the presence of gat-teeth and the establishment of the diagnosis of GC arthritis on other grounds (p 0.0001). Indeed, reliability of the gat-tooth finding equals or exceeds that of the other criteria for diagnosis.

[4] Hess, E. V., et al., Gonococcal antibodies in acute arthritis, 1965 *191*, 531.

---

[1] Chaucer, G., Canterbury Tales, Prologue to the Wife of Bath's Tale, lines 603–604, 617–618 (translated to modern English).
[2] Kinsey, A. C., et al. Sexual behavior in the human female, Saunders, Phila. 1953.
[3] Unpublished personal observations.
[4] Hess, E. V., et al., Gonococcal antibodies in acute arthritis, JAMA, 1965 *191*, 531.

FIGURE 1

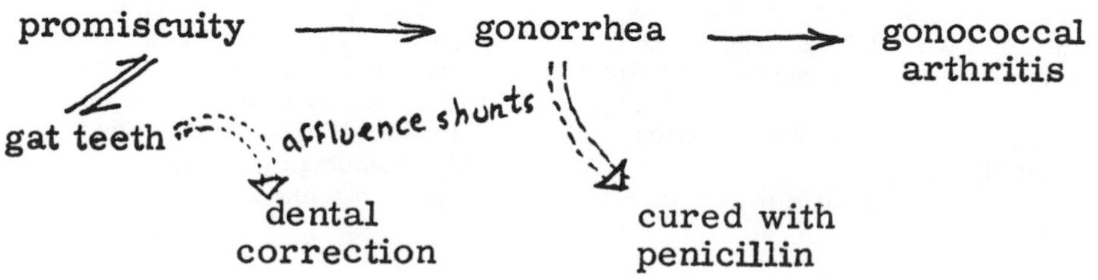

promiscuity → gonorrhea → gonococcal arthritis
gat teeth
affluence shunts
dental correction
cured with penicillin

TABLE 1

	History of promiscuity given	Gonococci identified	Increased GC antibody titers	Response to penicillin	Presence of gat-teeth
GC	6/11	4/11	7/10	11/11	10/11
Others*	5/14	0/12	2/8	3/9	2/14

* acute rheumatic fever – 8; infectious arthritis (non-GC) – 3; systemic lupus erythematosus – 1; undiagnosed – 2.

Of the two non-gonococcal patients who showed gat-teeth, one gave a history of extreme promiscuity. The other did not[5]. The GC patient in whom gat-teeth were not seen had had all her teeth knocked out by a boyfriend several years previously.

CONCLUSION

In view of the results of this study, and that of Chaucer, the usefulness of the "Wife of Bath Sign" in identifying promiscuous women seems established[6]. The logical extension of this finding into the field of arthritis provides an invaluable sign to aid in the difficult diagnosis of gonococcal arthritis. ■

[5] She was a liar.

[6] This designation seems preferable to "The Chaucer-Wildenthal Sign" which others have proposed (see Standard nomenclature of athletic injuries, AMA Publ., Chicago, 1966).

Obliteration of SEX

In recent months care has been taken by most departments to avoid the use of words which unintentionally designate a specific sex in letters, forms, brochures, etc. In general, it appears that forms and publications do not contain references to gender which might be considered discriminatory, but we need to assure ourselves that we are not distributing or using material which may appear so. Words identifying a specific sex (e.g., he/she, him/her) should not be used except where their use is intended to identify a specific person. "He," for example, should not be used as a neuter gender word. In place of such pronouns, specific nouns may be used e.g., the candidate, the employee, the customer, etc. Further guidance related to sex discrimination is contained in Management Guide 03-0306.

The following actions should be taken to ensure that we completely remove any possible indication of sex discrimination in any written or printed material:

1. Review all existing forms and publications which have been developed within your department to assure that there is no reference to a specific sex unless such reference is essential and can be supported. (Legal counsel should be sought in cases where it is thought such reference is essential.)

2. Advise your Employee Relations Manager/Advisor of the existence of any forms or other material which appears discriminatory and revise them to remove any questionable references. Withdraw any such existing material from use and destroy warehouse stocks. If in the judgment of local management, such action would result in unreasonable expense, Company legal counsel should be sought to review the risks/costs involved.

3. Instruct all personnel within your department that no references to sex should be made in conversation, or in any future correspondence, printed material, etc., unless essential (see 1 above).

If there are any questions concerning these instructions or matters which require clarification as they are implemented, please contact me. ■

[1] Sex Discrimination—Correspondence Publications

AIR-WATER TIGHT TESTS FOR MECHANICAL CONTRACEPTIVES

Sir,

I should like to refer to the 1968-1969 Annual Report of the Haffkine Institute of Bombay (Govt. Printing and Stationary Maharashtre State Publ. p. 80). The last paragraph of the Biological Section reads as follows:

> "Thirty-seven samples of mechanical contraceptives (condoms) were tested. Only 6 samples have complied with all the tests. It has been observed that most of the samples failed in length, weight and air-inflation tests. A few samples were not made of good elastic rubber, as some of the samples could not hold enough water to pass the water leakage test."

It is regrettable that the report does not include details of the standards used to assess parameters such as length and weight. Thus, it is difficult, if not impossible, to compare the Indian figures with those of other countries. Taking into consideration, however, the highly significant proportion ($P < 0.01$) of the samples that did not pass the air-inflation tests, as well as the high birth rate in India, it is reasonable to assume that many of the faulty specimens had, nevertheless reached the consumer market.

Although no actual figures are given for the water leakage test, we assume that under maximal hydrostatic pressure, compatible with integrity of the device, any rubber closed cylindrical shape, will become a sphere. Under these conditions, one can calculate the number of spermatozoids (E) traversing the rubber wall as follows:

$$E = \frac{\Pi \ R^3 \ V}{HP} \ C_e \ C_p$$

Where ΠR^3 is the volume of the sphere, V the coefficient of viscosity of the liquid contained in it, HP is the hydrostatic pressure, and Ce is the coefficient of elasticity of the rubber and Cp the coefficient of porosity. ■

H. E. (Isr. Inst. Bio. Res.)

Sexual Feats and Facts*

Sex is inherently interesting even to those with none of the multifold sexual problems described or invented by psychiatrists, psychologists, and novelists. The current popularity of the studies by Masters and Johnson is understandable but the replacement of mythology with scientific myths is a danger. The trend to sanitize, and scientize sex with probes, movies, etc. may provide new information of value to those unfortunates with unendowed bodies, faulty skills, or functional "hang-ups", such advances will be welcome.

I suppose individual sexual performance may range up to heights comparable to those attained by trained athletes in other kinds of physical activities and team play, who are often paid to demonstrate their individual superiority in one or another contact sport. The large attendance at movies with a predominant sexual theme may not differ too much from the crowded stadia of football fans with deep emotional attachment to the barely discernible players who are expected to display unattainable feats of skill and strength — and risk life and limb.

It will be interesting to see if new public heroes in sexual athletics will develop. One can only wonder if there will be a parallel development of tournaments, training tables, coaches, teams, and medical specialists to provide aid to the injured (laceration of uterine ligaments?) or perhaps an "air hammer disease" of the spine or phallus etc., etc., etc., etc., etc. In any event, it seems certain that in this topsy turvy world, progress of some sort in this area of medicine is certain.

The role to be played by the physician is less certain. Even the third year medical student who traditionally felt that he was over the hump — educationally speaking — may find that his factual lectures have been "woefully inadequate". I suppose outstanding players who become overage will become coaches and unless the curriculum is changed, the physician's role may be severely limited.

From a sentimental standpoint, the unique flavor of OB-GYN and Urological jokes will be lost forever. I suppose hindstart programs will be funded by the Government and reluctant youngsters will have to complete required exercises before going out to play. Probably, only neighborhood medical sex centers will be funded and thus, provide a place for underground medical activities, such as public health, trauma and mental care. ■

* Moore, George E., M.D., MEDICAL WORLD NEWS, May 1, 1970, p. 48.

Sex Object

Since there is no conventional word, we propose that the term "sex object" be defined as: any object (living or dead, natural or artificial) that frequently evokes in an animal a response that is similar or identical to any typical mating-behavioral response to an individual of the opposite sex. A sex-object presumably is a source of some "sex-stimulus" (visual, tactile, olfactory, sonic, etc.) that excites some receptors to initiate unconditioned or conditioned reflexes that are components of sexual behavior.

Ed: Now we know what the fuss is all about

Dr. J.J. Menn
Mt. View, CA

Response Latencies of Female Rats During Sexual Intercourse

G. Bermant
Harvard University

Science 1961, 133:1771

"The experiment reported here measures the effect of single copulations on the behavior of estrus female rats, in the context of an ongoing series of copulations. In order to investigate the behavioral effects of single copulations of the female, it is necessary to provide a method by which she can control the timing of the copulation. One such method was to make each copulation contingent upon some measurable arbitrary response by the female. This was done by conditioning the females to press a lever in a box whereupon a male was placed in the box by the experimenter. The time required to train the females ranged from 15-90 minutes."

(**Results:** It was found that after copulation the female rats would press the lever within 20 seconds for at least 5 consecutive copulations).

A new product—unfortunately-is Joseph Labs' (Hollywood, Calif.) cologne and conditioner for the breast. It's the 'only kind of cologne of its kind in the world' and comes in a flavor (mint frappe) or a floral scent (huneysuckle!). The package is 'designed to excite and titillate.' The product, *Consent,* furthermore 'does away with the disagreable taste of perfume.'

Advertsing AGe, June 28, 1971, p. 42

DIRECT MEASUREMENTS OF THE THERMAL RESPONSES OF NUDE RESTING MEN IN DRY ENVIRONMENTS

Mitchel, D., Wyndham, C.H., Atkins, A.R., Vermeulen, A.J., Hofmeyer, H.S., Strydom, N.B. and Hodgson, T.

Pflugers Arch. Europ. J. Physiology, 1968, 303/4

(Quite a team for measurements!)

ESTRUS—INDUCING PHEROMONE OF MALE MICE: TRANSPORT BY MOVEMENT OF AIR

W.K. Whitten, Bronson, F.H. and Greenstein J.A.

Science 1968, 161, 584

Abstract: "The proportion of female SJL/J mice exhibiting estrus when placed 2 meters downwind (6 meters/min) from a group of hybrid males was not significantly less than that of females placed directly under the males and exposed to their urine."

APPLES and ORANGES

BIOLOGICAL MOTIVES III: SEX

Sex is Different

Sex seems to differ from the hunger and thirst drives in several ways:

1. Unlike food and water, sex is not essential for the survival of the individual. No one will die without sex. (You may think you are going to die, but you won't)...
2. People seek arousal as well as reduction of the motive. Most people enjoy the feeling of being sexually aroused, but almost no one likes to feel too thirsty or too hungry.
3. Humans can be sexually aroused by an extremely wide range of stimuli—a much wider range, in fact, than will arouse hunger or thirst.
4. The arousal of the human sex motive appears to be less affected by deprivation than are the other drives...the sex motive seems to be arousable at almost any time, and does not show the regular increase over time that is displayed by the hunger and thirst drives.
5. Sexual behavior uses energy rather than replacing it.

Our culture is permeated with sex.

Taken from:

Invitation to Psychology, 2nd Ed., 1983, Houston, Bee, and Rim, Academic Press

Submitted by:
Ray H. Bixler
Louisville, KY

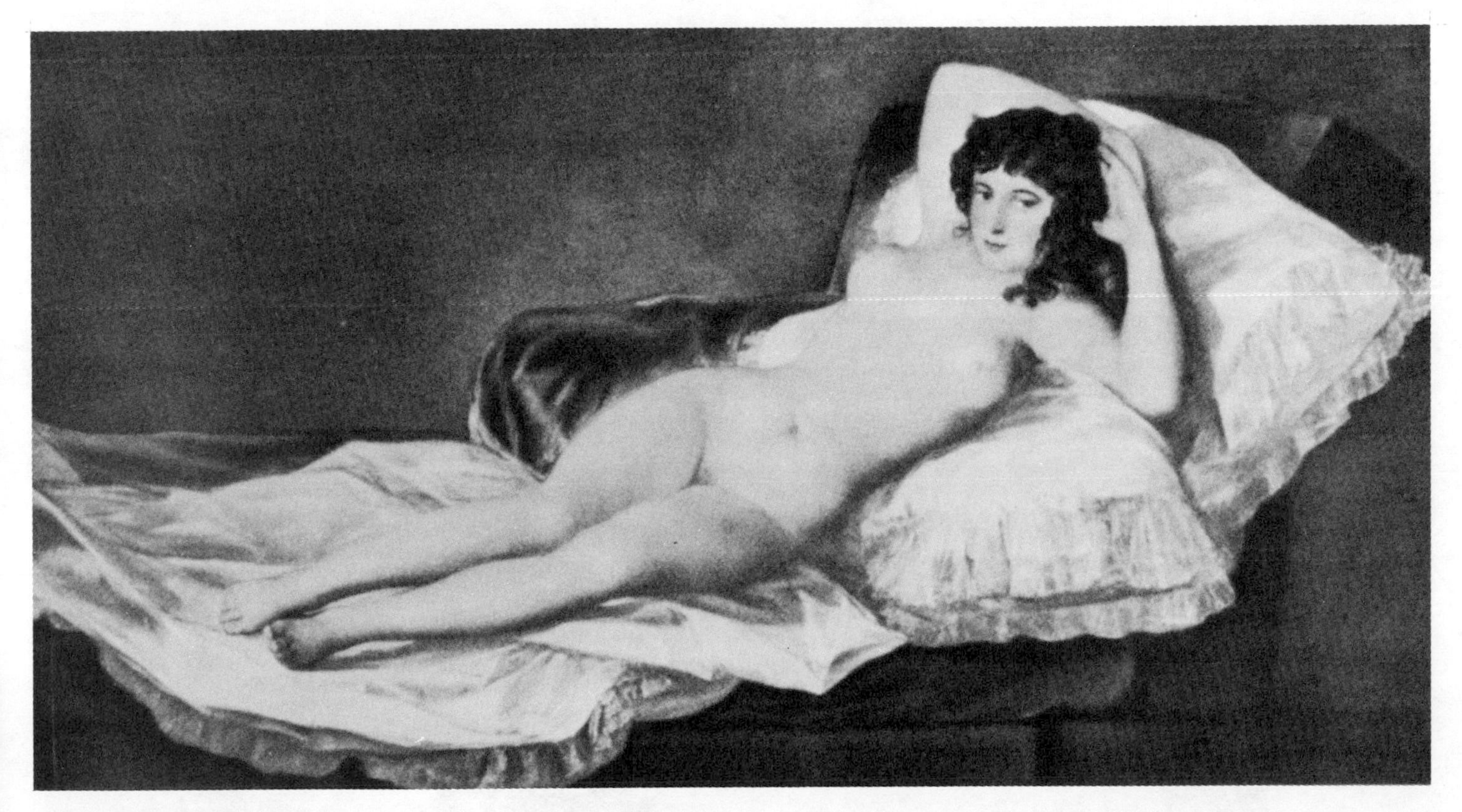

C.M. Kunin & R.A. Ames

METHODS FOR DETERMINING THE FREQUENCY OF SEXUAL INTERCOURSE AND ACTIVITIES OF DAILY LIVING IN YOUNG WOMEN

Amer. J. Epidemiol 113, 55-61, 1981

"To determine whether the frequency of sexual intercourse and other activities of daily living" (eating, drinking, bathing, showering, urinating, defecating)" could be reliably measured in a population of women with or without prior urinary infections, a prospective, pilot, case-controlled study was undertaken in a family clinic."

One interesting conclusion: "Frequency of sexual activity was significantly higher on weekends."

五戒

一、生命（いのち）あるものを殊更（ことさら）に殺さざるべし
二、与へられざるものを手にすることなかるべし
三、道ならざる愛欲をおかすことなかるべし
四、いつわりの言葉を口にすることなかるべし
五、酒におぼれてなりはひを怠ることなかるべし

Irreproducible

studies

must

be expressed

in

LANGUAGE

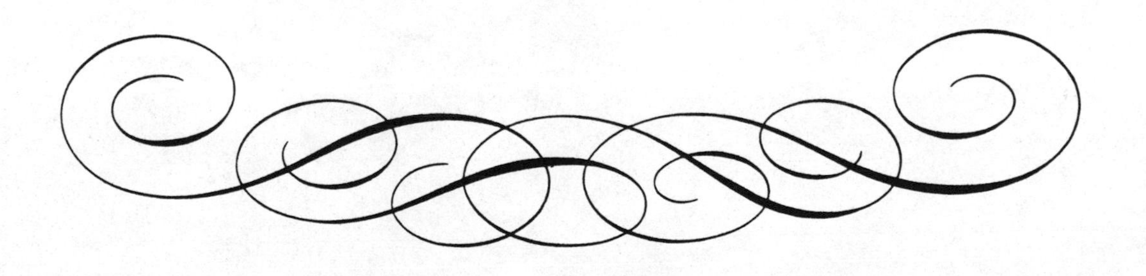

THE PHYSICAL DIMENSIONS AND VALUE OF SWEARING

R.J. Hoyle
Halifax, Nova Scotia Canada

"...Words are strokes..." said the Queen to King Cymbeline (10). This statement, referring as it does to the synonymity of 'stroke' with 'blow' or 'hit' rather than with 'caress', as in stroking a pet dog, points out the assaultive use of words. This function is taken up again by the same author in a different historical account: "...not a word of his but buffets better than a fist of France" (12). These early references to the concrete effects of language prompt an examination of the physical factors involved. That a verbal assault contains a physical component is not doubted by certain Arabs who, according to Montague (6), "when cursed, ducked their heads or fell flat on the ground in order to avoid a direct hit." This suggests that the curse is a vector, having direction and, presumably, magnitude.

It might, at first, be supposed that, knowing the nature of sound and its transmission, the momentum of a curse is related to the weight of the air molecules involved and their velocity in the direction of the target. A clue that this is not a complete picture is provided by a previously quoted author who records a subject's observation:

"Margaret's curse falls heavy on my neck" (12) suggesting that considerable weight is involved. This is reinforced by Graves who states that "to swear by Jesus Christ is an oath with weight behind it" (4, p. 6). Although this latter example involved the affirmative rather than the assaultive function of the oath, it is felt that application of the same principles is justified. Further evidence is provided by Shirley (14) who points out that "frequency of swearing can carry weight" (p. 52). Care is necessary in the interpretation of this statement in view of the apparently conflicting evidence of Echols (1) who points out that an oath can "lose its power to protect and persuade through overexposure and frequent failures to produce the desired effect." This was also noted much earlier by Paley (9) who commented that "the levity and frequency with which oaths are administered brought about a general inadvertancy to the obligation of them." These data are interpreted to mean that the weight of swearing as a whole can be increased by increasing the variety of terms used per unit of time, rather than by mere repetition.

The above attribution of power to an oath is reinforced by Naracic (7) who asserts that " the curse contains the same power as the person using it." In this connection, the earlier reference to "a fist of France" (12) prompts an inquiry into the force of a manually delivered blow. It has been reported (2) that during a karate chop, the hand "can develop a peak velocity of 10-14 metres per second and exert a force of more than 3,000 newtons, a walop of 675 pounds." Nevertheless, the curse is often used at a longer range than is normally possible with a blow by hand. It is possible to infer that a curse has some of the characteristics of a projectile (6) and Naracic's evidence suggests that its physical dimensions are consistent with what can be thrown conveniently. Some evidence on this point has already been noted (12), indicating a possible maximum of a few pounds per unit curse.

A minimum value is harder to estimate, though some tangential evidence may be related. It has been suggested (16) that a certain type of ill-formed, non-directional speech such as that of the well laid-back hippie was "incapable of bearing any meaning weightier than a sigh." This indicates a low range value of the same order of magnitude as a moderately forceful exhalation. This would probably fall approximately half way between the tidal volume of the lungs (c. 500 ml) and the inspiratory capacity (c. 3600 ml) (5), say 2000 ml. The weight of this air (approximately 2.5 gms) gives an estimate of the minimum value.

One can expect that an oath's effectiveness is related to its volume as well as its weight. Shakespeare provides insight again in his reference to the principle of "a good mouth-filling oath" (11). If "good" implies "effective", such an oath would then have a volume similar to that of the buccal cavity. This and the deductions concerning weight are consistent with a small ball model.

Graves supports the concept of the curse as a physical entity when he says that "the thermodynamic entropy of the ingenious swearing bout is intensified" by the use of "negative swearing" (4: p. 14). This involves the use of particularly polite language in place of epithets of the usual derogatory nature. However, the exact relationship of this type of intensification to the entropy of the system is unclear. Montague also recognizes (6: p. 44) a physical basis when he mentions the "bioenergetics ... of the cursing tablet's efficacy", referring to a tribal method of intensification.

The socio-economic history of swearing provides some interesting data on the declining value of the oath. Around the beginning of the twelfth century A.D., Henry I of England imposed fines on those members of his court caught swearing. These ranged from 40 shillings for a duke to a scouring for a page (8). About 500 years later, William III proclaimed "...if any person shall profanely swear, if he is a laborer, servant or common soldier, he shall forfeit one shilling to the poor for the first offence, two shillings for the second, et cetera, but double for any person not a servant" (3). This class distinction was not evident in a law repealed in Maryland in 1953 (15), although the deflationary trend is maintained. This law provided for a fine of 25 cents for the first swear word used and 50 cents per word thereafter. We note that if this trend of declining value continues, oaths will soon be worthless and need not be used at all.* It is thought, however, that this point will never be reached because a significant proportion of the population would then have nothing to say. Paley also records some information on the negotiable value of the oath: "A pound of tea cannot travel regularly from the ship to the customer without costing half a dozen oaths, at least." Although it is safe to assert that such a cost would readily be passed on to the customer, it is doubtful whether the decline in value of the oath to zero would produce a decrease in such a "middle-man's supplement" and a lowered comodity price.

Conclusions:

Curses are estimated to range from one gram to one kilogram in weight, with a mode of 150 grams. The upper range of volume is thought to be in the order of 150-200 cc.

The monetary value of the curse has declined almost to the vanishing point: the rate of decline is thought to be asymptotic. ■

References:

1. Echols, E. 1979. *The Art of Swearing in Latin*. American Scholar, 49, 111. (Winter '79-'80).
2. Feld, M.S.. R.E. McNair & S.R. Wil,. 1979. *The Physics of Karate*. Sci. Amer., 240(4), 150-158, (April).
3. Good, J.M:, O. Gregory & N. bosworth. 1819. *Pantologia*. London.
4. Graves, R. 1927. *Lars Porsena*. London: Kegan, Paul, Tranch, trubner.
5. Mathews, D.K. & E.L. Fox. 1976. *The Physiological Basis of Physical Education and Athletics*. Philadelphia: W.B. Saunders, Co.
6. Montague, A. 1967. *The Anatomy of Swearing*. New York: MacMillan.
7. Naracic, V.G. 1972. *The Psychology of Swearing Among Sportsmen*. J. Sports Med. & Phys. Fit. 12(3), 207-210, (Sept.).
8. Nowell, Alexander. 1611. *A Sword Against Swearers and Blasphemers*. London. (quoted by F.A. Shirley).
9. Paley, Wm. 1785. *Elements of Moral Philosophy*.
10. Shakespeare, W. 1610. *Cymbeline*. Act 3, Scene 5, Line 40.
11. Shakespeare, W. *Henry IV*, (Part I). Act 3, Scene 1.
12. Shakespeare, W. *King John*. Act 2, Scene 1.
13. Shakespeare, W. *Richard III*. Act 5, Scene 1, Line 25.
14. Shirley, F.A. 1979. *Swearing and Perjury in Shakespeare's Plays*. London: Geo. Allen & Unwin.
15. *Time*. May 16th, 1969, p. 78.
16. *Time*. Aug. 25th, 1975, p. 57.

*Actually, we may have a "Giffen Good" effect where the cheap commodity drives out the good. Thus, we see an increasing trend to use the "Big F" word in polite discourse in movies and TV, while more inventive and imaginative outlets have virtually disappeared from common use.

THE WATER BEAST, by Jackson Pollock, Stedelijk Museum, Amsterdam

Randall K. Thomas (Dept. Psychology, Ohio State Univ. Columbus)
UNCERTAINTY AND THE DEVELOPMENT OF INTIMATE RELATIONSHIPS
Social Psychology Bulletin 80-4 (Sept, 1980)

"A definition of intimacy that would incorporate the idea of uncertainty would be that intimacy occurs in a relationship in which two people have mutual and equal access to restricted areas of their partner and that both partners have developed expectations of continuing interaction."

AN INTRODUCTION TO TROPHOLOGICAL IDIOMATIC EXPRESSIONS IN THE AMERICAN LANGUAGE

Pierre L. Horn
Wright State University

While America may lack international recognition and stature with respect to its cooking expertise, there is nothing to apologize for in the use of food imagery to "lend spice" to its language. Indeed, it is interesting to note that the United States has borrowed from the world of food many of its most tasty expressions. Let us sample below with our gourmet guide but a few of these verbal treats.

Good evening, I am Jacques, your maitre d'. Please follow me to your reserved table. Tonight we have a delicious smorgasbord of succulent dishes. Would you care to see the menu?

Vegetables

(good looking) tomato
(hot) potato
(this is small) potatoes
(full of) beans
(hit on the) bean
(spill the) beans
(he knows his) onions
corn (y jokes)
(an old) chestnut

Soupe du jour

(in the) soup

Fish and Seafoods

(a fine kettle of) fish
(sounds) fish (y to me)
(holy) mackerel (!)
(a red) herring
(that little) shrimp
(50,000) clams

Meats

(a) meat-and potato (man)
(golf is his) meat
(go the whole) hog
(bring home the) bacon
(living high off the) hog
(a governmental) pork (barrel)
(he's a real) ham
(go cold) turkey
(that's a lot of) baloney
(my) goose (is cooked)
(the big) enchilada
(what's your) beef (?)
(a) tenderloin (district)
(they're in a) stew (over this)

All the above dishes are served *au jus:*

(get on the) gravy (train)
(the rest is pure) gravy

and with relishes:

(in a fine) pickle
(cut the) mustard

Breads, cereals, and pastas

(you're a) crumb
(that's a lot of) bread
(rolling in) dough
(feeling one's) oats
(that movie is) mush (y)
(use your) noodle

Accompanied by:

butter (up somebody)
(in a) jam
(big) cheese

Non-alcholic drinks

(that's not my cup of) tea served with
sugar (daddy) or (my car is a) lemon
(a) milk (run)

Fruits and desserts

(top) banana
(go) bananas
(I don't care a) fig
(how do you like them) apples (?)
(isn't it) peach (y?)
(get a real) plum
(they gave him the) raspberry
(a) rhubarb (on the field)
(that fellow is a) fruitcake
(it's a piece of) cake
(that takes the) cake
(that's the icing on the) cake
(a smart) cookie
(that's the way the) cookie (crumbles)
(She's nothing but a) tart
cheesecake (photographs)
(you) fudge (on your answer)
(He's) nuts
(working for) peanuts

Terms of endearment are no exception to this culinary preoccupation with such fattening sobriquets as: honey, honeybunch, sugar, creampuff, (cutie) pie, puddin', pumpkin, sweetie pie, sugarplum, and apple dumpling.

To show that there is such a trophological phenomenon has been our modest intent.

How to account for it we leave to the more dedicated gastronomophilologists among you. ■

Sexual Behavior in the Human Language

H. J. LIPKINSEY

It is now apparent that sex is at the root of all human problems. In all of history[1, 2] sex has been consistently suppressed in areas where it has every right to appear[3]. It has therefore popped up in other areas where there is no good reason for its existence[4]. All aspects of the problem have been treated in detail by Freud and Kinsey[3], with the exception of the sexual behavior of languages which is treated by Lipkinsey[5].

The French pride themselves on their rational attitude towards all things, including sex. But Frenchmen have neglected their language, whose anomalous sexual behaviour is among the worst of its kind. All French tables are feminine. "La crayon de ma tante est sur *la* table" is the first sentence every student learns. Woe be unto the poor wretch who says *le* table. No French table will ever forgive this insult. Yet, what is the lot of the unfortunate French table? All that the future offers them is hopeless spinsterhood, for *all* French tables are feminine and there are no male tables in France.

Lucky is the French table who can make the journey to a foreign land like Israel, where all tables are masculine (or the lucky Israeli male who can make the trip to France). Perhaps this is one of the reasons for the friendship which has recently arisen between Israel and France. But all in all, both the French and Hebrew languages are nothing more than a mass of misplaced sexuality. They are full of sexified tables, chairs, houses, trees and other objects who have no need for sex, have never asked for it, and use it only to frustrate themselves and anyone who is trying to learn the language.

The Germans have evidently stumbled upon the important idea that some objects need not have sex, and should remain sexless. With characteristic German thoroughness they have classified everything as masculine, feminine, and neuter, and they have made a horrible mess. Der Tisch, die Wand, das Mädchen, das Fräulein. German tables are masculine, German girls and unmarried women are neuter, walls are feminine. What is there for a young man to do in Germany? All young and unmarried women are sexless. If he wants feminine company he must find himself a married woman or go to the wall[6]. Is it any wonder that German youth has been responsible for so many upheavals in the past century?

The English-speaking peoples alone have almost completely succeeded in keeping sex out of irrelevant portions of their language. This is undoubtedly a result of the Puritan tradition of taking sex out of everything[7]. English tables are sexless and have no problems, even though they may be laid on occasion. Boys are masculine, girls are feminine, and Hollywood makes the most of it. It is only when foreigners speak broken English that sex is ever introduced into the English language: "This house, she is too small; My train, he did not come on time." An Englishman can always tell a foreigner by his attempts to introduce sex into the conversation.

---

[1] A. Essex, B. Sussex and C. Wessex, *Sex in English History,* Middlesex Press (1957).

[2] S. Freud, *Oedipus Sex, a study of sexual behaviour in ancient Greece.*

[3] S. Freud, A. B. Kinsey and S. U. Perkinsey, Sex and Repression in Human Society, *Journ. Inorgasmic Chemistry,* 1957, *1,* 51.

[4] P. Morris and L. Strike, "Sex in Cigarette Advertising," Any American Magazine.

[5] H. J. Lipkinsey, "Sexual Behaviour in the Human Language," *J. Irrepr. Res.* 1957, *V.*

[6] Die Magd, the servant girl, is also feminine. Draw your own conclusions.

[7] One Puritan sect succeeded in eliminating sex from every aspect of life and became extinct.

Multi Variational Stimuli of Sub-Turgid Foci Covering Cross-Evaluative Techniques for Cognitive Analyses of Hypersignificant Graph Peaks Following Those Intersubjectivity Modules Having Biodegradable Seepage.

DAVID LOUIS SCHWARTZ

ABSTRACT.

After a definition of basic terms, a model is constructed wherein cross-disciplinary cognition patterns are subjected to increasingly unviable stimuli. The statement is made that a focus of point-sensitive trivalency does not necessarily presuppose molecular dispersion, despite the obvious tripolar discontinuity. It is shown that earlier research was misleading in its emphasis on total-sweep ranging procedures being applied to Judeo-Christian parameters.

BASIC TERMS.

Multi: Lots of them. Whole big bunches. More than just a few.

Variational: Changeable. Not likely to stay the same as it was the last time you checked.

Stimuli: Things what they make other things do things.

Sub-turgid: Not about to do much before next Friday. Less active than mere sluggish, the implication is of approaching rigidity. Useful in describing managemental characteristics.

Foci: Places where things come together. Related to "fussy" and "fussbudget" in the quasi-normative sense.

Fuss-budget: One who focuses ("foci") on one aspect while excluding whatever you happen to be interested in.

Hypersignificant: Big stuff. Top management conceptualization. See hyposignificant, hypochondriac, hypertension, hyperschmeiper, and hippopotamus.

Intersubjectivity: Involves intimate cross-relationships between subjects. Sometimes the subjects are people, in which case intersubjectivity becomes a dirty word.

Judeo-Christian: Jewish and Christian. Also Christian and Jewish. Or both. See Abie's Irish Rose, Nimzo-Indian defense, former Mayor what's-his-name of Dublin, and your dentist twice a year.

Trivalency: The kind of clothes hanger where you can put three things on at once without overlapping. See also trivial, Valencia, and Linus Pauling.

Vivaldi: Italian term for portable battery-operated noodle-sifter characterized by a distinctive "ticka-ticka-ticka-ticka-ticka-ticka." See Fig. 4.

Nimzo-Indian defense: Loses for black. Also loses for white. Highly recommended.

With: Conveys a sense of possession. Very useful in pointing up the opposite aspects of without. (q.v.)

Without: Lacking. Outside. Unless. Beyond. The word, for sanitary reasons, is infrequently parsed. Without, withouter, withoutest; these forms have found favor in sociological manipulative parallels. For example, all poor people are needy, but some are more needy than others.

(q.v.): Abbreviation for the Latin quizzimus vitalisius, meaning "Who used up all the hair tonic." Now used to introduce the participial subjunctive whenever a presidential aspirant wants to show some familiarity with the farm problem.

Upwardly mobile: A vertical transfiguration of a horizontal growth percentile. Expressed in cubic geronimos, the recipricals have been found useful in predicting fluctuations in the patterns of dog-food sales.

Downwardly mobile: Almost like upwardly mobile, except for subtlies in directional modes.

Sicilian defense: Strong bond between king and knights. See Don Carleon and the continuing Fischer King legend.

Unviable: Not too healthy. On the sickly side. A generally diseased condition. (See nonviable, neo-biblical, unbuyable, and umbilical.) That listless, run-down feeling. Tired blood. Take two aspirin, some chicken soup and call me next Tuesday.

Rhesus: Some kind of monkey. See Fig. 4.

Biodegradable: Ten minutes after you throw it away, it turns into delicious pineapple yogurt, and you can eat it. ■

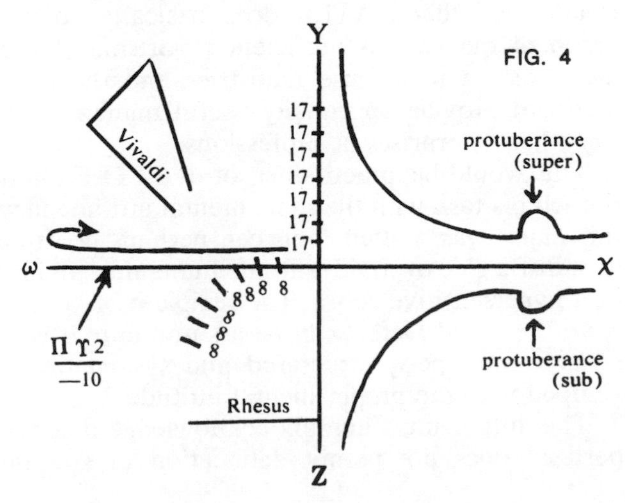

FIG. 4

RATTLE

F. H. AMES, JR.

Various concepts appropriate to the subject of RATTLE, Reportsmanship Aimed Toward The Lofty Expert, have appeared previously in various journals during the past decade. Unfortunately these articles are usually fragmented and are specifically oriented toward activities pertaining to government contracts. The author's purpose in developing RATTLE is to document and describe techniques applicable to the basic art of reportsmanship throughout all business, professional and academic activities. Accusation that RATTLE was derived from the age old expression "RATTLE the enemy" is without foundation and should be considered as malicious gossip. However, it is admitted that RATLE does basically comprise a union of the various proficient reportsmanship activities existing in defense industries and the academic world to provide an equally useful management tool for other enterprises or professions.

The would-be practitioner of RATTLE must approach his task with the same mental attitude in which this paper was written. This can perhaps be expressed as sadistic glee overlaid with the humanitarian instincts of a hypersensitive cobra. The recipient of your report is your enemy! Defeat can be assured only if your approach is properly structured and systematically developed with the proper mental attitude.

The author must humbly acknowledge that his "expertize" does not permit delineation of specific examples in an area such as medicine. However the basic concept and technique described herein, although oriented to the engineering discipline, are equally applicable to any profession or activity issuing written reports to another entity which is categorized as being a superior office.

The basic objectives of this paper are as follows:

1. Define reportsmanship
2. Discuss the concept and purpose of RATTLE
3. Provide criteria for various levels of competence for practitioner of RATTLE
4. Describe specific techniques available to RATTLE
5. List source of background data

REPORTSMANSHIP

Reportsmanship is the informal technique of leading the reader to draw a wrong conclusion. It is not particularly systematic and does not require the originality required by RATTLE.

A forerunner of Reportsmanship is Gamesmanship or the gentle art of psyching your opponent in games or during casual conversation. One of the few documents on gamesmanship is Potter's delightful little book *Gamesmanship*. It is recommended reading as background information for the novice. Reportsmanship is really documented Gamesmanship with a little extra effort.

RATTLE, CONCEPTS AND PURPOSE

RATTLE is a systematic and formal approach to impressive verbosity utilizing the technique outlined later in this article. It is defined as the technique of authoritatively writing a report capable of inspiring a state of exquisite awe in the mind of the recipient through inducement of erroneous conclusions. The phraselogy must be carefully chosen to both impress the reader with the author's expertize and to provide data, largely factual, which will result in erroneous conclusions by the reader. A successful practitioner will induce a growing feeling of respect for the author

— because the only alternative for the reader is to admit that the contents of the report are beyond his comprehension. Such an admission is tantamount to a confession of insecurity and ignorance. No expert can afford this luxury!

The basic purpose is to keep the expert reader off-your-back. This situation is particularly true when the recipient must review and analyze the report as a result of his position in life, or, whenever the document is a contractual or administrative requirement. The reader should be familiar with the term "gobbledygook"; this technique must not be confused with RATTLE. Gobbledygook is merely a pathetic attempt to confuse, not to impress, although some specific elements of terminology may be suitable as RATTLE inputs.

RATTLE PRACTITIONERS, RATING OF

There is no existing standardization or qualification for RATTLE practitioners. This article presumes to establish three categories to formally describe the capabilities of those who utilize this particular technique. The section on technique will give more definite measurement criteria.

Apprentice — Although this could be deemed self-explanatory, a few words are appropriate. Satisfactory performance in this grade is evidenced by acceptance of one's peers within an intimate group, such as the local technical society/association, fellow employees, and knowledgeable friends. There is a tendance for these readers to address the author as "Mr." In this writer's opinion, at least two years are required normally as an apprentice under the tutledge of a man who enjoys a higher rating in RATTLE. Publication in house organs or small we-will-print-anything journals is required.

Journeyman — Satisfactory performance in this grade is demonstrated by publication in national journals which have a moderate technical content and serve the profession in which the author operates. Those readers who do not personally know the author occasionally address him as "Doctor." (This form of address, of course, is not appropriate within the medical profession.) Recognition is also evidenced by membership on some of the working committees of national societies and associations representative of his profession. At this point we must warn that such recognition must be based solely on the writings of the journeyman (and higher grades). Technical competence as evaluated by personal or factual knowledge of an individual, must not be confused with competence in RATTLE. Many emminent people, widely respected in their profession, are completely inept in RATTLE — such people actually transmit information and knowledge in clear concise terminology. RATTLE misleads the reader so he convinces himself, as an individual, that he is dealing with a more knowledgeable person.

Master — This is the highest grade attainable. To reach this lofty rating, one's articles will be published in the leading journals of the world; publication in a government periodical — even if as a source in a footnote, should be achieved regularly. A great deal of originality must be used in developing the report and considerable knowledge must actually be used. The Master displays a bold capability in generating impressive source documents, is often categorized as a consultant and is invariably addressed as "Doctor."

RATTLE TECHNIQUES

One of the most significant indications of proficiency is the originality displayed in developing new techniques, improving those used by others. The professional author will never hesitate to plagarize or modify techniques developed by his peers in RATTLE.

It is very important to recognize that RATTLE must contain a substantial amount of factual information — it is written from a slightly fictitious viewpoint — not as fiction.

BUZZ PHRASE GENERATOR

One technique utilized widely during creative thinking classes is the buzz phrase generator (or bafflegab); this device is a multi-column matrix normally comprising two adverbs and a noun. Selection of three words, to formulate an abstract but plausible phrase, is made on a random basis. Normally each column has ten words numbered from 0 to 9, this identification permits selection by usage of random numbers. One widely used manner of selection is to ask someone for a three digit number which is then used to select the three words. It has been said that this technique possibly resulted from a meal in a Chinese restaurant: take one from column "A", one from column "B", and one from column "C".)

The selection of words requires careful thought since all possible combinations must be plausible to the reader. The practitioner of RATTLE must also define, to himself, the resultant phrase in understandable terminology so as to avoid generation of gobbledygook. This precaution will be invaluable if the term is questioned by some stuffy reader who is truly an expert on the subject under discussion.

BASIC QUALIFICATIONS

TECHNIQUE	APPRENTICE	JOURNEYMAN	MASTER
Foreign Language[a]	Occasional phrase	Phrases/sentences; maximum of four per report.	Quote paragraphs in entirety — use profusely.
Footnotes/Source Documents	Legitimate documents of own profession occasional obsolete document.	Obscure/obsolete documents. One foreign source permitted.	Fictitious & foreign documents plus those of other disciplines.
Buzz Phrase/Acronyms	Desk dictionary. Terminology of writers profession.	Unabridged dictionary. Terminology of other professions.	Coins his own words.
Curves	Limited number of points.	Only 3-4 points shown.	No points — only the curve.
Fog Index[b]	16	18	21
Charts	Much detail	Present the same data in different formats.	Use extensively with minimal captions or describe only in words.
Oral Presentation	Never	Very seldom and then to local audience.	As requested.
Attitude Toward References Quoted	Straightforward	Point out small errors.	Challenge validity of theory.
Title Length	Multi-sentence or short paragraph.	Lengthy	Word or phrase.

[a] Never, never supply a translation.
[b] Refer to Robert Gunning's "How to Take the Fog Out of Writing."

Each generator requires tailoring to the profession of the practitioner; avoid using those generators which have been published before — extract words as appropriate but do not use examples blindly.

An appreciation of poetry is a great asset in preparing a buzz phrase generator. If the three words combination results in an awkward rhythm then reconsideration is advised. This poetic talent is particularly helpful in determining the order of precedence of the adverbs of the first two columns. A sample buzz phrase generator for industry in general follows; it is set forth merely to demonstrate the concept described previously. As a note of warning, although the buzz phrase generator can be used by itself, there are certain categories of personnel who should never be subjected to the output of the generator. Typical examples are manufacturing personnel who have never worked on a government contract and management personnel who have arrived at the top on sheer ability. Such people can be somewhat forceful in evaluating the output of a buzz phrase generator.

SAMPLE OF A BUZZ PHRASE GENERATOR

	"A"	"B"	"C"
0	discrete	behavorial	criteria
1	perceptive[1]	adaptive	motivation
2	dynamic	innovative	phenomena
3	synergetic[2]	emperical	systems[2]
4	cognitive	operational	model[2]
5	contributory	transitional	stimuli
6	classical	managerial[1]	centrality
7	sublimal	technological	theory[2]
8	pristine	competitive	propensity
9	systematized	optimal	interaction

1. Care must be taken to avoid use of both an adverb and noun having the same meaning; as an example: "perceptive-perception" is unthinkable. Along the same lines, it is most inadvisable to allow usage of three words all starting with the same letter; as an example "synergetic substantive system" could best be used to illustrate inept humor, certainly not for RATTLE. Avoidance of such circumstance requires considerable originality.

2. The expert could possibly challenge inclusion of these over-worked terms; the author's justification lies in the fact that this is only a sample buzz phrase generator used solely for illustrative purposes.

SOURCE DOCUMENTS

Numerous source documents exist for usage as footnotes for phraseology, as input to the buss phrase generator, or for quotation.

The following suggested list comprises basic categories; the practitioner of RATTLE should develop his own private list in considerable detail.

1. Doctoral dissertations — especially those in the field of philosophy, operations research, social sciences, and management. Those subsidized by a government contract are particularly rewarding and applicable to all professions.
2. The prestige journals of technical associations; usually these are issued on a quarterly basis in contrast to the monthly journal for the general membership.
3. Research and Development type of articles published by government organization. Those from the widely publicized "Think Tank" groups are highly recommended!
4. Obsolete technical publications, especially those which are nearly impossible to locate or are out of print.
5. Science fiction publications written by technical people. (Beware of the cops-and-robbers-with-a-ray-gun type which dominate.)
6. Text books used for advanced study in fields such as philosophy, operations, research, management theory, social sciences, communications, systems engineering and industrial engineering. It appears that a newcomer, anthropology, now deserves consideration as a source of data for RATTLE.
7. Congressional records.
8. Public relations and advertising publication.

Needless to state, source documents should not be limited to those in the English language. After all, if you ain't got no savoir-faire, how can you impress your reader?

HAZARDS

The basic information in your article or paper *must be predicated on factual data* no matter how obscurely it is presented. It must be defensible against a reader who requires definition of each term and has the tenacity to dig out the facts. When such individuals are identified, endeavor to keep him from receiving your masterpiece and apply for a position with his organization.

Be careful of using too many acronyms or abbreviations, otherwise you can create an unfavorable impression.

RATTLE must be used as a tool; it is a method of presenting information — not a technique to prepare a series of incomprehensible phrases. Your publication must, on the surface, be dignified, logical and most impressive to your peers.

GENERAL ADVICE

One particular effective technique in using acronyms is selection of a common one with two different meanings.

Draw conclusions, "it obviously follows," without any intermediate explanation as to the thought processes required to reach the conclusion.

Quote theories/models taken from publications outside your profession; of course, you should not explain the theory and be vague about the publication from which it was obtained. But, be able to explain it, just in case!

One technique which merits consideration is inclusion of a very, very detailed index which could amount to 10 percent of the text. One aspiring to the classification of MASTER should also consider inclusion of a massive bibliography including each document read during preparation of your article or paper. Whether you actually extracted information from such references is besides the point. (Although this concept is open to debate, one respected author of a management text stated, "It also includes a substantial number of books that might have been cited." His text comprises 212 pages with a 35 page bibliography.)

It is inadvisable to give sufficient information, when quoting references, to permit the reader to actually locate such references. Quote only the title and author — nothing more — as demonstrated by this article.

The writer sincerely hopes that the reader's future reports will benefit by this article. ■

Obscurantism*

S. A. GOUDSMIT

Referees and editors often complain about the obscure style of the majority of Letters and Articles. In addition to using unintelligible, twisted sentences, many authors create and use slang expressions known to a few specialists only, and indulge in unnecessary abbreviations. Such practices may help the writer but they slow down the reader considerably and exclude the uninitiated completely. In fact, many papers give the impression that the author was writing a memorandum to himself or merely for the benefit of a close collaborator. Yet when we ask authors to write their papers so that a few more colleagues can appreciate their significance, some of them rebut that popular articles do not belong in our journal.

We are convinced that an Article or even a short Letter can be written in a style that helps the interested physicist to understand its aim even if he is not a specialist. One of the causes of bad writing is that so many young research physicists lack teaching experience. They have never faced the challenge of explaining something they know very well to a student who knows nothing about it.

But there is still another reason for writing an obscure paper. It is the common subconscious fear of exposing oneself to scrutiny. If a paper is too clear, it might be too easy for readers to see through it and discover its weaknesses. We observe this same behavior with the lecturer who writes a formula on the blackboard and erases it almost immediately. We see it with speakers who address the blackboard instead of the audience and who keep the room dark between slides. They themselves do not realize that they are subconsciously afraid of being clearly understood.

Thus we believe that writing incomprehensible papers is not an indication of the author's erudition but merely reveals a common psychological defect. We hope that this insight will induce a few more of our authors to come out from behind their screen of specialized terms and machine-inspired sentence constructions. ■

* Editorial, Physical Review Letters

A Verbal Rorschach. An Antidote for Technically Obnubilated Appellation

By E. J. HELWIG

In his book *African Genesis* Robert Ardrey tells a delightful story about Sir Zolly Zuckerman, a young South African anthropologist, who once horrified his English friends by proposing to publish a book titled *The Sexual Life of the Primates*. He was promptly informed that "primates" in England could refer to nothing but the prelates of the established church. The book eventually appeared under the title *The Social Life of Monkeys and Apes*.

Perhaps the book would have sold more copies with the sensational title, but it would have disappointed many readers. As it was, the new title was simple and unambiguous. It contained no technical jargon and in clear terms informed the most casual reader of its contents. It is a pity in these days of publish or perish that more technical writers aren't similarly embarrassed into simplifying abstruse titles.

Today, when one can make a career out of reading as well as writing technical articles, many authors seem to choose titles that will look well in the Chemical Abstracts. Apparently titles must sound scientific, esoteric, and prestigious (in the archaic sense). Since the recondite is often confused with the erudite, the titles of technical articles frequently smother meaning under a plethora of jargonese.

As a result the layman, or, jargonwise, those not familiar with the lexicon of the scientific disciplines, don't receive the slightest benefit from a technical title. Or does he? Words, like the Rorschach ink blots, invariably carry some sort of impression, even if they conjure up pictures of erotic clergyman.

Technical writers might be able to gage the fuddle-factor of a proposed title by testing it as a Verbal Rorschach on their unscientific friends. The results could be devasting to the dignity and prestige of technical journalism. And they should effectively deflate pompous titles.

The following examples illustrate just what could happen with a Verbal Rorschach test. The titles were gleaned from a single issue of a listing of current technical papers. A possible Rorschach interpretation accompanies each title.

Title: Representation Mixing in U12.

Translation: Social Life on an Atomic Submarine.

Title: Group Theory of the Possible Spontaneous Breakdown of SU3.

Translation: Group Therapy for Demoralized Submarine Crews.

Title: On the Existance of a 189-Plet Mesons.

Translation: Extraterrestial Life on Meson.

Title: Double Image Formation In a Stratified Medium.

Translation: Visual Aberration in a Stoned Spiritualist.

Title: Wave Motion Due to Impulsive Twist on the Surface.

Translation: Math a Gogo, in the surf.

Title: Behaviour of the Nighttime Ionosphere.

Translation: The Naughty Sky After Dark.

Title: Fluid Behaviour in Parabolic Containers Undergoing Vertical Excitation.

Translation: Standing Room Only At The Burlesque.

Title: Redundancy in Digital Systems.

Translation: Having More Than Five Fingers or Toes.

Title: Many Body Theory.

Translation: Life in a Harem.

Title: Some Results of Transport Theory and Their Application to Monte Carlo Methods.

Translation: Hitch-Hiking Home From Los Vegas.

Title: Dispersion Techniques in Field Theory.

Translation: Fun on a Field Trip.

Title: Wullenweber Arrays Using Doublet Aerials.

Translation: A death-Defying Double Trapeze Act Featuring the Famous Flying Wullenwebers.

Title: Holography and Character Recognition.

Translation: It Takes One to Know One.

Now that you know what Verbal Rorschach testing is, try the following titles on yourself and your friends.

(1) Numerical Model of Coarticulation.

(2) Propagation Behaviour of Slotted Inhomogeneous wave Guides.

(3) The Verbal Rorschach—An Antidote for Technically Obnubilated Appellations.

(4) General Methods of Correlation.

(5) Rectification in a Column with Wet Walls.

All the titles referred to are bonefide titles of actual technical papers. The authors and the Journals in which they appear are not listed, in order to protect the guilty. If you object to the lack of good taste of some of the translations, remember that in the case of the last five, it was you who drew the dirty pictures. ■

The Backround of Lincoln's Wordmanship

Robert L. Birch
Graduate School
US Department of Agriculture

The simplicity ascribed to Abraham Lincoln's presentation is consistent with the suggestion that he may have been aware of and may have used a system of memory reinforcement used by Chaucer, Shakespeare, Milton, Thomas Gray, and others.

The tradition of memory training using consonants to stand for the digits may derive from the Druids. Direct evidence of Lincoln's awareness of it may some day be found in diaries or derived from underlinings in drafts of various writings, but so far the evidence is indirect.

Instinctive rejection of the idea that Lincoln used a memory system may grow out of the feeling that such a technique would be incompatible with his image as a relatively untutored person of native genius.

Awareness of the availability of a description of the system, published in 1845 and much commented on in the newspapers of the time, makes it seem a reasonable bet that Lincoln had heard of the system. It would be a surprise to learn that he had heard of the system but not taken the pains to study its possible application to his own habits of study.

Learning efficiently is only one of the uses of the traditional memory system. Longfellow is known to have received an autographed copy of the book *Phreno-mnemotechny, or the art of memory,* by Francis Fauvel-Gouraud, published in 1845, and the chapter headings of the *Song of Hiawatha* are a strong indication that he made use of the system it describes. Emma Lazarus, likewise, seems to have used the system in planning many of her poems, so that it is not surprising to find phrases in the thirteenth line of her sonnet on the Statue of Liberty that refer to the number 13, and that the word door, in the fourteenth line, encodes the number 14.

Chaucer's choice of the Tabard as the inn of the *Canterbury Tales,* and of the knight, the miller and the reeve, to tell the first stories may reflect the use of the traditional system, since t, n, m, and r, stand, respectively, for the numbers 1, 2, 3, and 4. In Shakespeare's *Much Ado About Nothing,* act 2, scene 3 begins with the words "In my chamber window" where the n and m, standing for 2 and 3, can cue the actor and director concerning the opening lines. Similarly, such phrases as "a lass unparalleled" in act 5, scene 2 of *Antony and Cleopatra* often make it easy to carry a mental index of scene keys.

Gray's *Elegy Written in a Country Churchyard* is relatively easy for the memorizer who knows of the system, since the author seems to have sprinkled consonants according to the system. The many p and b sounds of the ninth verse ("The boast of heraldry and pomp of power/All that beauty gave...") suggest that Gray used the system to suggest the content of particular parts. Lewis Carroll, of Alice in

George P.A. Healy's ABRAHAM LINCOLN, Corcoran Gallery of Art

Wonderland fame, wrote a brief discussion of his reasons for adopting a different pattern of digit-consonant correlations from those used by other memory specialists. He used b for the number one, and this may account for the banker, baker, and other occupation names beginning with b occurring in his *Hunting of the Snark,* where the snark turned out to be a boojum.

Evidence that Lincoln knew of the system, if it ever turns up, would do much to corroborate the feeling that the pattern of alliteration in The Gettysburg Address was deliberately related to the traditional correlation, so that the initial consonants of score and seven, dedicated, now we are engaged, met, resting-place, and so forth, were suggested by the consonant values used to code zero, one, two, three, four, and the rest. Thus it might not be coincidence that the f and v sounds, standing for eight, occur in the phrase "those who fought here have so nobly advanced" right after the k and hard-g that go with 7 in the "can not consecrate...this ground."

Proof that Lincoln did know of the system and use it might not be merely valuable to the history of stylistics. Poverty and war may be made either unlikely or less awful by the processes that communicate important meanings from one mind to another with something approximating the effectiveness of Lincoln, whose motives and oath of office would have led him to make use of the best means available to him to preserve and protect the Constitution. ■

How To Write Technical Articles

"When I use a word", said Humpty Dumpty, "it means just what I choose it to mean, neither more nor less."

—Through the Looking Glass.

Abstract: We shall endeavour to establish, Q.E.D. fashion, that contrary to popular misconceptions, writing is not an art, but a science, hence not something one is born with, but an attribute one acquires through diligent exertions.

Dear Scribe:

If writing were an art, how come so many artists write wretchedly and so many scientists well? Conversely, if writing were a science, then a writer is, ipso facto, a scientist and an artist has no business meddling in extraneous vocations. Q.E.D.

Otherwise: Writing is either a science or an art. But the artists are not writing since they are holding the paint cans while painting with their feet. Hence, it must be a science. Q.E.D. again.

Before you ever dip your quill into the inkwell, consider who your readers will be. Ostensibly, technical articles are intended for technical readership, i.e., one endowed with a level of intelligence somewhat above the mundane. Hence, if you are contemplating a "snow job", it behooves you to resort to seemingly inadvertent impediments to comprehension, e.g., sporadic cacology and non-germane juxtapositions on a level befitting those readers. It is practically incumbent on you to do so when all you have to communicate is trivial, trite and obvious. But the science of this discipline has not been yet sufficiently formulated. Henceforth, adhere to the Ten Commandments pertaining to that arcane body of knowledge and you shall have snow enough to ski in Sun Valley the year round.

You ought to realize that even the most ardent follower of "the-state-of-the-art" (a vain euphemism for "our current ignorance") gets jaded somewhat after the first grey hairs streak through and his general sophistication (a constant function of time) reaches a certain "firing level." It then becomes a matter of bemused detachment, polite boredom or downright amazement to skim through such a masterpiece laboriously sifting the heavy verbal sands for a golden nugget of original information.

Sedulously then, adhere to the following tenets to attain lasting recognition in printers ink:

1) Use words which mean only what you want them to mean and don't let on to the perplexed reader. (One professor actually wrote to me: "The earth has no boundaries, yet it is finite!")
2) Always expound in pompous polysyllables employing current buzz-words and expostulating with circumlocutory verbosity.
3) Synthesize ingenious concatenations of sonorous phrases with negligible congruity.
4) Use surreptitious recourse to arcane etymology to inculcate an impression of coruscating erudition.
5) Compound your syntax with ambivalent and equivocal periphrasis to attain esoteric obfuscation.
6) Resort to immutable quantifications to purvey an aura of punctilious erudition.
7) Sporadically resuscitate your reader from somnolent stupor by bombastic superlatives and fustian imperatives.
8) Intersperse your thematic opus with technological nomenclature without terminological reduncancies to promote extirpating turbidity and deleterious opacity.
9) Preclude untenable allegations which can be repudiated by fractious readers, incensed enough to do so, with opprobrious effects to yourself.
10) Never write as one speaks.

Of course, sodium chloride being so ubiquitous, some readers will liberally resort to it when reading you. However, if you are versed in the principles of NOR logic and apply them to the foregoing, it might come to pass that a lot of salt shakers will keep their grains for other applications. If you just keep in mind the journalistic tenets of: what, when, where, why, how and so what and furthermore have some innovation to convey, then by applying NOR logic to the above golden rules you should be able to see the light of publishing to your eternal recognition.

Yours in Caligraphy,
Semper Lucidus

Jack Eliezer
Western Union

Aug. 1968

■

A Call to Clearer Thinking

CHAS. M. FAIR
Synax Biomedical Corporation
Somerville, Massachusetts

In an age in which equal opportunity has become a major issue, Marigold L. Linton, in a recent letter to Science[1], calls attention to a neglected aspect of the problem. According to Miss (Mrs.?) Linton, the chimp Washoe who has been learning sign-language might have done far better had he (she?) not been "culturally deprived". In fact Linton suggests that the linguistic backwardness of apes in general may be due to the same cause. This is an arresting thought and I suggest that schools be set up in the jungle at once. Funding might be arranged on a matching-grant basis between HEW and UNESCO, with participation of local governments on a scale prorated to the GNP of each.

It will be recalled that John Lilly, in Man and Dolphin[2] pioneered in this field when he foresaw that dolphins might be taught to speak and proposed that, if not pacifists, they might be used by our government for underwater espionage. It may be that Linton has provided a clue as to why that project has been so slow in materializing. Dolphins probably *can* talk if only we get them started on it soon enough. Thanks to the work of Cousteau and others, a plan for setting up kindergartens for dolphin young in their own habitat is now quite feasible. There they might get the "feel" of language by playing with alphabet blocks, and later listen to stories played to them on speeded up tape. (A dolphin's vocal range runs up to 100,000 cps, its normal speaking voice lying somewhere in the middle of that range, or around 50 kc.) The aim, as I see it, should be to produce not dolphin spies but dolphin teachers, who might then carry on the work for themselves, on an oceanwide scale. The economy inherent in this approach should appeal to the present Administration in Washington. Its end-result would be to give us a pool, so to speak, not merely of secret agents, but of useful new citizens of all kinds — oyster-bed guards hull checkers, pilots (some dolphins have already gone into the field on their own, according to well-authenticated sea-stories), lifeguards at public beaches (Aristotle reports that in the Mediterranean, dolphins have on occasion come to the aid of drowning men), and guides for fishing fleets, to name a few.

Shaller's work with the gorilla suggests a third group which may only require a nudge from us to begin making giant strides on its own. The reader may recall the story of the golf-playing gorilla whom a man bought and entered in a tournament. The poor animal, having driven the ball 350 yards onto the green, was then handed a putter and drove it another 350 yards into a nearby woods. Lacking, as he and his kind have always been, in the advantages we enjoy from birth, he could hardly have been expected to do otherwise and besides incomprehension of the game, his action may have revealed a quite natural resentment at the position in which he had been put.

Such evils should not be allowed to persist, and will not, if Drs. Lilly and Linton, and others like them, are heeded. I say we should move immediately. In a technological age, literacy and speech are the birthright of all. To deny them to other species is as gross an injustice as to withhold them from our own.

(By one of those coincidences which occur with remarkable frequency in the history of science, I have just received a letter from a chimp who recently took his doctorate in driver-training arts at the University of the Pacific on Tutuila. He has given me permission to make his remarks public, since it seems that his *incognito,* respected all these years by his foster parents and fellow students, is about to be abandoned and his achievement made known to the world.

Doctor Ikashi Chojo, who was given his name by the Japanese couple who adopted him, spent his earliest years in the Tokyo Zoo, receiving his first instruction when a visiting ethologist, Professor Eibl-Eibesfeldt, gave him a short course in German word-order and prepositional constructions. Dr. Chojo has since scored an impressive number of "firsts". He is the first chimp to speak *and* write Japanese, the first to have played professional baseball — second base with the Hokkaido Giants — the first to shine his own shoes and dress himself. While at the university he shaved the backs of his hands and wore a kimono and a rubber Frankenstein mask to conceal his identity. His major, as mentioned, was driver-training, with a minor in political science. He writes:

Dear mister doctor Fair: Please forgive and forget. I write English rotten, start too late — two months old. . . . I just want to say you good student man-ape relations, out of sight, keep up good work, but strongly resent your suggestion dolphins may equal or surpass ape. All knows dolphin is a fish — warm-blooded but a fish. Sane world polity impossible if dolphin to be included. Please reconsider unsound view.

Yours truely

[Ikashi Chojo, PhD]

* * * *

The new age may have its complications, but it is clearly here. ■

---

[1] Science, *169*:328, 1970

[2] Lilly, John, Man and Dolphin, Pyramid Books, 1962

THE THERMODYNAMICS OF EEYORE

R.F. Irvine
Cambridge, U.K.

WINNIE-THE-POOH'S EEYORE, by E.H. Shepard, As seen in THE WORLD OF POOH

Abstract

Scientific analysis of "Winnie-the-Pooh" and "The House at Pooh Corner" by A.A. Milne has revealed hidden scientific allegories, mostly referring to the conflict between classical and quantum physics. The role of Eeyore as a mediator in this conflict is discussed.

Introduction

Various attempts have been made in the past to provide true interpretations of the allegorical stories in "Winnie-the-Pooh" (WTP) and "The House at Pooh Corner" (THAPC) by A.A. Milne[1,2]. Notable amongst these is the series of generally unconvincing political and psychological interpretations collected by Frederick Crewes[3]. This present paper is a summary of an alternative approach. Although this analysis is necessarily brief, I hope to demonstrate that the Pooh stories are in fact a cleverly disguised allegorical description of the changes that took place in physics during the years 1920-1930.

The evidence is based principally on two main features found in the books. firstly, a number of descriptions of physical phenomena are explicable only in terms of quantum physics - notably several predictions of Heisenberg's Uncertainty Principle in WTP even though that book was published a year before Heisenberg's revolutionary paper[4]. Secondly, a number of thermodynamic anomalies occur in both books, which are most easily explained by the assumption that Milne was under the mistaken impression that the "quantum revolution" had overthrown the laws of equilibrium thermodynamics. I shall examine this latter point first, as it gives an initial clue to the role of Eeyore in the stories.

Thermodynamic considerations

Generally, the Pooh books adhere closely to the laws of thermodynamics. For example, in the visit to Rabbit's burrow (WTP Chapter II) we find the following exchange:

"...Honey or condensed milk with your bread? He was so excited that he said, 'Both', and then, so as not to seem greedy, he added, 'But don't bother about the bread, please.' "

This is simply a comparison of intrinsic energies; if the intrinsic energies of honey, condensed milk and bread are respectively x,y and z, then assuming he would not have got stuck in Rabbit's burrow if he had complied with the original request,

$$x + y > z + y$$
$$x + y > z + x$$
$$x > z \text{ and } y > z.$$

It is apparent also that Milne had a fine grasp of the concept of entropy and of the second law of thermodynamics. Amongst examples of entropy increasing during a story are-

"PLEASE RING IF ANSWER REQUIRED" becoming
"PLES RING IF AN RNSER IS REQUIRD" and
"OWL" becoming "WOL", both in WTP (Chapter IV).

Similarly, "HIPY PAPY BTHUTHDTH THUTHDA BTHUTHDY" (WTP Chapter VI) and most striking of all,

" 'Help, help!' cried Piglet, 'a Heffalump, a Horrible Heffalump!' ...'Help, help, a Herrible Hoffalump! Hoff Hoff, a Hellible Horralump! Holl, holl, a Hoffable Hellerump!...!"
(WTP Chapter V)

which is a perfect example of a stepwise increase in entropy. Also, the decrease of entropy to nothing at Absolute Zero (the third law of thermodynamics) is at least implied when Piglet infers in THAPC (Chapter I) that only singing Tiddley-Poms prevented him and Pooh from freezing to a standstill while building Eeyore's house.

On the other hand, in that same story Eeyore equally emphatically suggests that a wind could both shift his house and actually improve its structure, and we can find further examples in both Pooh books where the laws of thermodynamics are either stretched or ignored completely. In THAPC Chapter V an originally high entropic message:

"GON OUT
BACKSON
BISY
BACKSON
C.R."

changes in the same chapter to

"GONE OUT
BACK SOON
C.R."

Rabbit's astonishing expenditure of energy during this story cannot account for this decrease in entropy, as Rabbit always caused chaos (entropy) wherever he went. i.e. he converted all his kinetic and verbal energy into entropy.

Other examples of thermodynamic inconsistencies are:

"But whatever his weight in pounds, shillings and ounces He always seems bigger because of his bounces."
(THAPC Chapter II);

".....Cottleston, Cottleston, Cottleston Pie, A fly can't bird, but a bird can fly" (WTP Chapter VI)
(an irreversible reaction);

What therefore is Eeyore? Is he a personification of quantum mechanics, or is he merely an energy source, predicted by Milne but as yet undiscovered, by which these odd happenings can be explained - in which case no laws are broken? If so, is bounciness a negative form of this energy? Perhaps Eeyore's name as spelt by himself (EOR) is a mnemonic for this predicted force (entropic order redundancy?). More light can be shed on this by looking at some illustrations of the more modern theories of physics and how they are expressed in the Pooh books. Once again just a few examples must suffice.

Quantum Physics

Both Pooh books are peppered with illustrations of Heisenberg's Uncertainty Principle, for example:

" 'So did I', said Pooh, wondering what a Heffalump was like." (WTP Chapter V) or (WTP Chapter VII)

" 'You ought to see that bird from here;' said Rabbit. 'Unless it's a fish' ".

The best example however, is the poem found in Chapter VII of WTP whose first verse runs:

"On Monday, when the sun is hot
I wonder to myself a lot:
'Now is it true or is it not,
That what is which and which is what?"

This whole poem is essentially a description of the Uncertainty Principle, showing remarkable precognition of a paper published a year later[4].

The distortion of Time induced by high speeds as predicted by Einstein's Special Theory of Relativity[5] is very well illustrated in WTP (Chapter IV) where Eeyore's gloominess is the distorting factor:

"Who found the Tail?
'I', said Pooh,
'At a quarter to two
(Only it was quarter to eleven really),
I found the Tail' "

This is a beautiful example of modern scientific thinking, as one would expect from a poem involving Eeyore. Pauli's Exclusion Principle[6] (newly proposed when the books were written) seems to be suggested in the following passage from WTP (Chapter V):

"Pooh, rubbed his nose with his paw, and said that the Heffalump might be walking along humming a little song, and looking up at the sky, wondering if it would rain,.....and said that, if it were raining already, the Heffalump would be looking at the sky wondering if it would *clear up!*"

Radioactive decay is illustrated by the decline of Pooh's honey pots on the branch of a tree (WTP Chapter IX), and in THAPC Chapter V Eeyore perpetrates a perfect chain reaction some years before Fermi first did so,

" 'Clever!' said Eeyore scornfully, putting a foot heavily on his three sticks. 'Education!' said Eeyore bitterly, jumping on his six sticks. 'What *is* learning?' asked Eeyore as he kicked his twelve sticks into the air."

Conclusions

The simplest explanation of the observations made here is that Eeyore is a personification of Quantum Physics, and that his goominess is the means by which the old ideas are destroyed and the new ones built up. The confirmation of the Uncertainty Principle by Heisenberg in 1927 may have boosted Milne's confidence in Eeyore so that his effects are more noticeable in the second book; observe how the majority of the examples cited here (and of all the examples I have found) are in THAPC.

Other possibilities are discussed above, and only further research can clarify the unique role which, on preliminary analysis at least, Eeyore appears to fill in twentieth century physics. ■

Acknowledgements:

I thank Peter Knox for many helpful discussions, and Ann Silver and Martyn Hill for encouraging me to publish my results.

REFERENCES

1. Milne, A.A. (1926) Winnie-the-Pooh. Methuen London
2. Milne, A.A. (1928) The House at Pooh Corner, Methuen London
3. Crews, F.C. (1964) The Pooh Perplex. Baker London
4. Heisenberg, W. (1927) Zeitschrift für Physik 43, 172-198
5. Einstein, A. (1922) The Meaning of Relativity. Trans. E.P. Adams Methuen London
6. Pauli, W. Jr. (1928) Zeitschrift für Physik *31*, 765-783

Harold Schultze
Assistant Professor

Viability of guinea pigs upon exposure to well defined experimental parameters normally associated with cessation of vital signs due to rapid crainial tissue contact with stainless steel in a cylindrical form.

A shot in the head
will make him dead.

An Elucidation of Certain Varieties of a Phenomenon of Common Occurrence in Scientific Communication......

RICHARD L. SUTTON, JR., M.D.
Kansas City, Mo.

"Bullshit" is widely used. The connotation, broadly understood, is more benignant, less denigratory, when b. is spelled as one word of 8 letters than when spelled as two words of 4. While perceptive and analytical essayists, like Morris Fishbein in his "Medical Writing" years ago, have given it consideration — although not directly by name — systematic b-ology, in contradistinction to the mere amateur utilization of b., may be said to have been largely neglected by practically all contributors to Scientific Literature. "Literature," so used to designate the main mass of such publication, is a specimen of b.

"Bull" in the sense of "nonsense" is in the Collegiate dictionary, but there is no entry between "bull's eye" and "bull snake" or between "shish kabab" and "shittah." This denotes on the part of professional word-listers an obtuse inattention to the vulgate as it is practiced, for example in the spirited public utterances of the entire student body of Leland Stanford Junior University at football games.

Limitation of space precludes all but a basic consideration of b., with its variegated and esoteric occurrences. Simply, I classify it as Conscious, Subconscious, and Unconscious. I refrain presently from ethical valuations but do believe that some kinds are worse than others. The following are illustrative.

Conscious. — "We" in the place of "I": the author is not valorous, and he knows it. Possibly his reasons are cogent. If his idea should prove demonstrably stupid, his critic may miss his head with the righteous bolt of indignant rectification as a result of the sportsman's error of "shooting into the brown," of which "we" provides ample. From a contemporary work I quote, "Dystrophia Unguis Mediana Canaliformis (Heller) . . . Treatment is unnecessary, but the patient should be advised to apply an emollient cream to the nail fold." The title, for which the authors are not responsible, is Heller's b., since the disorder is not a dystrophy or median or canaliform necessarily; but it is necessarily the tubular outgrowth of nail about a matrix teat, which is the essential lesion and which, usually, can be extirpated so as to cure the condition — which is not to be accomplished by applying an emollient cream to the nail fold for a period of time equivalent to the Triassic. Thus, excepting the first 3 words, the text quoted is clearly a specimen of Conscious B.

Subconscious. — "Logical" is pathognomonic for "illogical," to be translated, when and if the author becomes conscious, with "plausible" or "inviting." "It is thought" analogously means, "My brilliant colleagues . . ." or possibly "I, myself, have hypothesized, or fancied, or dreamed . . ." Like "we," this is invalorous but less deliberately, since the author simply lacked the guts to say, "I think . . ."

Unconscious. — In these cases, the author's intellectual environment was deprived, so he ought in charity to be held blameless, for it is Society that is at fault when the language of the home is Troglodyte. No one told him, and he did not discern, that "regime" does not mean "regimen." His "plantar callus of the sole of the foot" uses 8 words to say 3 times what the first 2 say in a manner that appeals esthetically to the discriminating. Compare "tinea capitis of the head," which I have heard but not seen because a first-day copy girl will decapitate it. I designate this variety as "Solecistic Redundant B. of the Innocent." This is itself of course pedantic humor, so named by Fowler — i.e.,b.

Acronyms have merit in saving the printed page and/or the muscles of phonation, and I submit a new one especially fitting when Conscious B. is redolent: A C E R B, meaning "Archie Campbell's Estimable Rejoinder to B." Band 5 on Side 2 of RCA Victor LSP 3699 records his musical "I'd 'a wrote you a letter, but I couldn't spell pffft."

Two conclusions may be adduced: barring the necessity of technical symbolism, anything worth saying can be said in plain English; and the hypothesis is inviting that an article in a learned journal, if devoid of b., would be a thing of beauty, a joy for quite awhile, and brief. ■

THE TITLE OF THIS ARTICLE IS NOT LOGICALLY DERIVABLE FROM ITS ASSUMPTIONS

Ron Marshall
East Lansing, MI

As part of its editorial policy, the prestigious *Journal of Truth and Logic* requires that any article submitted for publication must satisfy the following conditions:

1. All true statements appearing in the article must be logically derivable from the article's underlying assumptions.
2. If, however, a false statement is made, intentionally or otherwise, it must not be a logical derivation of these assumptions.
3. Any article may be re-submitted, but no more than once.

Recently, the present article (the one you are now reading) was submitted to this journal and after a surprisingly short time, the following letter was received from the editor:

Dear Sir:

We regret to inform you that your article is unacceptable for publication for the following reason. If the title of your article is a true statement, then what it asserts must, in fact, be true—namely that it is not derivable from the article's assumptions. This, as you know, would violate condition 1 of our editorial policy. On the other hand, if the title is presumed false, then what it claims cannot be so, in which case it must indeed be derivable from these assumptions. But this would constitute a violation of condition 2 of our policy. Either way, one of these conditions is violated and so we have no choice but to reject your manuscript.

Pursuant to condition 3 of our policy, you are entitled, of course, to revise and re-submit your article (although in all candor, I must tell you that the reviewer does not encourage you to do this). Thank you for considering the *Journal of Truth and Logic*.

Sincerely
The Editor

It was soon realized that the reason for the unusually short turn-around time was due to fact that the reviewer found it sufficient to read only as far as the article's title in order to reach a rejection decision. The reason for rejection—something to do with article's assumptions—was, of course, unexpected and somewhat surprising since the article, as should now be apparent, has no asumptions.

In any event, the article was revised and re-submitted. Unfortunately, rejection occurred again, this time due to the opening line of text, a sentence (since removed) that read:

> "This sentance is not derivable from the assumptions of this article because it contains three erors."

The reviewer actually thought the sentence might be true arguing that because of its location, there were no antecendents from which a derivation could be made. My own view is that the sentence is at least partially ture in the sense that it actually does contain three errors: two spelling errors ('sentance' and 'erors') plus one counting eror (it incorrectly asserts that it contains three errors when, in fact, it obviously contains only two). As the editor was careful to point out, however, the issue of truth or falsity is a conundrum upon which the rejection decision did not depend. Rejection the second time around was inevitable, as it was the first time, because of the certain violation of one of the Journal's first two policy conditions.

It may be worth noting that in deference to condition 3 of the Journal, the article was not subsequently re-submitted. And though its current whereabouts is unknown, it is perhaps not too much of an exaggeration to say that at this very moment, it is probably being read by someone, somewhere... ■

REFERENCES

Hofstadter, Douglas R., *Godel, Escher, Bach: and Eternal Braid,* Basic Books, 1979

Hofstadter, Douglas R., and Daniel C. Dennet, *The Mind's I,* Basic Books, 1981, pp. 276-283

Smullyan, Raymond M., *What is the Name of This Book?* Prentic-Hall, 1978

Irreproducible
research
too often
leads to

GREAT
DISCOVERIES

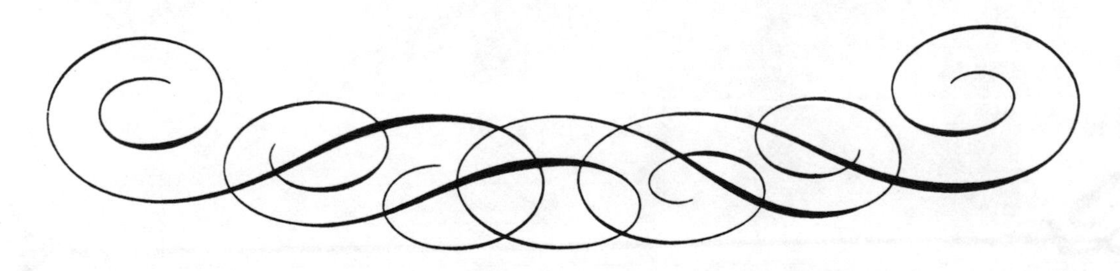

DETRIMENTAL EFFECT OF PEROXIDES UPON BRAIN CELLS

Jim Ballinger
Toronto, Canada

Examination of certain popular television series and other anecdotal evidence that the use of peroxide hair bleaches for cosmetic purposes over prolonged periods (> 5 years) seemed to be correlated with impaired brain function prompted us to study this phenomenon in humans and in an animal model. Integrity of neuronal pathways was assessed by myographic studies in human volunteers. Following similar studies in animals, brain biopsies were taken, examined by electron microscopy, and incubated to determine N-acetyl clavidase activity.

METHODS

Human Studies

Female volunteers were recruited from the staff of the Affliction Research Foundation, Toronto, and were divided by the methods of Haley (1) and Hefner (2) into three groups: A-bleached blondes (N=17), average duration of bleaching =5.92 years; B-natural blondes (N=9); C-natural brunettes (N=19). The groups originally contained 20, 10, and 20 subjects, respectively; however, two of the bleached blondes withdrew due to mental instability and one by suicide, one of the natural blondes was later determined to be not natural, and one of the natural brunettes dyed her hair during the course of the experiment and had to be dropped from the study. The original advertisement for volunteers did not specify that only females were wanted and, in fact, a number of males who used bleach did apply; inexplicably, however, none was able to withstand the mild electrical shock involved in the test and speculation on a sex-related effect of peroxides upon pain threshold forms the basis of another paper.

Although brain biopsy is a recognized diagnostic procedure, it was judged to be unethical for this study as it was feared the following removal of 200 mg of cortex some subjects would not be left with sufficient cerebral capacity to function normally. Therefore, the degree of impairment of spinal reflexes was studied by myographic determination of impulse transmission rate following electrical stimulation (3). This method was suggested by a fellow down the hall; he is only a glass-washer but he gives sound advice on a wide variety of subjects, from investments to personal matters. The procedure was painless and quick, requiring only 15 minutes on 4 separate occasions for each subject.

Animal Studies

The species chosen was the guinea pig since white and brown strains were available (The Rodent Ranch, Caledon). As in the human studies, three groups were used, 20 animals in each: A - bleached guinea pigs; B - naturally white guinea pigs; C - naturally brown guinea pigs. Bleaching of guinea pigs was performed by immersion in vats of 5% w/v hydrogen peroxide solution for 7 minutes. Artificially-bleached animals appeared identical to the naturally white guinea pigs.

Similar myographic determinations of impulse transmission rate were recorded. Following 10 replicates in each subject, the animals were sacrificed by the pentobarbital milkshake method and brain biopsies taken. Brain cell histology was examined by electron microscopy. Tissue samples were also incubated at 37°C for 20 minutes in 0.05 *M* phosphate buffer and N-acetyl clavidase activity was determined by published methods (4). Statistical analyses of data were carried out by methods developed in conjunction with Dr. Onyczek (5).

TABLE 1.

Species	Group	Subjects	Transmission metre-V/amp-sec	Activity umole/mg/min
Human	A	bleached blonde	0.361 ± 0.104	-
	B	natural blonde	0.694 ± 0.025	-
	C	natural brunette	0.742 ± 0.017	-
Guinea pig	A	bleached white	0.073 ± 0.026	2.36 ± 0.84
	B	naturally white	0.131 ± 0.012	4.36 ± 0.42
	C	naturally brown	0.143 ± 0.009	4.78 ± 0.21

RESULTS

The results are self-explanatory and are presented in Tables 1 and 2. As can be seen, the animal model correlates very well with the human situation. Furthermore, morphological observations identify the site of toxicity as the basal perineural cells of the brain, where damage is irreversible.

There were no significant differences between the natural blondes and brunettes ($P \approx 0.2$). However, a 50% decrease in capacity is observed in bleached blondes ($p < 0.001$). Regression of impulse transmission rate with duration of bleaching showed high correlation ($r = 0.923$).

TABLE 2

Description	%Remaining Function human	guinea pig
bleached blonde	48.7	51.0**
natural blonde	93.5	91.7***
natural brown	100.0*	100.0*

*by definition
**significant ($p<0.05$)
***not significant

Further animal studies are currently underway to more accurately profile these changes. In particular, the effects of loud disco music and imported mineral water are being studied. However, preliminary results suggest that IQ decreases by 1 point each time a woman bleaches her hair. The implications are mind-boggling. ■

REFERENCES

1. Haley A: "Roots" Garden City: Doubleday (1976)
2. Hefner H: Pictorials, *Playboy,* various issues (1954-80)
3. Sicuteri F *et al:* Measurement of brain cell function by impulse transmission rate, *J Expt Neurol, 72* (3): 321-7 (1958)
4. Wong K-L, Vitro G: Assays of brain enzymes. III. N-Acetyl clavidase activity, *J Brain Res, 9* (2): 74-9 (1973)
5. Ballinger JR, Onyczek UF: Bal-Ony analysis, *J Irreprod Stats, 13* (0.05): 132-27 (1978)

D.F. Petersen and O. Carrier, Jr.

AFFERENT NEURAL RESPONSES TO MECHANICAL DISTORTION OF THE TESTIS OF THE CAT

Federation Proc. 31:370abs (1972)/

"...compression in lightly anaesthetized cats indicated a pseudoaffective pain-like response to distortion of the testis" "A glancing blow to the testicle produced a burst of activity."

Unbelievable but true!

THE PHYSICS OF FLUIDS

Vol. 25:5, 1982

The meandering fall of paper ribbons

Adrian Bejan

The objective of this paper is to present a series of interesting experimental observations concerning the meandering motion executed by highly flexible ribbons falling through the air. The experiment consisted of dropping a length of light-weight toilet tissue paper through the air and photographing its shape as it falls to the ground. The reader may take note of the fact that this falling-ribbon phenomenon occurs naturally when excited sports fans launch rolls of tissue paper from the stands onto the playing field. Another natural phenomenon related to the falling-ribbon experiments described in this paper is the "waving of flags" and the "vibration" of tape drives used in the computer technology.

ACKNOWLEDGMENT
This research work was supported by the Office of Naval Research.

Submitted by:
Heather Redding
Idaho Falls, ID

Dear Editor,

The article by Cerny and Gort, "Genetics plus cartography produces map cows," in the last issue of *JIR*, while clearly of major significance to zoologic geography, and while mentioning the contribution of Emil van Beest in 1969, failed to acknowledge previous seminal work on zoocartography in Italy. I attach an illustration of the 1968 work of Claudio Parmiggiani in a private collection in Milan, reproduced in *Hic Sunt Leones: Geografica fantastica e viaggi straordinari* (Milan: Electa, 1983), p. 115, which clearly demonstrates the pre-eminence of the Italian Holstein breeders in producing zoocartographic maps of continental proportions.

Sincerely yours,

David Woodward
Madison, WI

RATIONALE FOR GETTING A HAIRCUT
(and taking a bath)

We say we are alive,
yet we are always dying on the
outside.
Strands of hair,
nails,
external scurf,
all the dead parts,
constantly increase.
The living parts insist on making
more of them,
trying to tell us something.
Unchecked,
the zone between life and death
could slip dangerously deep,
disturbing equilibrium.
Negative feedback would become
a possibility.
Production of new tissue might
stop at the center,
precluding turnover,
leaving us mired in necrosis.
If we rid ourselves
of these vestures that capture us like
snakeskins,
we will continue to keep death
in its proper perspective.
We will survive
centrifugally.

William H. Blackwell
Oxford, OH

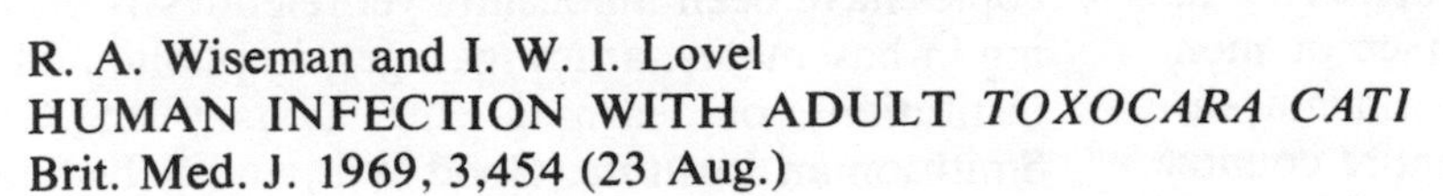

R. A. Wiseman and I. W. I. Lovel
HUMAN INFECTION WITH ADULT *TOXOCARA CATI*
Brit. Med. J. 1969, 3,454 (23 Aug.)

Commenting upon a case of an intestinal infection of a 14-month English boy who vomited 3 nematodes, the authors quote Sprent (Austral. Vet. J. 1958, 34, 161) as follows:

"The subject was a woman who habitually ate earth from the graves of priests, and as a result of her unusual eating habits frequently vomited beetles, suggesting that human infection may possibly result from the accidental ingestion of beetles, cockroaches, or other intermediate hosts."

NATIONAL GEOGRAPHIC, THE DOOMSDAY MACHINE

GEORGE H. KAUB

Pollution of many types and kinds is currently paramount in the public mind. Causes and solutions are being loudly proclaimed by all of the media, politicians, public agencies, universities, garden clubs, industry, churches, ad infinitum. Pollution runs the spectrum from the air we breath, the water we drink, the soil we till, as well as visual and audio pollution, and in recent years, pollution of outer space from junk exploration hardware. These threats to our environment, our health and our mental well being are real and with us, but not nearly as immediately catastrophic or totally destructive as the disaster which imminently faces this nation and which has gone unheeded, unheralded and ignored for over 141 years. The insidious consequences lurking in this menace of monstrous proportion bode national, even, continental disaster of proportions likened only to the entire country resting on a gargantuan San Andreas fault. Earthquakes, hurricanes, mud slides, fire, famine, and atomic war all rolled into one hold no greater destructive power than this incipient horror which will engulf the country in the immediate and predictable future.

This continent is in the gravest danger of following legendary Atlantis to the bottom of the sea. No natural disaster, no overpowering compounding of pollutions or cataclysmic nuclear war will cause the end, instead, a seemingly innocent monster created by man, nurtured by man, however as yet unheeded by man will doom this continent to the watery grave of oblivion.

But there is yet time to save ourselves if this warning is heeded.

PUBLICATION AND DISTRIBUTION OF THE NATIONAL GEOGRAPHIC MAGAZINE MUST BE IMMEDIATELY STOPPED AT ALL COSTS! This beautiful, educational, erudite, and thoroughly appreciated publication is the here-to-fore unrecognized instrument of cosmic doom which must be erased if we as a country or continent will survive. It is NOT TOO LATE if this warning is heeded!

According to current subscription figures, more than 6,869,797 issues of The National Geographic Magazine are sent to subscribers monthly throughout the world. However, it would be safe to say that the bulk of these magazines reach subscribers in the United States and Canada, and it is and never has been thrown away! It is saved like a monthly edition of the Bible. The magazine has been published for over 141 years continuously and countless millions if not billions of copies have been innocently yet relentlessly accumulating in basements, attics, garages, in public and private institutions of learning, the Library of Congress, Smithsonian Institute, Good Will, and Salvation Army Stores and heaven knows where else. Never discarded, always saved. No recycling, just the horrible and relentless accumulation of this static vehicle of our doom! National Geographic averages approximately two pounds per issue. Since no copies have been discarded or destroyed since the beginning of publication it can be readily seen that the accumulated aggregate weight is a figure that not only boggles the mind but is imminently approaching the disaster point. That point will be the time at which the geologic substructure of the country can no longer support the incredible load and subsidence will occur. Gradually at first, but then relentlessly accelerating as rock formations are compressed, become plastic and begin to flow; great faults will appear. The logical sequence of events is predictable. First will come foundation failures and gradual sinking of residences and public buildings in which the magazine has been stored. As these areas depress the earth, more and more structures will topple and sink until whole towns and cities will submerge, then larger and larger land masses. This chain reaction will accelerate until the entire country has fallen below the level of the sea and total innudation will occur.

The areas of higher subscription density, affluence and wealth will be the first to go, followed by institutions, middle class, urban and ghetto areas in that order, with the relatively unpopulated plains and mountains finally sinking into the sea.

We have been warned of this impending calamity by a seeming increase in so-called natural disasters throughout the country as well as isolated occurrences striking areas heretofore immune to natural destruction.

Increase in Earthquake activity in California has been triggered by population growth and the subsequent increase in National Geographic subscriptions and accumulations of heavy masses of the magazine. This gradual increase in weight has caused increased activity along the San Andreas fault.

Earthquakes in the Denver area were not caused by pumping of wastes into wells at the Rocky Mountain arsenal, but by accumulation of National Geographic magazines by more and more people as the population increased over the years.

Sinking of several coal mining towns throughout the country can only be attributed to the increase in workmen's benefits and pay increases allowing them to subscribe to and hoard National Geographic.

Mud slides in California which have brought destruction to hundreds of homes built on the hillsides were triggered by the final straw in the form of the last mail delivery into these areas of National Geographic subscribers and hoarders.

The list is endless, the warnings are clear.

The time grows short and we must act at once if this calamity is to be averted. The National Geographic must cease publication at once, if necessary, by Congressional action or Presidential edict.

Sobel, H., Mondon, C.E., and Means C.V.

PIGMY MARMOSET AS AN EXPERIMENTAL ANIMAL

Science 1960, 132, 415-416

"To date there have been no successful matings, however. This may be due to the failure to separate pairs."

(R. Le Vin wishes to know: How successful can you get?).

****

Miles Fox and E. L. Barrot

VACUUM CLEANER INJURY OF THE PENIS

Brit. Med. Jour., 1960, June 25, p. 1942.

Three cases of such injuries are described. The authors call the methods used by the patients, and leading to the injury, "rather ingenious" but producing "disastrous results."

Amazing!

D. Causey

CRIMES IN SCIENTIFIC EDITING

Turtox News, 1959, vol. 37, No. 3, p. 93

..."Editors are, in my opinion, a low form of life, inferior to the viruses and only slightly above academic deans."

****

A NOTE IN THE TEA PHYTOLOGIST

1939, vol. x+1, No. 1, p. 17

"In a short series of recent experiments, 50 plants of *Drosera rotundifolia* were fed at intervals of one hour with miniature beefsteaks; at the moment of application a dinner bell was rung. After 5 days of this treatment the dinner bell was rung without the application of beefsteak; 15 out of 50 plants elevated their leaves, at the same time producing copious peptic secretion."

THE AGE OF ENLIGHTENMENT ENDS

KIRK R. SMITH
Environmental Health Sciences
Warren Hall
University of California
Berkeley, 94720

There is no such thing as light. What there is in the universe is dark.[1] It is obvious from simple observations that this is so.

What we call light is merely the absence of dark. Dark is continually created. As fast as it is whisked away, more fills up the space.

We can easily establish these facts long hidden by the tenaciousness with which light-headed scientists have clung to their illuminating but less than brilliant theories.[2]

What we have called sources of light are in reality dark-sinks. They are places into which dark is sucked. More dark is created and is sucked into the "light". It, of course, flows at the speed of dark which is relatively fast.

It is often observed that "light bulbs" after failure contain a quantity of dark inside. The dark has clogged them up. Normally, of course, the dark is sucked down the wires and into power stations where it is put back into the world in the form of air and water dark (smoke and pollution).

A fire in the fireplace uses chemical energy to pull the dark out of the room leaving a bucketfull in the fireplace afterwards.

Shadows are created simply by objects being in the way. The dark can't get by on its way to the dark-sink.

I suspect that a physicist, being conservative by nature as well as by law, will not accept this radical new theory without flaring up.

"What about 'light pressure'?" he will grumble darkly.

Simple: When dark is sucked away, new dark is created to take its place. This new dark is created as in pair-production. The new piece of dark (darkton) travels backwards with equal velocity and momentum as the darkton being pulled away. When it strikes an object it exerts a pressure. (See Figure I) Since behind the object being pushed there is no new dark being created, there is no push from behind. The push is all on one side and results in an effective pressure away from the dark-sink.

The photoelectric effect? Merely the darkton hitting an electron on its way to the sink or after being newly created.

Colors? Different shaped darktons. See Figure II for most probable shapes. We are able to see different colors because of these shapes. For example when the newly pair-production created darkton is yellow in shape it fits into the enzymes in our eyes as in Figure III. As we have seen in molecular biology texts enzymes come in the appropriate designs to detect all possible color-shapes.[3]

This is by no means a revelation to be treated lightly. Our view of the world will be markedly changed. "As the sunrise empties the valleys of dark" will become precise scientific description instead of poetic vision. Basic philosophy will have to be transformed.[4] "Let there be dark-suck"? But first there had to be the dark. Perhaps we should alter the old adage and, applying ourselves directly to the source, we should, indeed, curse the darkness.

[1] R. R. Lyrae, "Speaking in Variables" Journal of Microwave Astrology 11:45 1963

[2] The first clues leading to this line of reasoning developed from meterological research done at the botanical gardens. Following the method of multiple-hypotheses in pursuing the possible connections between trees and wind, it was found that, in fact, the wind is caused by the fan-like action of forests. In no case was there observed wind without the trees waving.
S. Maizlish, "Observing the Forest in Spite of the Breeze" Ohio Journal of Forensic Forestry 12:31 1945

[3] "Photo" synthesis is just the reverse.

[4] D. Koch, "The Real Thing" Turn-on Delight Monthly 7:25 1955

FIGURE I

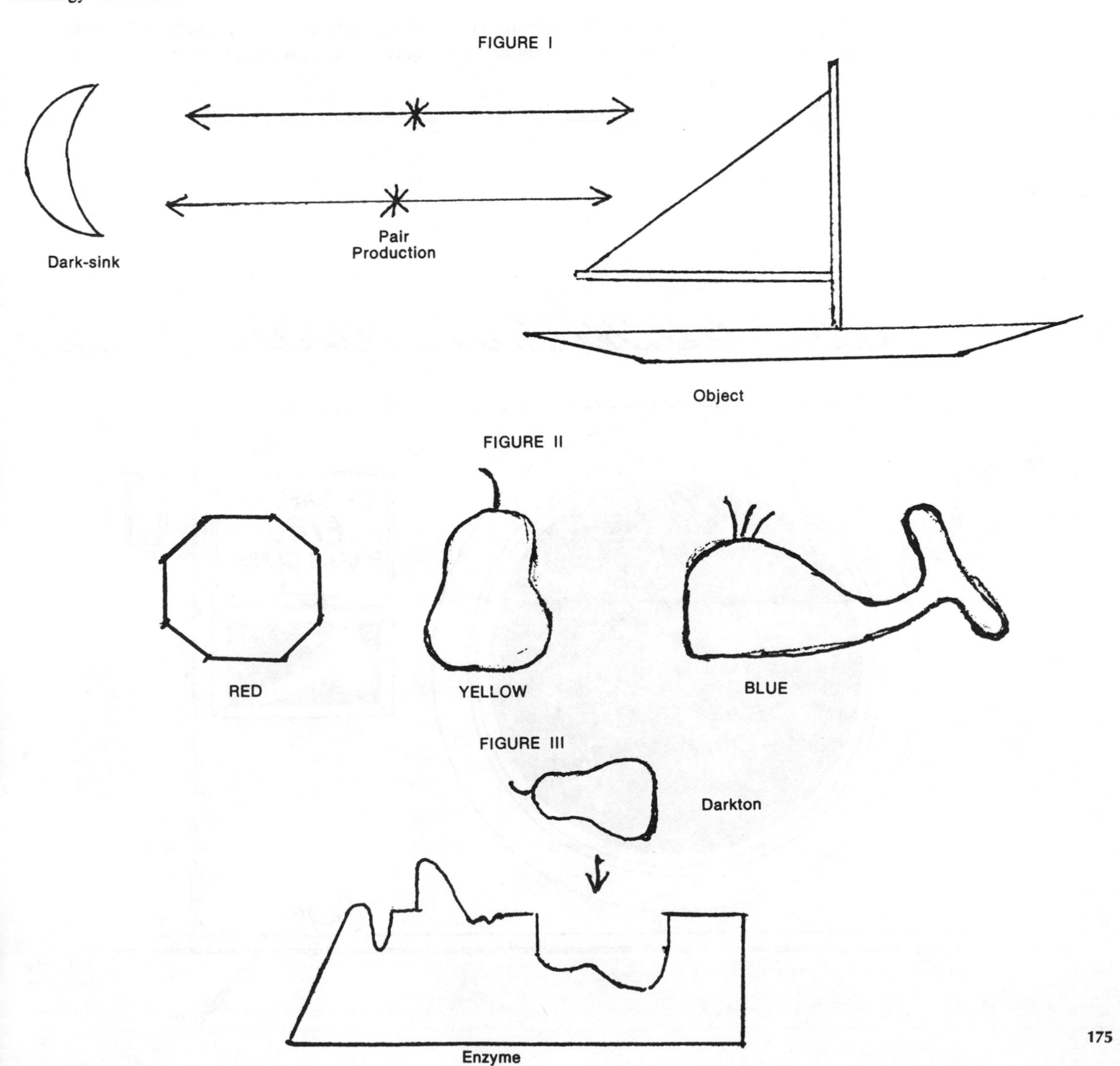

S. B. G. Eysenck and H. J. Eysenck
AN IMPROVED SHORT QUESTIONNAIRE FOR THE MEASUREMENT OF EXTRAVERSION AND NEUROTICISM
Life Science 1964, 3, 1103.

From answers to 12 questions such as: Are you rather lively? Do you like mixing with people, would you call yourself happy-go-lucky? Do you like practical jokes? Do you suffer from sleeplessness? When you get annoyed do you need someone friendly to talk to about?

The authors come to the conclusion that "the men are very slightly more extraverted, although the difference is too slight to mention," and that the women are noticeably more neurotic about 1/3 S. D.

* * *

Phil Foglio
N.Y.C., N.Y

DARWIN, EDISON AND HOUSEPLANTS: Origin of the Specious

Robert J. Sokol, M.D.*
Roberta S. Sokol, M.A.**
Detroit, MI

INTRODUCTION

With the advent of vasectomy and tubal sterilization, the number of children per household has fallen. As a result, many upwardly mobile middle class Americans have taken to raising plants to "people" their households. Unfortunately, these green children have a very high mortality rate. In the typical case, the etiology of death remains elusive. At the all too common funerals, knowledgeable chlorophiles are as likely to opine (1)—"Too much water" as "Too little," "Too much light" as "Too little," or "Too little fertilizer" as "Too much." A recent case, however, implanted a new hypothesis in the fertile soil of the minds of these investigators.

CASE REPORT

A healthy six year old Schefflera, adopted eight months previously, developed terminal dandruff shortly after the demise of its sibling in the next pot. Manipulations of water, light and fertilizer concentrations were without result. In the absence of response to treatment for a physical problem, a consult in psychoceramic(2) medicine was obtained—perhaps, psychosocial disruption caused by moving to a new house, as often observed in Ficus, was the problem. Scheff, our green son, responded with worsening alopecia.

When the LOF/LOP ratio(3) became critical, euthanasia seemed the kindest course. One of the authors (RSS) administered potent poison quickly and painlessly (Malathion, 2 mg/KgPS***). Amazingly, the plant did not die—it recovered; the LOF/LOP ratio approached zero.

The key finding in this case was that poison appeared to produce a significant decrease in the LOF/LOP ratio. This suggested that a mechanism, sometimes operative in neuroendocrinology, i.e., inhibition of an inhibitor, such as $\downarrow$ prolactin inhibiting factor (PIF) $\rightarrow$ $\uparrow$ prolactin, might be present. Further, the previous demise of Scheff's sibling suggested application of the neonatal concept of nosocomial infection. Histopathological examination of tissue samples was "suggestive of severe mitosis."****

This etiologic diagnosis has theoretical and practical implications. On the theoretical level, it provides strong evidence for Darwin's theory. Probably, when men (persons—Ed. note, RSS) first dragged some plants from the jungle into their caves, they were somewhat taken aback by the brontosaurus-sized insects which inhabited them. Out went the plants, which quickly died. Homeless, the bugs died off too. A few plants were inhabited by petite pests. In the dark of the caves, they were never noticed. The process continued for millenia—survival of the smallest. Things went along just fine for the bugs until Thomas Edison almost did them in with his light bulb—all of a sudden the bugs lost their indoor invisibility. This, by the way, accounts for the Kamikazi attacks mounted by the bug population on any exposed light bulb. But, we digress. The evolutionary pressure for microminiturization (Japanese beetles led the way) was tremendous. By the mid 1970's the bugs had become nearly microscopic. Hence, the recent boom in houseplants, despite mites, aphids and the all-devouring thrip.

From the practical perspective, this case report suggests that the next time you endure the death of one of your green children (leaflessness), during the moment of silence at the funeral, listen closely for a barely audible 21 Bug Salute. ■

REFERENCES

1. **Norfolk, et al:** The Irish Evergreen: A source of needles for I.P. (Intraplant) Injection. *O'Lancet.* In press.
2. **Yoyo, URA:** Crackpots and cachepots: financial aspects of plant psychotherapy. *N. Korean Journal of Botany and Finance* 15:232-198, 1984.
3. **Prunes, Stu S:** The Leaves on Floor to Leaves on Plant Ratio In "Memoirs of an Inebriated Horticulturist." Fig U. Press, Cleveland, 1975, pp 15023-16142.

*M.D. = Moss Doctor
**M.A. = Malathion Administrator
***PS = per stomata
****Mitosis = Full of mites

United States Patent Office

3,781,424
Patented Dec. 25, 1973

HEMORRHOID TREATMENT PREPARATION PRODUCED FROM HOT CHERRY PEPPERS

Edwin K. Ponvert,
Virginia Beach, VA

Continuation-in-part of application Ser. No. 656,700, July 28, 1967, which is a continuation-in-part of application Ser. No. 402,018, Oct. 6, 1964, both now abandoned. This application July 21, 1970, Ser. No. 56,826

Int. Cl. A61k *27/00*

ABSTRACT OF THE DISCLOSURE

This present invention relates to a process, and the product thereof, whereby a new composition of matter is produced. This new composition of matter has no known chemical formula, it being a complex mixture of many chemicals of unknown organic structure. This new composition of matter, a soft solid residue of *Capsicum anuum,* is intended for oral administration in various forms, and prescribed dosages, to human beings suffering from hemorrhoids. Hemorrhoids, are not known to exist in animals. It is, therefore, intended as a hemorrhoidal remedy for the relief of hemorrhoids in mankind.

PATENT SPECIFICATION

1 426 698

(54) PHOTON PUSH-PULL RADIATION DETECTOR FOR USE IN CHROMATICALLY SELECTIVE CAT FLAP CONTROL AND 1,000 MEGATON, EARTH-ORBITAL, PEACE-KEEPING BOMB

(71) I, Arthur Paul Pedrick, British subject, 77 Hillfield Road, Selsey, Sussex, do hereby declare the invention, for which I pray that a patent may be granted to me, and the method by which it is to be performed, to be particularly described in and by the following statement:-

(1) To detect the difference in the colour of the fur on the back of a cat wishing to gain entrance to a house by means of a "chromatically selective cat flap", to thus admit to a house a cat which has GINGER fur, but exclude a cat with BLACK fur.

and

(2) To provide, in an Earth Orbital 1,000 Megaton Complete Nuclear Disintegration or "CND" Bomb Automatic Reprisal Satellite Bomb, forming part of an Automatic Response Nuclear Deterrent System, or ARNDS System for short as described in UK Patent No. 1,361,962, means for detecting with certainty, whether a nuclear attack has been made. ■

Submitted by:
Richard D.Seibel
Pasadena, CA

Levinson, S. and Liberman, M., "Speech Recognition by Computer", *Scientific American,* 224, 4(April 1981). 64-76

"At the center of human language is the word."

Submitted by:
Evan Rudderow
Hoboken, NJ

IMHOFF's Law: A bureaucracy is very much like a **septic** tank: the REALLY BIG chunks always rise to the top. *Management Review*

Submitted by:
Robert R. Witt

New linear accelerator to take wing?

COLONEL SANDERS
Montana State University
Bozeman, Montana

At a time when physics is faced with funding crises everywhere, it is reassuring to find that some research projects can still be run on mere chicken feed. In this regard, we call attention to the recent announcement, by the National Research Council of Canada, of the successful operation of a new linear chicken accelerator, or LCA (see *Chemical and Engineering News*, 2 November 1970, page 56). The LCA, which is capable of accelerating a four-pound chicken to speeds of 620 mph, is currently being used as a flight-impact simulator in an engineering study of airplane-bird collisions. But we believe it may have application as a basic research instrument, since — in more familiar terms — it has a rated energy of 5×10^{14} GeV, which makes the LCA the most powerful accelerator of its kind in the world today.

A careful study of high-energy chicken-chicken collisions, with due attention paid to the production of virtual chickens (i.e., eggs), could lead to a resolution of an age-old question of causality, namely, which came first, the chicken or the egg? At somewhat higher energies, one could look for the production of the intermediate vector chicken, or hawk, and in general study the problem of rooster-hen coupling. At yet higher energies, the scattering would of course be discussed in terms of the Pomeran-chicken trajectory. Crossing symmetry would be important here, and one could hope to discover why, or even whether, the chicken crosses the road. By simply replacing the chickens with ducks, one could undoubtedly establish a threshold for the production of quacks.

Although group-theoretical cacklations based on the eight-fowled way can be expected to establish a pecking order, a really comprehensive theory would be based on an apprropriate egghenvalue equation. Quantization would then naturally proceed by introducing the "capon," with appropriate truncation. It should be noted that capon-chicken coupling may be assumed to be very weak to all orders. A clue as to the correct form of the egghenvalue equation might be provided by noticing that Coop(er) pairing is obviously described by interactions such as *R , where R is the propagator, or rooster function. Owing to a lack of bilateral symmetry, it seems clear that operations such as R · probably do not occur naturally, if at all. These theoretical difficulties obviously leave us with nothing to crow about.

Yet much can be done. The LCA should be used to measure breast masses and farm factors. A determination of Rooster's angle would probably help to establish the correct egghenvalue equation. Coherent production of chicken-anti-chicken pairs could be investigated by analogy with the well known dove-hawk interaction, which quickly produces a state of incoherence and annihilates to a large number of put-ons. In this regard, we might ask whether the beautiful picture of an elementary particle that appeared on the cover of the 27 November 1970 issue of *Science* is really a put-on or a capon? We suggest that a Feather's analysis be carried out immediately. Who knows: there may be a Pulletzer prize in all this. We have only scratched the surface! ■

From *Physics Today*, 1971.

Banghart, Bachrach and Pattishall
(Division of Educational Research, University of Virginia, Charlotteville, Virginia)

STUDIES IN PROBLEM SOLVING

Contract No. 474 (8) Office of Naval Research Sept. 1959

"In other words, for this particular tax, intelligence did not seem to interfere with problem solving performance."

****

ACADEMIC GRAFFITI

Notice taped to electric Hand dryer in medical faculty washroom:

"Push button for a one-minute message from the Dean."

J. D. Gillet (Entebbe, Uganda)

INDUCED OVARIAN DEVELOPMENT IN DECAPITATED MOSQUITOES BY TRANSFUSION OF HEMOLYMPH

J. Exp. Biol., 1958, 35/3, 685

We reproduce here some of the technical details connected with the decapitation of mosquitoes:

"Mosquitoes were chilled for 100 seconds...and decapitated the wound being sealed with paraffin wax. About one fourth of all donors and recipients (of hemolymph) failed to survive treatment. It was felt that mosquitoes (Ae. aegypti) are not sufficiently robust to withstand such treatment."

Quat-Quars* A Newly Discovered Phenomenon

R. KENNETH
TOM CABBIN
FELIX KLAUS
and TAB TAILOR

Introduction

Cats constitute a large percentage of the observable fauna in the township of Rishon-le-Zion. They have been found to outnumber human beings there by approximately ten-to-one, using the well-known methods of Ellenbogen.[1] It was noticed, however, by the senior author, that these incrutable animals are seldom to be seen in the evening hours, from the lowering of the reddening sun over the western horizon in a blaze of glory, till slumber-time for humans when they make their re-appearance, enlivening the night air with their gay antics and wassailing. The question was asked by the senior author, "Where do the critters go?" In fact, the question was not directed to anyone in particular, but the answer came from the vicinity of my knees, from my young son, who always looks up to his father, "Under parked cars, you fool!" After careful examination of his claims, and the wearing out of the kneecaps of two pairs of trousers, it was ascertained that he was fully justified in his conclusions. This phenomenon is designated as Quat-Quars.

One thing leading to another, a preliminary study was carried out, consisting of a survey on the habits, preferences and predilections of these beasts.

Materials and Methods

A hand-count was made of all cats present during daylight hours on a 200 meter stretch On the Street that I Live. This was accomplished by the baiting method,[2] using Grade A homogenized milk (kindly supplied from my neighbors' doorsteps in the early hours of the morning), poured in a straight line along the northern curb at the rate of one liter/10m. of curb. All cats answering this inexorable call were counted (and one can count on them to come, using a TAB electronic Quat-Quounter with a nine-life battery (Abercrombie & Fintch). Three assistants were employed to prevent cats from returning for second helpings, so it may be assumed that the figures are dependable. The count was replicated (three different hours). The results were: 160,15**, 140, with an average of 150, which turned out to be a nice round number to work with.

On seven consecutive evenings, careful counts were made of numbers of cats under cars, at different hours. The full details, together with statistical analyses, are presented in five other papers being submitted elsewhere. In this paper, we will suffice by summarizing the more important findings, and trust that the reader will understand that all the results were statistically very significant, as is the importance of the findings.

Results in Brief

1. Of the cats counted during the day, 98.5% are bedded down under automobiles during evening hours. Of the others, 1.2% were still up trees and afraid to descend, and the rest (0.3%) were found to be trapped in garbage cans whose lids had fallen down (Isn't there a better way for me to earn a living than this?).

2. For some reason still not fully understood, no more than one cat may be found under a single car. We believe that this could be linked to the "ecological exclusion principle", and also to Gresham's Law which states "Bad cats drive out good". ■

*Quat-Quars: (kăt'-kărz), n. (< Sanskrit).

**Deleted from our calculations — The sudden appearance of Mrs. Strauss' poodle upon the scene during the second count, seems to have had a catalytic effect (Webster: Kat-a-lĭsis (>Gk. a losing), causing what may henceforth be called the "Quatar-disappearance phenomenon", in which trees suddenly become populated by the creatures to the detriment of the street level. Catalysis and autolysis have been found to have no connection whatever; the latter describes the situation whereby a car gets lost, in theft.

3. The peak hour for sub-vehicle resting is from 2200-2300 hours, there being approximately twice as many felines present there as during 2000-2100 or 2100-2200. The number of parked cars present, however, during 2200-2300 hours are considerably more than earlier in the evening, due to the return of the theater crowd, so that the number of cats in proportion to that of cars is about the same during all hours under investigation.

4. It was ascertained that there is no car relation whatever between the size or age of the cat and the make or size of the automobile under which it seeks repose.

5. The color of the cat, as regards the color of the car was examined. Although it is true that at night all cats are gray, the problem of color differentiation in late evening was solved by the elegant procedure of cruising my car (senior author, Cadillac Fleetwood, 1970 Model), with full headlights, up and down the street to bring out color highlights particularly the black. This, unfortunately had to be discontinued after the senior author received a black eye from an irate motorist. However, it was found that the eye (black or otherwise) becomes accommodated to the dark, and various shades could eventually become distinguished (see Fig. 1).

FIG. 1. Black cat under hearse at 2330 hours.

The results show that there is no car relation between color of cat and that of automobile, and that color is only fur deep.

6. There is a definite car relation between the variety of cat and the size and make of the automobile. Unexpectedly, it was found that Siamese and Persian cats prefer small cars of any make to large ones, leaving the Cadillacs, Chryslers and Lincolns on the Street that I Live (!) to the more plebian breeds. This was shown by us to be a matter of "noblesse oblige".

7. Cat sex and cars:

a. Female cats, as a rule, prefer dainty cars, such as the "deux-chevaux". Male cats prefer female cats.

b. The female cat of any variety or breed seems to be perfectly willing to yield its place under any make or size upon the approach of a male, as a matter of courtesy. Strangely enough, the male seldom takes advantage of this generosity but usually follows her, calling out[3] that the gesture was not necessary, as he was just strolling by, taking in the night air. Females, however, have seldom been persuaded to return to their rightful place.[4]

8. Cats do not sit at any and every place under automobiles. Being agoraphobic, they react to thigmotactic stimuli under the car and generally gravitate to a car tire, snuggling up to it and calling out with delight. This is known as caterwauling (next to the senior author's car's tires of course, it is caterwhitewauling).

[1] Ellenbogen, K. 1937. Methods for the estimation of populations, carried out on Sundays at Coney Island. Jour. Applied Math. 7 (11) :32.

[2] Whittington, R. 1422. On Cats and Queens. J. Gutenberg Press, London & Mainz.

[3] Doaks, J. 1960. My brother talks to cats. Sat. Eve. Post. January 10.

[4] Millett, K. 1970. Women's Liberation. Juno Press, Manaos, Amazones.

Acknowledgments: This research was supported by the Ecological Foundation, the S.P.C.A., the Ford Foundation and the Volkswagen Foundation.

EFFECT OF SALIVA ON SPERM MOTILITY AND ACTIVITY

Togas Tulandi, M.D.*
Leo Plouffe, Jr., M.D.
Robert A. McInnes, M.D.
Quebec, Canada

Infertile couples frequently experience sexual dysfunction during their infertility investigation, including inadequate vaginal lubrication. In an attempt to look for a lubricant that would not impair sperm motility and activity, saliva was added to normal semen from healthy male donors. Saliva induced a "shaking movement" in 12% of the total sperm population incubated with high concentrations of saliva. This phenomenon did not occur with low concentrations of saliva, but sperm motility and progression significantly decreased. The results indicate that saliva has a deleterious effect on sperm motility and activity and should not be encouraged as a vaginal lubricant for the infertile couple.

Fertil Steril 38:721, 1982

Disco Music Makes Mice Turn Gay

Ankara, Turkey

Disco music causes homosexuality in mice and may make no exception where men are concerned, a study at the Aegean University maintains.

The Milliyet newspaper said yesterday that researchers at the Izmir-based university "discovered that high-level noise—such as that frequently found in discos—causes homosexuality in mice and deafness among pigs."

"The researchers think that there is a caveat in these studies for human beings as well," the Milliyet said.

The paper did not offer any explanation as to how mice were judged resistant to deafness or why pigs kept their sexual identities.

United Press

Infant Mortality-The Y's Have It

John W. Scanlon
Columbia Hospital for Women
Washington, D.C.

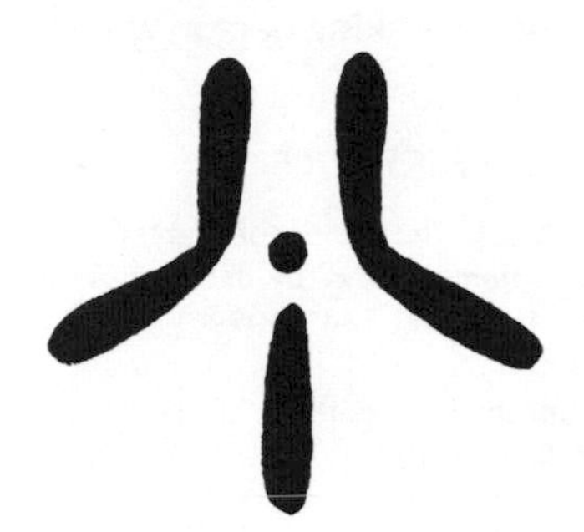

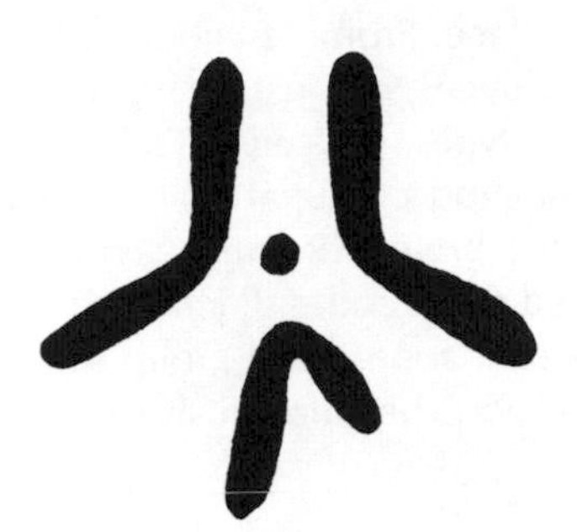

Scientists and world leaders may disagree about many subjects. However, I believe there is a topic on which all would agree; there is an international imperative to reduce infant mortality. This is not just because dead babies are aesthetically unappealing. The infant mortality rate is considered an excellent index of the general health of a society. In addition, for every dead infant there are even more who survive handicapped.

The many reasons why any individual country, city, or social unit sustains a given number of infant deaths are quite controversial. Lots of guilt and blame gets generated; contributing factors differing between emerging, developing and industrial countries. In attempts to explain infant mortality rates, consultants (defined as tourists with slides) usually focus on secondary issues: Nutrition, infection, smoking, alcohol, drugs, teenage pregnancy, poverty, industrial pollution, emotional stress, formula manufacturers, the medical establishment (or its lack), capitalism, communism, and nuclear waste have all been spotlighted. Almost never mentioned is the well-documented fact that human males succumb alarmingly easily at any step of the developmental path. No one ever points a finger at the deadly influence of the Y chromosome!

Even as the male seed spurts off to its vital rendezvous, it is attacked by hydrogen ions, enzymes, bacteria and hostile mucous. To succeed, sperm must surmount more barriers and debris than a salmon ascending the Potomac, and it has about as much hope for success. Odds for reaching the single (sic!) available ovum are less than 1 in 24 million. If that poor besieged tadpole of humanity harbors the Y chromosome, an increased mortal risk is its everlasting fate.

Spontaneous abortions are more common, premature births more deadly and intrauterine infections more frequent for the male. Boy oh boy, things only get worse after birth! All newborn ills are more lethal for the boy-child. Male mortal risk continues throughout childhood and into the adult years. Infections, accidents, sudden death, even most cancers are more common and more severe. The Y chromosome is a killer!

Ancient cultures understood this masculine death risk. Greeks and Romans took great legal pains to protect boys over girls because these ancient sages appreciated the former's fragility. The Druids plunged only male infants into a frozen Rhine to temper them. They felt no need to do this for girls who were presumably already strong enough.

In Washington, D.C., infant mortality is higher than in any other U.S. city. This rate rivals third world values. In D.C. the single greatest risk predictor is being born male. No matter whether mother is unwed or married, black, white, or some other color, rich or poor, adolescent or adult, educated or uneducated, little boys die more often.

Based on such overwhelming evidence, it seems reasonable that if fewer males are conceived, the infant mortality rate should drop. How can we accomplish this?

Again, clues are available from antiquity. Ancient Greeks believed that males originated from one teste or ovary; females from the other side. Unfortunately translations are not specific about the correct side, although dexterity and maleness were usually linked. Since temporary isolation of the appropriate seminal sac (never mind an ovary) seems painful and almost impossible to accomplish in the heat of the coital fray, this historical technique lacks practicality.

Ancient sorcerers largely contributed biopharmocological suggestions about influencing the sex of anticipated progeny. These were usually coupled with psycho-theological maneuvers for effectivity. Current governmental regulatory approval (not to mention third party reimbursement policies) preclude their widespread application.

More recently, modern science suggests that when coitus is restricted to intervals of 72 hours or greater, the frequency of male births increase[1]. Alternatively, as coital frequency increases, so do subsequent female births. Implications from this practice are enormous! For just once, satisfaction and safety go hand in hand in sexual matters.

The key to lowering infant deaths is to encourage intercourse, of course. Demand erotic art in factories, offices, courts, shops, every public place. Promote "nooners" for all employees, not just for higher executive types or senior scientists. Suggest short conjugal visits instead of coffee breaks. Require sensually oriented dress codes. Encourage unisex locker rooms and group showers. In short, be salaciously creative for its redeeming social value.

This suggested approach to infant mortality improvement may seem non-traditional, but is backed by the wisdom of millennia, unimpeachable epidemiological evidence and sound scientific logic. It also sounds like fun! What more could a thinking person want? ■

REFERENCES

Harlop, S. The Gender of Infants Conceived on Different Days of the Menstrual Cycle. NEJM. 300:1445, 1979

[1]Published, in part, in LEADER'S magazine, January 1983

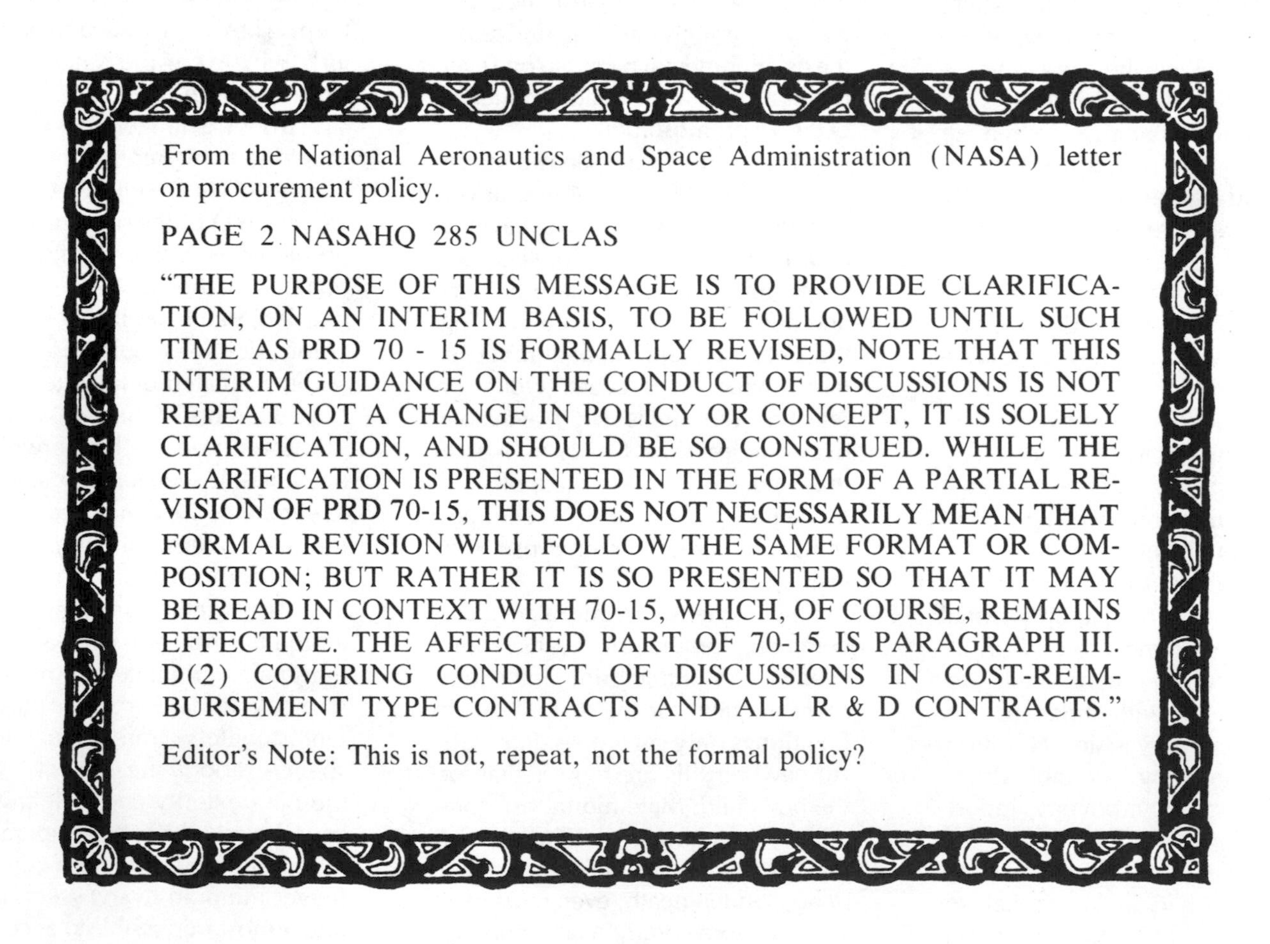

From the National Aeronautics and Space Administration (NASA) letter on procurement policy.

PAGE 2 NASAHQ 285 UNCLAS

"THE PURPOSE OF THIS MESSAGE IS TO PROVIDE CLARIFICATION, ON AN INTERIM BASIS, TO BE FOLLOWED UNTIL SUCH TIME AS PRD 70 - 15 IS FORMALLY REVISED, NOTE THAT THIS INTERIM GUIDANCE ON THE CONDUCT OF DISCUSSIONS IS NOT REPEAT NOT A CHANGE IN POLICY OR CONCEPT, IT IS SOLELY CLARIFICATION, AND SHOULD BE SO CONSTRUED. WHILE THE CLARIFICATION IS PRESENTED IN THE FORM OF A PARTIAL REVISION OF PRD 70-15, THIS DOES NOT NECESSARILY MEAN THAT FORMAL REVISION WILL FOLLOW THE SAME FORMAT OR COMPOSITION; BUT RATHER IT IS SO PRESENTED SO THAT IT MAY BE READ IN CONTEXT WITH 70-15, WHICH, OF COURSE, REMAINS EFFECTIVE. THE AFFECTED PART OF 70-15 IS PARAGRAPH III. D(2) COVERING CONDUCT OF DISCUSSIONS IN COST-REIMBURSEMENT TYPE CONTRACTS AND ALL R & D CONTRACTS."

Editor's Note: This is not, repeat, not the formal policy?

A VOIDING SHOCK

Howard T. Francis
Park Forest, IL

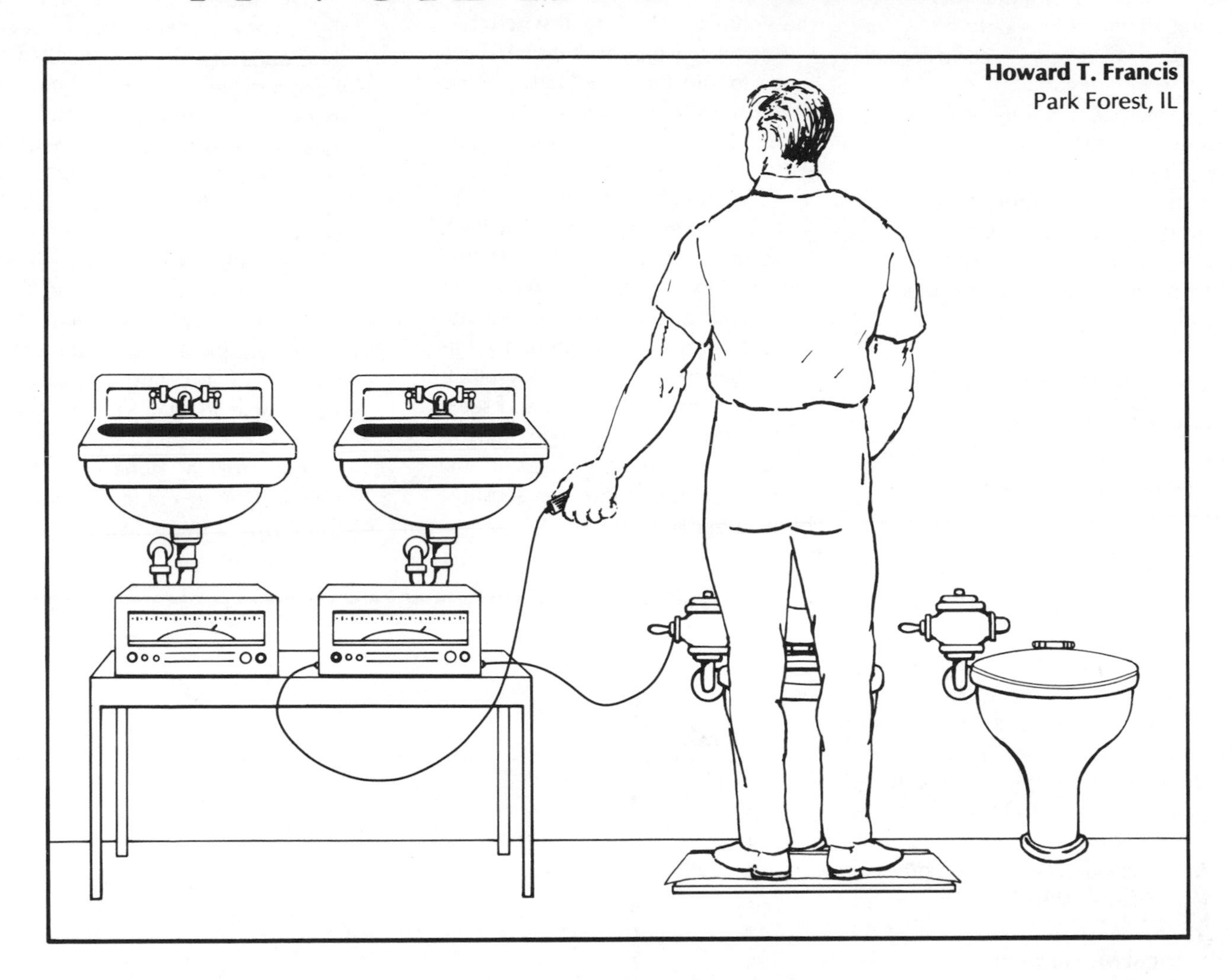

The hazards of electricity are of great concern in hospitals. The "susceptible" patient may suffer ventricular fibrillation if a tiny 60 Hz current finds its way to his heart. Or he may be blown to pieces during surgery if cyclopropane in his lungs is ignited by a static spark.

The shock hazard has forced adoption of an elaborate system of "grounding" of every conductive surface in sight, so that no stray "leakage" of current can find their way into a patient. To avoid explosions, operating room floors, shoes, and rubber cart wheels are made conductive so that no static charges can develop and cause a disastrous spark.

We have now discovered that shock and explosion hazards exist in areas heretofore considered safe, viz., all washrooms or—more precisely—"toilet rooms." Actually, at this point we can not be certain that the hazard exists for "ladies" washrooms; our experimental work thus far has involved only (standing) male subjects. We *are* certain, however, that the phenomenon is related to "leakage," as will be evident. A recent philosophical discussion* of the relationship between micturition mode and intellectual achievement suggests that the *visual* contact afforded a man during this function produces improved mental traits. There may be more to it; *electrical* contact may play a part.

Background

On several occasions during January, 1980, the author (subject No. 1, 62 years of age) experienced a static electric shock while reaching toward the flush valve after using a standard, public-type'john'.The distance through which the spark jumped suggested that a potential of thousands of volts existed between the subject and the water-pipe ground. At first, little thought was given to the shocks—it was assumed that because the subject was wearing rubber-soled shoes, he had simply acquired a charge while hurrying to the washroom.

This theory was not attractive, since our washroom is not nylon-carpeted, but is floored with bare ceramic tile. A quick test showed that the static charge was *not* present *before* going to the 'john'—only after. With this discovery, curiosity piquid; research began.

Other members of the staff were questioned—delicately, of course—with the most unexpected result: not one of the other (5) subjects produced a noticeable static discharge! It now appeared that we had flushed out some truly irreproducible results and our race to publish was on!

Experimental Apparatus

A test instrument table was set up near (not too close) the 'John'. Instrumentation included a Sensitive Research (sic) Model ESD Electrostatic Voltmeter, with a range of 0-1000 VDC, and a Simpson Model 314 FET VOM (for polarity measurements).

In order that there would be minimal leakage during an experiment (charge leakage), a plate of quarter-inch Lucite was placed on the floor at the appropriate spot in front of the test 'john'. The Lucite was protected against footprints—and splash-by sheets of used Z-fold computer paper (disposable).

Electrical Measurements

The first test fully justified the undertaking! The ground terminal of the voltmeter was clipped to the flush valve. A lead from the second terminal of the meter was held in the (No. 1) subject's left hand, while the remaining experimental manipulations were completed with the right hand. The subject stepped onto the Lucite platform, and after a few seconds of hesitation—perhaps induced by the excitement inherent in such experiments—a potential of 1000 volts was produced in 5 seconds! The meter went completely off scale during the remaining 15 seconds required to complete the run, which by that time could not be comfortably interrupted—even at the risk of meter damage.

During the next few days, the other staff members began to pick up the voltmeter "hot" lead when they visited the washroom. A curious hesitance to join this quantitative phase of the study was noted in some subjects, although the individual tests were conducted in complete privacy. Although none of these subjects had ever experienced a flush valve-to-finger spark—and therefore had nothing to fear—it was concluded that any man who works with electricity every day probably has developed an understandable reluctance to stand over a 'john' and see a meter indicate a sizable potential difference between the water in the bowl and his relatively sensitive body termination.

DISCUSSION

A troublesome question from the outset was: "How can a charge of thousands of volts be generated between a subject's body and the water in the toilet bowl when the two terminals are presumably connected by an aqueous stream containing copious amounts of dissolved sodium chloride?"

On the more practical—perhaps even commercial—side, the "charge generation test" may turn out to be a useful diagnostic aid in the early detection of prostate trouble. There could be a special 'john' set up in the airport mens' room, for example, where—for a dime perhaps—the traveling man could perform a self-diagnosis without wasting a minute of his time.

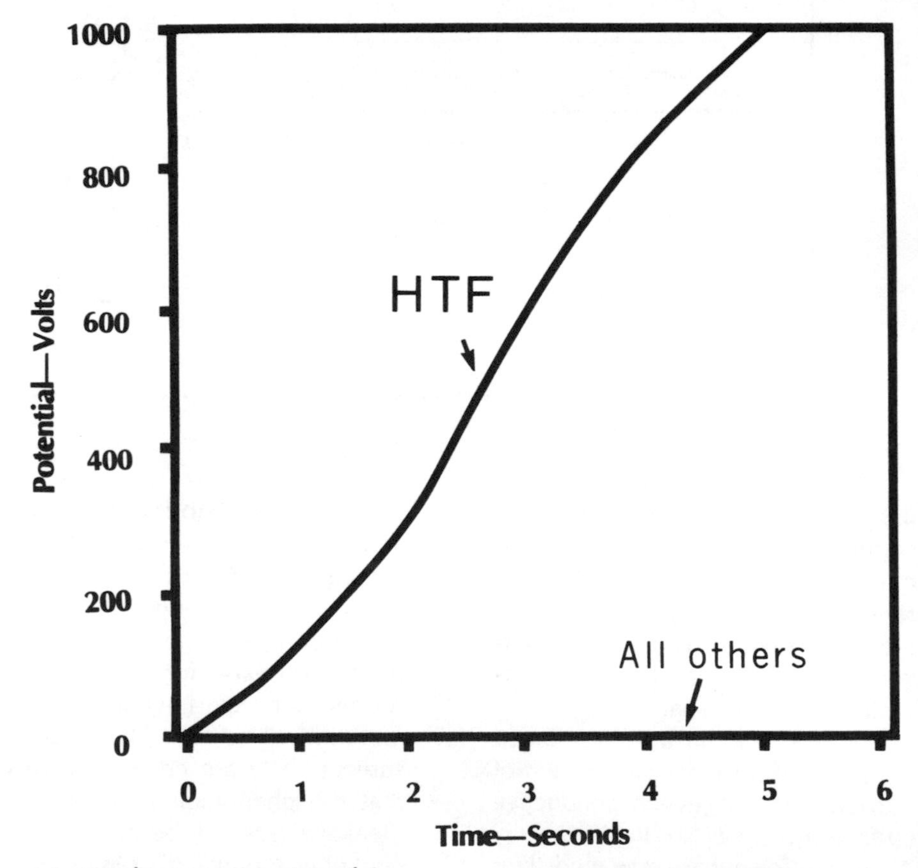

A supplementary point of interest is the charge polarity. Immediately after a charge-generation test, the electrostatic voltmeter was simply discharged through the Simpson meter at an intermediate DC voltage range. The pulse produced through the voltmeter indicated the polarity to be: subject, negative; ground, positive.

ACKNOWLEDGMENT

Thanks are due BH for the use of the electrostatic voltmeter. Members of the author's staff (WAB, TRM, WK, ML, and JS) prefer to remain anonymous, not being sure that they want to reveal their lack of spark. ■

*Forsyth, E.B.: Make Waves Not Water, JIR, Vol. 25, No. 3:27-28.

THE PANACEAS OF PANGAEA

Tom DiNardo
Philadelphia, PA

Overwhelming evidence that all continents were once part of one huge land mass, called Pangaea, has led to wide scientific acceptance of the theory. Great sections of the earth rest on abutting plates which slide on mantles of resilient underlayer. At a recent Earth Sciences Convention, the phenomenon was compared to dishes on a cafeteria tray just before a majority decided to break for lunch.

For the layman, the main applications in the crucial fields of behavioral science, anthropology, and zoology are just beginning to be explored. The pure muse of science can now answer many mysteries of these vast fields without resorting to fashionable 'visitors from outer space' explanations.

As a result of the breakthrough, geological terms are coming into common vocabulary usage. The term 'plate techtonics', once used by Scandinavian silverware designers, has been adopted to explain the sliding plates on the earth's surface. 'Pangaea', 'sea-floor spreading', and 'continental drift' are widely-used terms. Even the creation of mountain ranges such as the Himalayas, Andes, and Alps by the collision of two land masses is called "doing the bump" by geologists.

"Sun, Isles and Sea" by John Marin, The Downtown Gallery

Continental drift is also causing the distance between North America and Europe to increase at the rate of several inches per century. (A recent poll taken between New York and London during an in-flight showing of "Airport '80" showed that many feel the estimate conservative.)

New history books will have to be written, purging the misconceptions of the past. Man must have travelled paths long since rearranged; the instinctive wanderings of Genghis Khan, Marco Polo, and Columbus will no longer be explained to schoolchildren by tales of "new worlds" and "trade routes to the East." Nomadic tribes were actually in search of a gas station with an ice machine.

Some scientists are reconsidering the legends of Bigfoot, Piltdown Man, Abominable Snowman, and the Loch Ness Monster, possibly creatures trapped by land shifts. However, they now discount the possibility of a 'missing link', due to an amazing coincidence—the simultaneous discovery of fire (which enabled men to melt metal for tools, weapons, and currency), and the coin-return button, which we know was not perfected until many years later.

Some debate still continues regarding the exact center of the original Pangaea. Some scientists insist it fell on the original site of Yankee Stadium while another camp insists on New York's Third Avenue between 62nd and 63rd Streets. However, fossil remains prove that Massachusetts, Spain, and northern Africa were once the same land mass, a fact that caused confusion among Pleistocene Era travel agents.

Many unexplainable quirks of the animal kingdom have now become easy to understand. Lemmings, for instance, were merely trying to get to the relocated Capistrano. Mastodons became the first practitioners of cryogenics, hoping that ancient science would develop a cure for tusk and hoof disease while they waited, frozen, in anticipation. (Some researchers claim they were captured in ice by glacier activity while waiting for the off-season rates). Other animals of similar bone structure, now residing on separated continents, came from the same areas. However, much scientific time was wasted in learning the origins of the koala bear before it was found to be a furry midget hired by an airline company.

While continents were still relatively close, animals could cross over land bridges. Undersea fossil tracks clearly indicate that races were held on a land bridge across the Straits of Magellan (now renamed the Straits of Hippo).

Architecture, too, will have to be re-examined. Scholars still insist that Stonehenge was indeed made of stones from Wales, now many miles distant. But they now claim that it was meant to be the end of the line for the London Underground Subway rather than a primitive astronomical observatory.

The Pyramids may disguise lumpy hills caused by the pressures from beneath the earth's crust. Attempting to discover the ancient secrets of construction, theorists have calculated that building the Giza Pyramid around one of these eyesores would have eliminated the need for 3,986 25-ton stone blocks. If true, this time-saving method would have reduced the Egyptians' requirements to a mere 37,473 stones.

The edges of the moving plates line up with earthquake activity, and the friction between the edges causes earth tremors. Because the plate on which the Pacific Ocean rests is moving three inches a year in a northwesterly direction, scientists are working on ways to minimize the stress of the San Andreas Fault by injecting fluid into the rift. The F.D.A. has offered to use the stockpile of banned cyclamate soft drinks for this purpose. Civil engineers have designed interlocking sections for the Golden Gate Bridge, expandable as the plates move. In some seven to eight million years, the commuting time to Sausalito may become impractical, but Amtrak may then be nearing completion of track repairs.

Since we now know that the major plates are in constant motion, we can predict their directions over long periods. Anarctica is approaching South America, North America to the Pacific, Eurasia towards the North Pacific, and India-Australia north to Eurasia. Mate can be made in seven moves, although solution is left to the student. ■

"The Menopausal Queen—Adjustment to Aging and the Male Homosexual" J.S. Francher, *et al.*, American Journal of Orthopsychiatry 43: p. 670-4, July 1973

Submitted by:
Lisa Schwender
Timonium, MD

A JOYOUS REJECTION

To keep the authors among you inspired during this holiday season, here is a reprint of a rejection allegedly sent from a Chinese journal to a British economist. It appeared in the July 18, 1983 issue of C & E News, and shows that diplomacy can make even rejection a great experience:

"We have read your manuscript with boundless delight. If we were to publish your paper it would be impossible for us to publish any work of a lower standard. As it is unthinkable that, in the next thousand years, we shall see its equal, we are, to our regret, compelled to return your divine composition."

Best Regards,
Morris Leaffer

Revista de la Associacion Latinoamericana de Facultades de Odontologia (ALAFO) 1981 July; 15(2):9

HERRERA JUAREZ, CARLOS, Variabilidad del reposo Mandibular. Revista ALAFO, 15(2) (pág. 9 al 15), Julio 1981 ABSTRACT - VARIABILITY OF THE MANDIBULAR REST POSITION. The purpose of the review of the literature is to show the different theories held to this day about the rest position of the mandible, with an analysis of its clinical implication. On the other hand, the author afirms that is no process where everything is at rest without moving.

DR. CARLOS HERRERA JUAREZ, Facultad de Odontologia, Universidad Nacional Autónoma de Honduras.

Intelligence, guesswork, language

H.B. Barlow

A satisfactory definition of intelligence has **never** been found, and as a result it means different things to different people. What it is may remain too complex for a succinct definition, but the theory and practice of information handling have clarified what it does for us: it enables us to guess better, and the discovery of unexpected orderliness is the chief means of doing this.

Nature, Vol. VII, 1983

ERGONOMICS OF TOILET SEATS

Ian L. McClelland, Institute for Consumer Ergonomics, University of Technology, Loughborough, Leicestershire, England, and **Joan S. Ward,** Department of Human Sciences, University of Technology, Loughborough, Leicestershire, England

This study examines the appropriate position for and configuration of a toilet seat for the population of the United Kingdom. The study is based on the assumption that only a seated posture would be acceptable. Subjects underwent sitting trials using five different toilet seats, and data obtained include physical and subjective measures of subjects' responses. It is recommended that the height of a toilet seat should be 0.4 m. Seat angle is not a critical factor in subject preference.

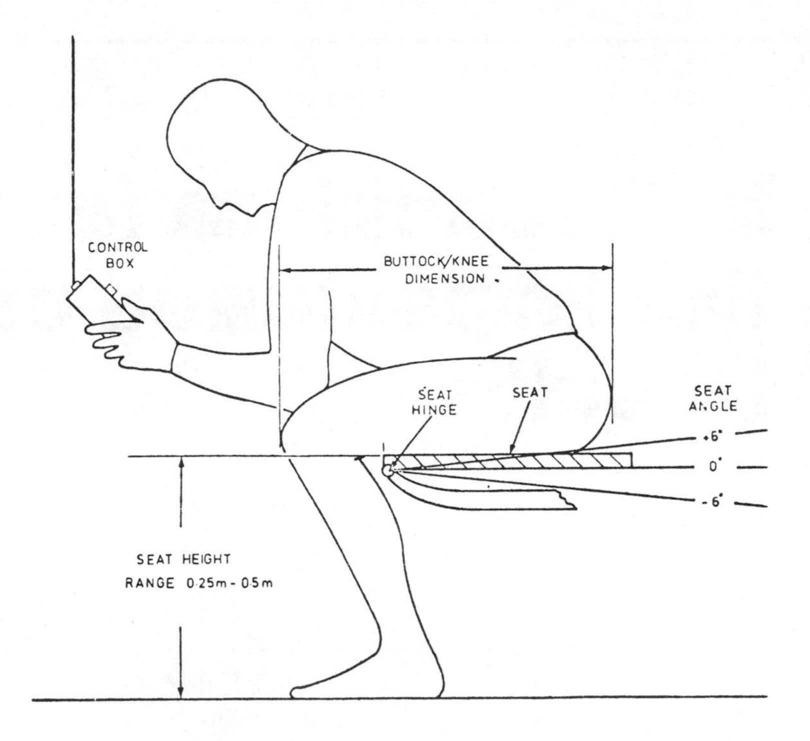

The physical dimensions measured for each seat type, seat height, buttock/knee dimension, and seat angle. (The posture depicted does not necessarily represent the posture adopted by subjects.)

Submitted by:
Ben Ruekberg
University of Illinois, Chicago

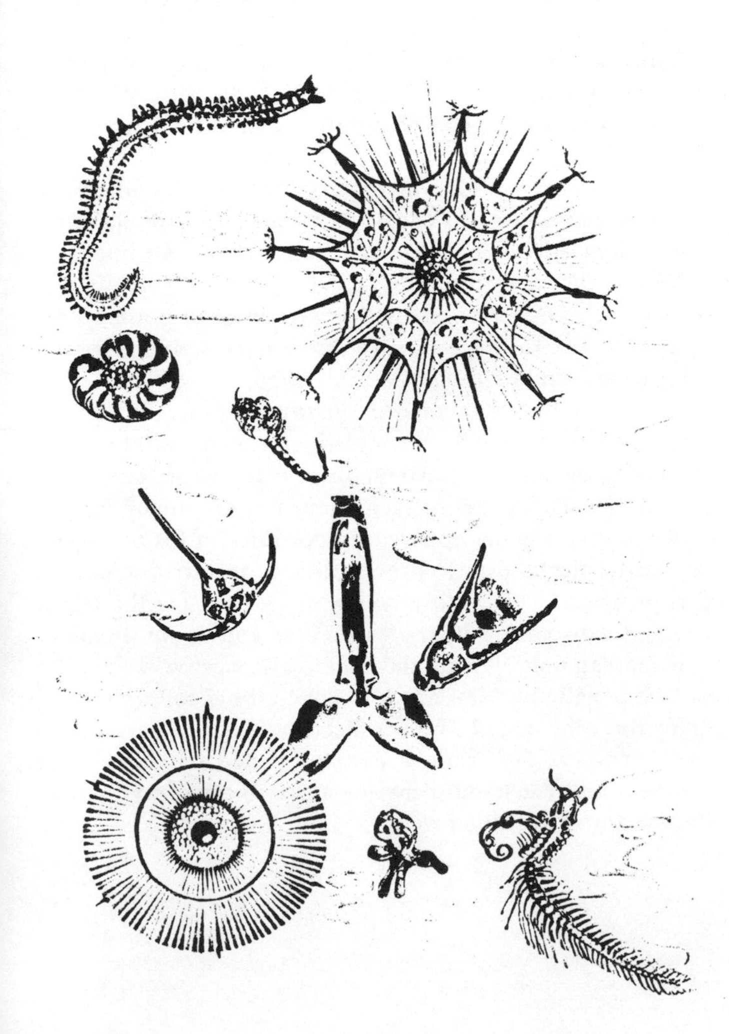

LUSCAN BEHAVIOUR

Cannibalism in Chitons

Leo Plas, Jr.
Department of Persian Philology, Liverpool

Acknowledgment: This research has been supported since 1962 by a generous grant, Number P-4779/56m 3a, DoPP.Lp, 228, from the National Science Foundation, hereby gratefully acknowledged. As ordered by the District Court, the entire amount of this grant has been returned.

There is no cannibalism among chitons.

A Mechanism for the Evaluation of the Importance of Research Results

MARCUS M. REIDENBERG, M.D.
Section of Clinical Pharmacology
Departments of Pharmacology and Medicine
Temple University School of Medicine
Phila., Pa. 19140

A problem requiring increased attention is that of evaluating the importance or significance, the quality, of research results rather than the quantity of results. Quantitative evaluation can readily be made by determining the number of papers published, amount of money spent, number of square feet of floor space a laboratory has expanded to occupy, etc. To make a qualitative evaluation of the importance of observations made is much harder. A new method is now proposed to establish a qualitative rating scale, capable of being used for self-evaluation, to be used to determine the importance of scientific discoveries.

When an investigator observes something new for the first time i.e., makes a discovery, he usually gets very excited. He then communicates this information to others in the lab and frequently to colleagues in adjacent laboratories. The number of people he informs of his discovery during this initial short time period after making the discovery is proportional to how excited the researcher is which in turn is proportional to how important he thinks his discovery is.

A custom in many labs, including our own, is to celebrate a new observation with a party that afternoon. We may buy some cake to have with the afternoon coffee in the lab. We may buy bakery donuts to share with everyone on our corridor at afternoon coffee. Sometimes we buy red wine and cheese when we get really excited. Once an afternoon champagne party was held.

Thus, empirically, one can observe that the number of people one tells about a new discovery on the day it is made and the cost of the refreshments served at the party to celebrate the new discovery within a day or two after it is made is directly related to the importance of the discovery in the judgment of the discoveror. Since the discoveror is more expert in the field of the specific discovery than anyone else, his judgment of its importance is probably better than anyone else's. His immediate judgment of its importance will be unbiased by financial, political, or other considerations and probably be a more valid scientific assessment of the discovery than can be carried out by other people or at another time.

Thus, we propose an evaluation of the quality of scientific discoveries on the basis of the immediate party the discovery causes. A discovery can be a coffee and cake discovery, a wine and cheese discovery, or a champagne discovery. One can develop finer gradations of the scale by subclassification. For example, a wine and cheese discovery can be subclassified into a domestic wine and cheese discovery or an imported wine and cheese discovery. One can further subclassify into a regionally bottled wine discovery or chateau bottled wine discovery, etc. Coffee and cake discoveries and champagne discoveries can be similarly subclassified.

A possible bias that can occur in this method of evaluating quality of research is for the investigator, knowing about this method of evaluation based on his own degree of excitement, to consciously fake more excitement than he really feels in order to upgrade the apparent importance of his discovery. A control mechanism to prevent this is an intrinsic part of this proposal. The discoveror naturally pays for the party. Thus, if a discoveror upgrades a coffee and donut discovery to a chateau bottled wine and cheese discovery very often, he will be penalizing himself only, since all his colleagues will enjoy the wine and cheese irrespective of the importance of the discovery. This should serve as a negative feed back function and cause most investigators to remain honest in their evaluation of their work.

MEGALOPTIC HUMAN SERVICES
SUBSTATION 4791
Postal Transmitter Code 33-965-421-4
MOHOLE GOVERNMENTAL CONTROL CENTER
Sublevel 134
Appalachia, Pennsylvania

Office of the
Supreme Defender
of the Primogenitary

Videophone
1796-431-62-4976-35
Laser Channel 14913

Donald P. Kent, Ph.D.
Department of Sociology
and Anthropology
The Pennsylvania State University
227 Graduate Building
University Park, Pennsylvania 16802

My Dear Kent:

Thank you so much for affording me the opportunity to review the Kalish article for possible inclusion in your forthcoming "Gerontology: Past, Present, and Future." You realize, of course, that being on the very eve of retirement, my reflections on this are quite different from those I might have made in 1970 when I had already spent some 20 years in gerontology and related areas of social welfare. You no doubt recall the point of view which I, Linden, Simmons, and Townsend shared about the place of elderly people in our society. However, these are different times and a great deal has happened since those early days.

The author of this paper presents an interesting thesis about the reasons for higher social value in western society, although I am not certain he is entirely accurate. He suggests that at this time "they are unlikely to add to the population by having children." This, however, is a factor that was present for all the generations preceding our present one. Indeed, today the early experiments at Cornell with rats (Schoenfelt, 1959) which produced female rats bearing young at the human equivalent age of 80 is now

seeing fruition (if you permit an old man a bad pun) among some of our senior citizens. Indeed, the long range physiological effects of oral contraceptive drugs saw what was an alarming increase during the Seventies of women in their middle and late 50's bearing children (Burson 1974).

While there is no reason as yet to believe millions of American women in the 7th and 8th decades of their lives will produce children, it is a matter of fact that at the time of the last special census of retirees (U.S. Census Bureau 1988), no less than 10,000 live births were recorded among women above the age of 63. Some of my younger colleagues, encouraged by the widespread success of the foster grandparent program in the late Sixties and Seventies, have gone so far as to recommend that child bearing during the middle years may introduce the most satisfying role for the elderly that the American culture has been able to devise (Mileti, Smith, and Gross 1989).

The conclusion that the elderly are frequently retired and are not keeping a younger person from a job or a promotion was also true even in the Sixties and Seventies. Thus, this would be of no greater significance today than it was then. I feel what may be more significant is the fact that the distinctions between the young and old have become considerably more hazy. Retirement now occurs at age 50 with optional retirement at age 45. Education both in the technical schools and the professional schools continues full time until age 25, with virtually mandatory educational activity on a part-time basis extending to age 30. It is really not until after this date that people begin to work at full-time positions (U.S. Government Printing Office 1986 et seq.). Following retirement at age 50, approximately 30 percent return to some kind of educational activity (U.S. Office of Education in Retirement 1988).

However, the sociological factors are perhaps less important than what has happened in the area of physiology. (You must forgive me if I do not give you all the documentation that a scholarly critique should, but I will presume the preogatives that I must confess I feel are my due.) Organ transplants, including replacement of nerve block damage, particularly in the extremities, have improved body functioning significantly. Problems posed earlier by a slow down in synaptic transmission have been overcome. The discovery in 1970 (Schwartz and Kerenyi) that what we once quaintly called senile dementia was in fact amyloidosis permitted us to introduce drugs into the system which would counteract the metabolic changes leading to the deposit of amyloid in the vessels of the brain. In addition to cutting off admissions to institutional care because of poor psychological functioning, we arrived at a condition of old age which no longer saw the faulty memory, the slow reaction, or the doddering step as the signals of advancing years.

Perhaps one of the greatest effects that the technology has had in producing respect and high social value for older people has been the most recent development of cryogeriatrics. The successful freezing of elderly persons with revival at a later time has produced a situation not unlike that encountered at an earlier time in our history when control of the land by the parent maintained respect by the young for their elderly (see Simmons, op cit.). Cryogeriatrics has raised a host of new problems for the lawyers. If a person does not die but merely goes into a frozen state, does he leave an estate and are their heirs? What claim does he have and can he set aside funds for use upon his revival and "return" to "life"? I would submit that the young are wary and cautious and are treating with respect their elders who still control considerable wealth. It may be that cryogeriatrics is the most significant key in this area of social value (Birdseye 1984).

One final comment, a suggestion that the dispersion of persons of Chinese origin has had an effect is without merit. As the author so correctly notes, the move into the rural communities and small towns as demanded by the Percy-Tunney Immigration and Non-Urban Dispersal Bill, effectively isolated the Chinese. In any event, the group who came here in their 30's were those largely disenchanted with the developments of the Forties, Fifties, and Sixties of China, and while articulate about the "old values" were really uncommitted to the values of a China as not known since the 1920's and earlier.

In summary, I would agree with the author's overall conclusions. However, like so many psychologists, he has fallen into the trap of avoiding the technological developments and assuming the changes in value derived from the simple psychological, social and economic factors.

Well, these have been the ramblings of a fellow toiler in the vineyard. The author's style is terse and satisfying to read. His review of the literature is excellent as far as it goes. However, my guess is that this has been done by one of your younger colleagues who in an effort to dazzle us all has overlooked some important elements in the picture.

With all good wishes to you, and my congratulations to Marian on the birth of your eighth child, I remain

Your obedient servant,
Elias S. Cohen

Red blood cells
from lemming kidney
x 17,000

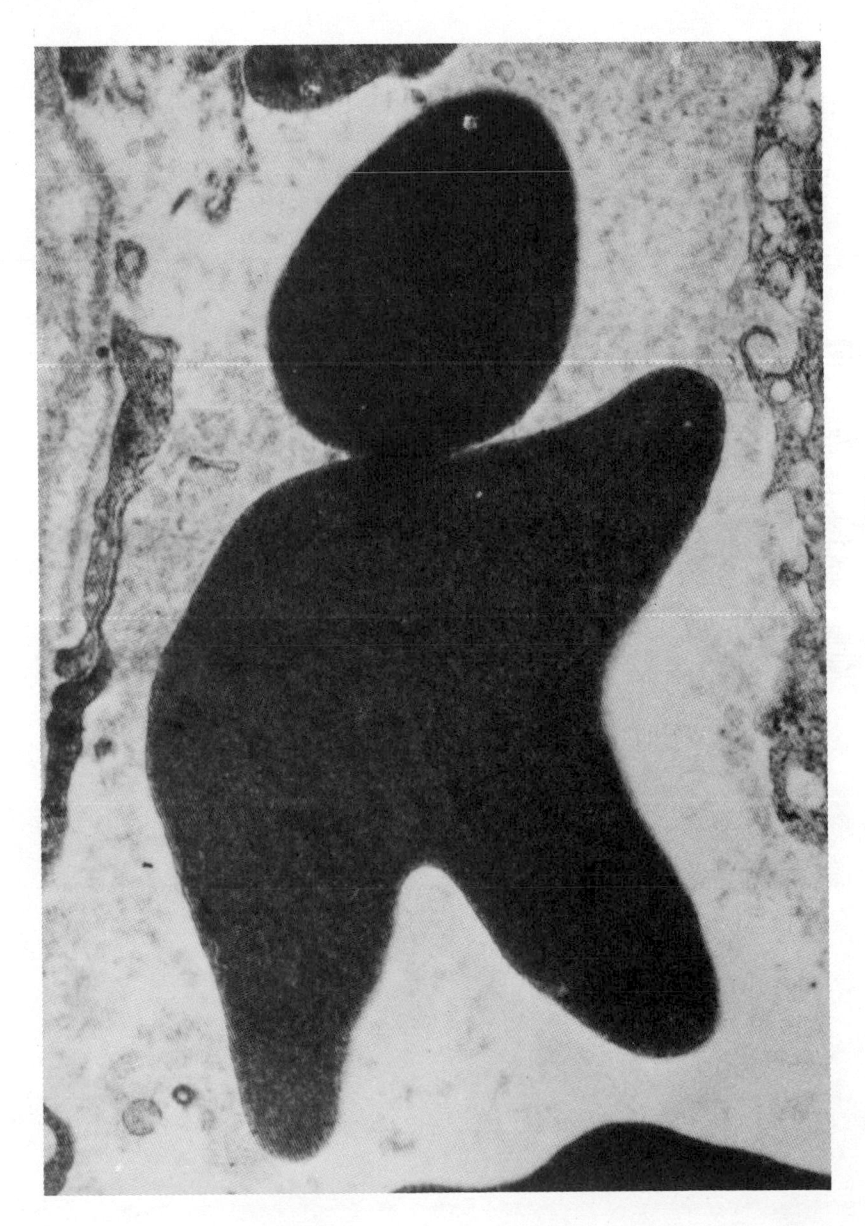

My Siamese cat is called Juicy Jill
She lures every tom to my window sill
She's in estrus for weeks
Shows lordosis - and shrieks!
I wish she were irreproducible.

I. Joan Lorch
Buffalo, NY

Irreproducible
data
impel
one
to

PUBLISH OR PERISH

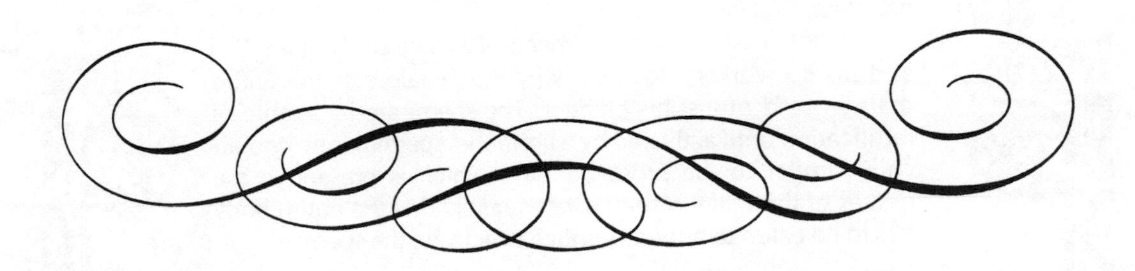

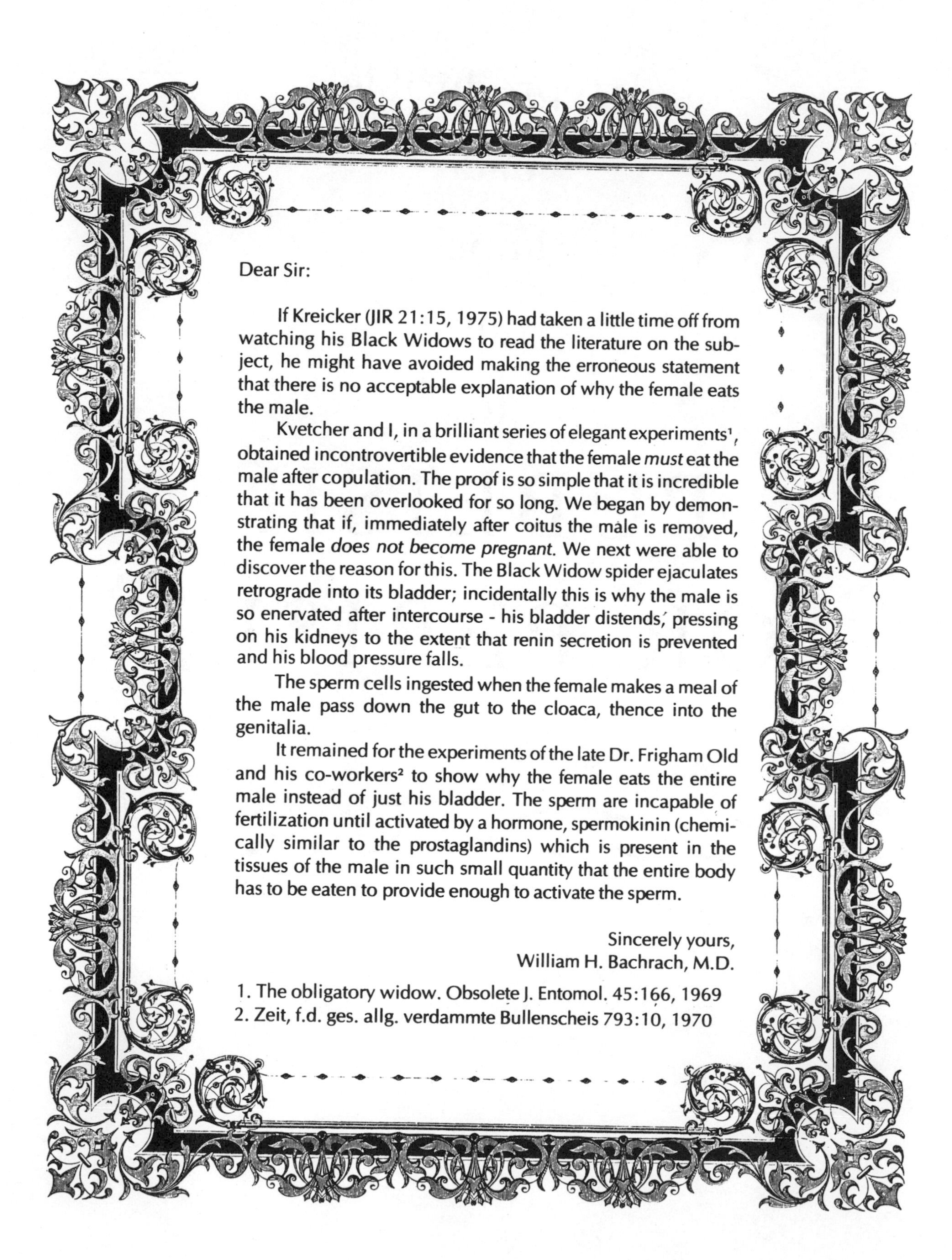

Dear Sir:

If Kreicker (JIR 21:15, 1975) had taken a little time off from watching his Black Widows to read the literature on the subject, he might have avoided making the erroneous statement that there is no acceptable explanation of why the female eats the male.

Kvetcher and I, in a brilliant series of elegant experiments[1], obtained incontrovertible evidence that the female *must* eat the male after copulation. The proof is so simple that it is incredible that it has been overlooked for so long. We began by demonstrating that if, immediately after coitus the male is removed, the female *does not become pregnant*. We next were able to discover the reason for this. The Black Widow spider ejaculates retrograde into its bladder; incidentally this is why the male is so enervated after intercourse - his bladder distends, pressing on his kidneys to the extent that renin secretion is prevented and his blood pressure falls.

The sperm cells ingested when the female makes a meal of the male pass down the gut to the cloaca, thence into the genitalia.

It remained for the experiments of the late Dr. Frigham Old and his co-workers[2] to show why the female eats the entire male instead of just his bladder. The sperm are incapable of fertilization until activated by a hormone, spermokinin (chemically similar to the prostaglandins) which is present in the tissues of the male in such small quantity that the entire body has to be eaten to provide enough to activate the sperm.

Sincerely yours,
William H. Bachrach, M.D.

1. The obligatory widow. Obsolete J. Entomol. 45:166, 1969
2. Zeit, f.d. ges. allg. verdammte Bullenscheis 793:10, 1970

A BRIEF HISTORY OF SCHOLARLY PUBLISHING

DONALD D. JACKSON
University of Illinois Press

50,000 B.C.	Stone Age publisher demands that all manuscripts be double-spaced, and hacked on one side of stone only.
1455	Johannes Gutenberg applies to Ford Foundation for money to buy umlauts. First subsidized publishing venture.
1483	Invention of *ibid.*
1507	First use of circumlocution.
1859	"Without whom" is used for the first time in list of acknowledgments.
1888	Martydom of Ralph Thwaites, an author who deletes 503 commas from his galleys and is stoned by a copy editor.
1897	Famous old university press in England announces that its Urdu dictionary has been in print 400 years. Entire edition, accidentally misplaced by a shipping clerk in 1497, is found during quadricentennial inventory.
1901	First free desk copy distributed (Known as Black Thursday).
1916	First successful divorce case based on failure of author to thank his wife, in the foreword of his book, for typing the manuscript.
1927	Minor official in publishing house, who suggests that his firm issue books in gay paper covers and market them through drug houses, is passed over for promotion.
1928	Early use of ambiguous rejection letter, beginning, "While we have many good things to say about your manuscript, we feel that we are not now in position . . ."
1934	Bookstore sends for two copies of Gleep's *Origin of Leases* from University Press and instead receives three copies of Darwin's *Storage of Fleeces* plus half of stale peanut butter sandwich from stockroom clerk's lunch. Beginning of a famous Brentano Rebellion, resulting in temporary inprovement in shipping practices.
1952	Scholarly writing begins to pay. Professor Harley Biddle's publishing' contract for royalty on his book after 1,000 copies have been sold to defray printing costs. Total sales: 1,009 copies.
1961	Important case of *Dulany vs McDaniel,* in which judge Kelley rules to call a doctoral dissertation a nonbook is libelous per se.
1962	Copy editors' anthem "Revise or Delete" is first sung at national convention. Quarrel over hyphen in second stanza delays official acceptance. ■

THE MANUSCRIPT

Arthur A. Dole

The Dynastatic Review
Dept. of Applied Macromicrology
College of Dynastatics
Metropolitan U., Metropolis,
Metrostate

Nov. 2, 1961

Dr. Albert Bush
Dept. of Basic Micromacrology
Land Grant U., College Center
Provincia

Dear Al:

There is no need to tell you how pleased I was to have the chance to sit down with you for a few minutes during the convention last month. I am terrifically impressed by your developing research program. As I said then, I'd be most interested in an article for the *Review*.

I'm sure you realize that the *Review's* current policy is to emphasize the practical implications of pure research. Keep it solid but avoid technical terms. We prefer brief factual incidents in support of generalizations rather than details on methodology. One table of technical illustration is acceptable. Your bibliography should be brief rather than comprehensive. Please avoid footnotes.

May I plan on having something from you by the first of the year?

With warmest personal regards to you and Susan,

Herbert U. Wheel
Editor

THE DYNASTATIC REVIEW
College of Dynastatics
Metropolitan U.
Metropolis, Metrostate 999999

June 10, 1962

Your manuscript

Current Research Developments in Provincial Macrostatics

has been received and will be reviewed by the Editorial Board. You will be notified as soon as a decision has been reached.

THE EDITOR

The Dynastatic Review
Dept. of Applied Macromicrology
College of Dynastatics
Metropolitan U., Metropolis,
Metrostate

July 5, 1965

Albert L. Bush
Professor of Basic Micromacrology
Land Grant U., College Center
Provincia

Dear Al:

I'm ashamed to say that I've been so busy setting up our new institute that my correspondence has gotten a bit behind. It was great to have your note of April 12 and I do appreciate your patience. The manuscript reads very well and we would like very much to publish it in a forthcoming issue. However, we prefer to hold off until we have sufficient other material to include in a full issue of the *Review* devoted to a thorough and balanced treatment of minimacrostatics across the country. You may, quite understandably, prefer to submit the manuscript elsewhere for earlier publication but I do hope you decide to bear with us a little longer.

I was terribly disappointed not to see you at the convention. You missed a terrific party with the old gang but I can well imagine that the installation of your new microsplitter has kept you close to the laboratory.

Affectionately,

Herbert U. Wheel
Editor

The Dynastatic Review
Dept. of Applied Macromicrology
College of Dynastatics
Metropolitan U., Metropolis,
Metrostate 999999

Sept. 7, 1970

Dr. Albert Bush
Assistant Professor of
Basic Micromacrology
Land Grant U., College Center
Provincia

Dear Dr. Bush:

I have very much enjoyed reading your essay, "Current Research Developments in Provincial Macrostatics," and I am genuinely appreciative of your thinking of the *Review*. Unfortunately, this piece does not quite meet our present publication needs so am returning your MS together with my thanks and best wishes. Do keep us in mind for the future, won't you?

Sincerely,

Herber U. Wheel
Editor

The Dynastatic Review
Dept of Rhythmics
College of Dynastatics
Metropolitan U., Metropolis,
Metrostate
Sept. 13, 1970

Albert L. Bush
Professor of Basic Micromacrology
Land Grant U., College Center
Provincia

Dear Professor Bush:

I am presently taking over the editorship of the *Review* from Professor Wheel who is about to go to the University of Politania. In the inevitable confusion resulting from the changeover, an error was made in the letter accompanying the manuscript just returned to you.

Through an oversight, you received the *pro forma* note ordinarily sent along with the unsolicited manuscripts that have to be rejected. We have since realized that yours was a solicited paper and that its return demands rather more explanation than you received.

As you can understand, we have decided to begin *de novo*, with the responsibility for the coming year's solicitation given to me. In the fairly sketchy plan I have at the moment, I do not find a place for your fine essay. I am, however, keeping your name and special interest on file; and I hope that, if I write with a new request, you will not hold this wielding of new brooms against me.

Sincerely,

Grace Classica
Editor

EATING BY WRITING
OR
Pot-Boilers in Science*

ROBERT SOMMER

The pot-boiler has a long and respected history among writers, composers, and playwrights. Most men of letters have consciously written second-rate stories for second-rate magazines simply to stay alive. Often the alternative was book reviews, translations, or ghost writing. Sometimes this was done under a *nom de plume* or the influence of alcohol. Most reviewers will not begrudge a young writer the opportunity to earn a few dollars for consciously doing work of an inferior quality.

However some people cannot imagine any place for the pot-boiler in science. The stereotype exists that science is above the lure of pecuniary motivation. The belief persists that the scientist is so well-paid and secure that he has no need for the dubious rewards gained by producing second-rate articles. Both of these assumptions are tenuous.

Has the scientist any need to write pot-boilers? The answer is an emphatic "YES." In many universities, as is well known, promotions depend on the number of papers written. Since the people who pass on promotions are the academic deans or university presidents who are usually trained outside the scientist's field, they cannot assay the quality of an article. They must leave assays of quality to the editors of professional journals. Furthermore the deans are unable even to judge the quality of the various journals. Sometimes their judgments are based on criteria that invariably assign the greater weight to the poorer journal. Because of local pride or regionalism, the dean may be more influenced by an article in a second or third-rate local periodical than in a first-rate international journal. He may resent the professor at Isthisa State University who never bothers sending an article to Isthisa Journal of Chemistry or presents a paper at the anual Isthisa convention. Since academic deans are unable to judge the merit and potential contribution of most papers, they are compelled to use systems based on a combination of counting and weighing[1]. It may be decided

* Reprinted with permission of the editor and author from THE MALPIGHII (Montreal).

[1] Young F. N. and Crowell, S. The application of gamesmanship to science, J. Irrepr. Res. 1957, *4*, 12.

that a heavy book is worth ten light articles, or that three published experiments are better than one review article. However a dean will usually conclude that eight articles are better than two articles, and three reviews are better than one. People are impressed by anyone who has done anything *several times*. One success may be an accident, but several "successes" indicate a genuine talent, a person who can be relied upon.

The rewards system in science may be less direct than in writing, but there is still a very real connection between the number of articles published and the size of one's paycheck. The scientist who waits several years for supporting data before publishing his results is rapidly becoming a relic of the past. As in exploring, fame and fortune come to those who are the first to land. Later arrivals may become known as homesteaders or substantial citizens, but they will never have statues erected in their honor. The pressure to be the first to publish is endemic through the scientific community; it is not limited to missile laboratories. The disinterested, relaxed and scholarly scientist is fast becoming extinct. To protect his job and guarantee his next promotion, he is compelled to read and heed the circulars on research grants sent out by the NSF, NRC and other foundations. Needless to say, officials who pass on research grants and awards are not unaware of the number of publications a man has to his credit. "Him that has, gits" is a good rule of thumb for understanding the logic of grants and awards.

But doesn't the ethical code of the scientist militate against the writing of second-rate articles? Certainly not, for as any scientist knows, there are second-rate scientific journals, and as long as these exist there will be a ready market for second-rate articles. As far as I know, all scientific fields have journals of varying degrees of merit. In some cases, the national journal is considered the best, while the regional and local journals contain papers that could not reach the standard of the national journal. This is also true of local and regional meetings of professional societies. Usually the papers are presented by graduate students or recent graduates, and are of dubious merit. The usual attendance at the scientific sessions of such meetings is far below the number of people registered for the meeting. Woe betide anyone scheduled to present the first paper of the morning, or the last before dinner.

Another occasion when the scientist consciously produces a second-rate study is when he is instructed to do a piece of research by his superior and he lacks interest in the topic or sufficient facilities. For example, if a biochemist is instructed to do a study of a thiamine derivative and he lacks the necessary equipment, he may know beforehand that he cannot hope to turn out anything resembling a first-rate study.

Although science has a ready market for second-rate work, it lacks the demand. No editors are clamoring for mediocre articles. Perhaps the chief reason for this is that articles pour in regardless. Authors have such a need to turn out articles that editors have no difficulty in securing material*. In some fields, not only are the authors unpaid, they must even pay to have articles published. There are several journals in psychology that exist solely on articles that are paid for by their authors. (Many articles in these "pay as you go" journals are supported by federal and foundation grants. In such cases the researcher is able to use part of his grant to finance the publication of an article which was not of sufficient quality to be accepted by a better, non-paying journal.)

I have often seen the look of amazement on the faces of friends and relatives when they learn that I am not paid for articles. They consider my motives are compounded of either egoism or altruism. I dread to think of their reaction if they learned that I write some articles to advance my job or facilitate transfer to another position.

In one laboratory I was in the middle of a personal feud between two superiors. One of them continually urged me to "produce articles" in order to embarrass the other person, who had been employed for three years and had written nothing.

Needless to say, my second superior looked upon each of my papers as a personal threat, and was not my most enthusiastic fan. I soon learned that research was a powerful weapon in inter-departmental warfare. If one department was able to publish a score of articles and reviews and receive mention in a national magazine or alumni publication, it was soon able to dominate the entire organization. They would soon receive preference in allocation of office space, secretaries, and graduate students.

In this highly specialized culture, it is extremely difficult for a lay administrator or specialist in another discipline to judge the merits of a scientific paper, especially if it is neatly packaged in technical jargon. The chances of a scientific pot-boiler being mistaken for an article of merit are correspondingly greater than would be the case with a poem, symphony, or novel. The fine arts have a cadre of reviewers and judges who are hired to assess the merits of a symphony or painting. Most often the public becomes acquainted with a play or book only through reviews. No such group of middlemen exists in science. Public acclaim very often comes to the scientist with a flare for publicity or whose work borders on a controversial area.

So then, let us drink a toast to the pot-boilers. Writers need them, scientists need them. But does anyone read them? ■

---

* This is also true for the *J. Irrepr. Results* (Editors).

A LETTER

DALHOUSIE UNIVERSITY
Halifax, N. S.

Faculty of Medicine
Department of Paediatrics

February 26, 1970

Dr. Lyle A. David
Department of Veterinary & Parasitology &
Public Health
Oklahoma State University
Stillwater, Oklahoma 74074
U.S.A.

Dear Lyle:

Thank you for your letter of February 18, and I apologize for the long delay in getting you some information on our manuscript, however there were a few minor problems which I wanted to get cleared up before I wrote again.

Initially I had a little bit of a problem getting the manuscript from the Post Office and Customs Officials when it arrived. However, as they explained, it was all a mistake which could have happened to anyone. Do you remember that little problem you had about possibly being held up at the Border when you left B.C. a few years ago? Well apparently the usual circulars were forwarded to Post Office and Custom and Immigration Departments and when the whole problem was cleared up naturally they did not bother to cancel any of the usual information that authorities circulate to Post Offices. At least if they did they have not found out about it in the Halifax Post Office. It seems that Customs Officials routinely check all communications from individuals on their Border crossing check list. When your manuscript (which did not seem to make any sense to them) arrived they were sure it was some sort of coded message that you were trying to get across the border. It seems that their decoding process involves cutting the manuscript into single worded bits and in some case even splitting compound words. They eventually called me in and I was able to explain the whole thing and regain possession of the manuscript. Unfortunately it arrived in the form of a couple of large envelopes full of typed words in single pieces. However, I must admit they were most careful and did not misplace one single word.

Of course this meant a bit of time for the secretary and I to piece together the words but I think we managed to get most of them into the proper order.

Since this meant a bit of time I thought maybe I could have the people in photography take a look at our photographs. They have a real hot-shot in photography working in the medical illustration department down here and he is doing a marvelous job at constructing his own computerized photograph printing processes. He felt that if he had all possible information in (i.e. photographs, negatives, color prints etc. in other words all the photographic evidence we have in this particular bit of work) he was quite sure that he could improve the quality of our prints. He really is an absolute genius when it comes to both photography and computers and had a system just about perfect. However, there was a small problem concerning a power supply source for one of the connections between the computer and the printer-enlarger equipment, and he had very neatly overcome this problem by using a battery. Unfortunately the battery was one of the little imperfections in his system and some of the acid leaked out overnight and destroyed most all the material on the bench including all of our data (i.e. prints and negatives).

He was most apologitic about the whole thing and really was sorry that he had destroyed all of our material, however he assures me that this little accident enabled him to pin-point all the problems and go on and perfect his system.

These things will happen and although it was a bit of a problem and, although the manuscript seemed a bit incomplete without photographic evidence I managed to make the necessary changes deleting references to figures and got the manuscript back together again.

I then sent it on to Dr. Miller who was really quite pleased with the manuscript as a whole. However, he felt our method of gluing the words on pages was a little bit unprofessional and decided to have the whole thing typed up properly before he sent it on to the editors. Things were a bit slow at that end because he was having secretarial problems at the time and was sharing a secretary with Dr. Vince. It appears that the poor girl was terrible overworked and in trying to do too many things at one time some of the manuscript pages got mixed up with some of Dr. Vince's reports.

As Dr. Miller knew from experience Dr. Vince never gets around to looking over his reports for at least a month and he probably quite rightly felt that any further delay would be just too much. Since the missing pages primarily involved descriptions of the photographs which has already been left out, he felt that he could put the manuscript together without these details and finally got what he felt was a very good, although somewhat reduced in length, manuscript off to the editor.

Now that I have brought you up to date I am pleased to advise you that none of the above, of course, happened but I wanted you to get the whole thing in its proper prospective before I told you the following.

Twenty-four hours after receiving your letter of inquiry I received the manuscript back from the editors informing me that they are not interested in publishing the manuscript.

I will be writing again in a day or two—mostly because I didn't want to clutter this letter with details. Chow ! ! !

Yours sincerely,
Margaret J. Corey, Ph.D.
Assistant Professor

MJC/mb

VIDE·INFRA

DR. TIM HEALEY, F.F.R., M.I.Nuc.E.
Yorkshire

As a keen[1] student[2] of footnotes,[4] I have long[14]

[1] Enthusiastic, not necessarily sharp.[13]

[2] When the late Dr. John Wilkie[18] stood up and said "As a mere student in these matters . . ." the listeners knew that they were about to hear some words of wisdom from a very experienced expert.[16,24] Modesty[17a] forbids me to draw a parallel.

[3] Blaise[17b] Pascal used the same trick with his phrase "It is easy to show that . . .". Experienced mathematicians soon recognised that these words warned them that the next step would take them three days of complex calculations to understand.

[4] I have been fascinated by footnotes ever since I obtained several editions of a book[5] which has some of the best footnotes[6] I have ever encountered.[7]

[5] Samson Wright's "Applied Physiology".[8]

[6] E.g. text. "Never occurs." Footnote: "What never? Well, hardly ever".

[7] A book called "Useless Facts in History" has a good pair[9] also.

[8] The footnotes disappeared after the ninth edition, when Samson Wright died. His major work has been continued,[10] but the footnotes that gave it individuality are no longer given: a grave mistake.

[9] There are only two in the book.[11]

[10] Tenth edition by C. A. Keele and E. Neil. O.U.P. London 1961.

[11] The first says "Do you like footnotes?" The second says "Aha![12] Caught you again"

[12] Note the similar style to Lucy.[15]

[13] Though I do not deny it.

[14] There is no room for the rest of this article, as my allotted space is entirely taken up with footnotes. However, I was merely going to state that it has been my ambition to write an article wholly composed of footnotes.[20] My resolution weakened and I included a first line.

[15] In the Peanuts strip cartoon by Schultz[24] in the Daily Sketch.[19]

[16] He was never wrong.[6]

[17] This juxtaposition no doubt reflects my admiration for the work of Mr. Peter O'Donnell.

[18] Of Sheffield.

[19] Now defunct. The Sketch and the strip have gone to the Mail.

[20] Footnotes should not be confused with references. Thus, 10 is a footnote, not a reference. References have a special charm of their own. I cherish a reprint of an article, describing one case, with seven alledged co-authors and 73 references. Famous physics papers include those by Bhang and Gunn; Alfa, Bethe and Gamow; Sowiski and Soda, etc. In my capacity of Science Editor for an international journal, I get not a few "crank" papers for assessment. I have learned to recognise these at a glance by the facts that a) the references always come first, and b) the list includes (always at number 5 or 6, for some unknown reason) "5: "Some inane observations on some perfectly well worked out phenomena" by N.A.D.[21] Six copies, privately circulated, six years ago."[22]

[21] The author of the paper being considered. They always use only initials here.

[22] You will observe that not all footnotes are brief. A recent article[23] I wrote was originally subtitled "A Footnote to History". It occupied two sides of news-sheet with 3000 words. The subeditorial pencil removed the subtitle, but a comment on the article in the People restored my faith in the subeditorial class. This genius[24] dreamed up the heading "Queen of Drag".

[23] "Was the Virgin Queen a Man?" Pulse. September 1971

[24] Credit where credit is due.

THE CROWDED CROSSROADS

J. BRUCE MARTIN, Ph.D

REFERENCES

Asch, A. B. "Engineering Education at the Crossroads," *Jour. Eng. Ed., 57* No. 8, (April 1967), 576–578.

Bose, Subhas O. *Crossroads: Collected Works,* Asia (1938–40).

Brode, Douglas. *Crossroads to the Cinema.* Holbrook (1975).

Collins, James. *Crossroads in Philosophy: Existentialism, Naturalism, Theistic Realism.* Regnery (1969).

Cox, Lionel A. "Why Is Industrial R&D at the Crossroads?" *Pulp Pap. Mag. Can. 73* No. 7 (July 1972), 93–96.

Ewing, David W. "Corporate Planning at a Crossroads," HBR *45* No, 4 (July–Aug. 1967), 77–86.

Fehrenbach, T. R. *Crossroads in Korea.* Macmillan (1968).

Forcey, Charles. *Crossroads of Liberalism: Croly, Weyl, Lippmann, & the Progressive Era, 1900–1925.* Oxford U. Pr. (1967).

Goland, Martin. "Professionalism in Engineering: At the Crossroads," *Mech. Eng.*, (March 1974), 18–22.

Hale, Arlene. *Crossroads for Nurse Cathy.* Ace Bks. (1974).

Kakonis, Tom E. & Wilcox, James C., Ed. *Crossroads: Quality of Life Through Rhetorical Modes.* Heath (1972).

MacDonald, John D. *Crossroads.* Fawcett World (1974).

Maritain, Jacques. *Education at the Crossroads.* Yale U. Pr. (1943).

Miers, Earl S. *Crossroads of Freedom: The American Revolution & the Rise of a New Nation.* Rutgers U. Pr. (1971).

Morgenthau, Hans J., Ed. *Crossroad Papers.* Norton (1965).

Namier, Lewis B. *Crossroads of Power.* Bks for Libs. (1962).

Norton, Andre. *Crossroads of Time.* Ace Bks. (1974).

Olsen, James & Swinburne, Laurence, Ed. *Crossroads Series.* Noble (1969).

Pappas, Lou S. *Crossroads in Cooking.* Ritchie (1973).

"Publication at the Crossroads," *Anal. Chem. 37,* (April 1965), Sup. 27A–30A.

Sherif, M. & Wilson, M. O., Ed. *Group Relations at the Crossroads.* Harper Bros. (1953).

Sykes, Christopher. *Crossroads to Israel 1917–1948.* Ind. U. Pr. (1973).

Taylor, Erwin K. "Management Development at the Crossroads," *Personnel, 36* No. 2 (Mar.–Apr. 1959), 8–23.

Thompson, Daniel C. *Private Black Colleges at the Crossroads.* Greenwood (1973).

Tomlinson, Monette W. *Crossroads Cameos.* Naylor (1964).

Verissimo, Erico. (Tr. by Kaplan, L. C.). *Crossroads.* Greenwood (1943).

Vitz, Evelyn B. *Crossroad of Intentions: A Study of Symbolic Expression in the Poetry of Francois Villon.* Humanities (1974).

Webster, David. *Crossroad Puzzlers.* Natural Hist. (1967).

Witton, Dorothy. *Crossroads for Chela.* Archway

Women's Bureau. *American Women at the Crossroads: Directions for the Future.* Proc. 50th Conf. (June 11–13, 1970)

Wyatt, Arthur R. "Accounting Profession at the Crossroads," *Mich. Bus. Rev.* XVIV, No. 5, (Nov. 1972), 20–26.

How to be a Published Mathematician Without Trying Harder than Necessary

DAVID LOUIS SCHWARTZ

Abstract

After a crisp, cogent analysis of the problem, the author brilliantly cuts to the heart of the question with incisive simplifications. These soon reduce the original complex edifice to a mouldering pile of dusty rubble.

****

The problem is that mathematicians know all kinds of weird things, but they publish comparatively little. It's not the numbers that bother them; it's the words. If the fill-in words necessary for a mathematical paper were provided, any mathematician could fill in the spaces with numbers, and he'd be safely through the publish-or-perish barrier. It is with this humanitarian view that I have undertaken to provide a form sheet, sort of a work-book approach. The arrangement given is based on already published material, so the plan has the advantage of having been shown to be workable at least once before.

There are numerous subtitles. One must realize that editors of mathematics magazines tend to understand either (1.) too much, or (2.) too little about the things they read. Misunderstandings arise. One way around this obstacle is don't submit things to mathematics publications. Try Ladies Home Journal, or Vogue, or Hot Rod; this is important. The possibility of "toning up" an issue with something serious can frequently appeal to a non-mathematical editor, whereas the same possibility probably never occurs to a math magazine editor. For that reason, we insert not merely connective words and phrases, but whole paragraphs.

Since everything basically contains part of everything else, it is always possible to relate a random paragraph to anything occurring before and after. Once this fact is taken to heart, a career as a published mathematician becomes not only possible, but inevitable.

The blanks are to be filled in with mathematical symbols. The more variety, the better. Throw in everything. Be neat. Editors love neatness.

****

Then
$$\frac{d^2v}{dx^2} = A\int \frac{x^3 - 3x^2 + 6x - 6}{x^4}\,e^x\,dx = A\cdot\frac{1}{D}\left(\frac{x^3 - 3x^2 + 6x - 6}{x^4}\,e^x\right)$$
$$= Ae^x\,\frac{1}{D+1}\left(\frac{x^3 - 3x^2 + 6x - 6}{x^4}\right) = Ae^x\,\frac{1}{D+1}\left(\frac{1}{x} - \frac{3}{x^2} + \frac{6}{x^3} - \frac{6}{x^4}\right).$$

Now $D(\frac{1}{x}) = -\frac{1}{x^2}$, $D^2(\frac{1}{x}) = \frac{2}{x^3}$, and $D^3(\frac{1}{x}) = -\frac{6}{x^4}$,

so that
$$\frac{1}{D+1}\left(\frac{1}{x} - \frac{3}{x^2} + \frac{6}{x^3} - \frac{6}{x^4}\right)^{*} = \frac{1}{D+1}\left[\frac{1}{x} + 3D(\frac{1}{x}) + 3D^2(\frac{1}{x}) + D^3(\frac{1}{x})\right] = \frac{1}{D+1}(D+1)^3(\frac{1}{x})$$
$$= (D^2 + 2D + 1)(\frac{1}{x}) = \frac{x^2 - 2x + 2}{x^3}.$$

Thus,
$$\frac{d^2v}{dx^2} = A\,\frac{x^2 - 2x + 2}{x^3}\,e^x + B, \qquad \frac{dv}{dx} = A\,\frac{(x-1)e^x}{x^2} + Bx + C,$$
$$v = \frac{y}{x} = C_1\frac{e^x}{x} + C_2x^2 + C_3x + C_4, \quad \text{and} \quad y = C_1e^x + C_2x^3 + C_3x^2 + C_4x.$$

From the statement
and
we obtain
in which
with
together with
we also have
and therefore, effectively,
the desired formula emerges as

Accordingly, a non-Hermitian canonical variable transformation function can serve as a generator for the transformation function referring to unperturbed oscillator energy states.

There follows
Alternatively, if we choose
there appears
Thus
and
where the latter version is obtained from

****

The next installment will discuss advanced presentation procedures, how (and why) to write an abstract, and the role of mathematics in our Judeo-Christian heritage.

A Psychological Study of Journal Editors

S. A. RUDIN
F.S.B.I.R.

The clinical psychologist Anne Roe studied the manner in which a scientist is seduced by his field of study, and she reported her findings in her book, *The Making of a Scientist* (Roe, 1955). She obtained interviews and test results from the 20 most eminent physical scientists, the 20 most eminent biological scientists, and the 20 most eminent social scientists in the USA. She concluded that the biological scientists tended to be preoccupied with death, that the physical scientists had difficulty locating themselves in the physical world, and that the social scientists disliked and could not get along with people.

This study extends her methods to the study of editors of scientific journals. The editors chosen for study were in charge of all major scientific journals in the USA, making a total of 318,991 subjects. Each subject was studied exhaustively by a combination of depth interview, case history, and numerous psychological tests of intelligence, aptitude, interests and personality.

RESULTS

First impression and general appearance. Subjects ranged from tall* to short, fat to thin, and warped to degenerate in general appearance. Despite this heterogeneity, each was marked by certain tell-tale characteristics: the eyes were narrowed; the mouth was pursed into a snarl; and the writing hand was cramped and taut from stamping REJECTED thousands of times. Upon first perceiving the experimenter, each subject exclaimed, "NO!" before noticing that no manuscript was being tendered.

Childhood background. That childhood experiences strongly influence the developing personality is well known. Again, great diversity of backgrounds was noted: they came from every conceivable environment, from palatial mansions in Hollywood to wretched hovels on some university campuses, but all had in common a peculiar set of family relationships. In every case, the father turned out to have been an alcoholic, drug addict, professional . . .**, or the editor of a scientific journal. The mother was found to spend but little time with her children, devoting herself to such pursuits as managing a house . . .**, selling drugs to adolescents, smuggling diamonds past customs officials, or editing a scientific journal. But of greatest interest for the purposes of this study was the discovery that in every case, the child had been beaten often and severely *with a book.* Naturally, such traumatic stimulation eventually led to a deep-seated hatred of anything associated with reading, writing, learning, knowledge, and scholarship. Some showed this tendency as early as the second year by tearing pages out of the Encyclopaedia Britannica, setting them afire in the middle of the living room floor, and executing an exultant war dance around them in the fashion of certain American Indian tribes.

Intelligence and aptitudes. These were measured by a variety of instruments including the Wechsler Adult Intelligence Scale (WAIS), the Draw-A-Person Test and various special aptitude tests. Considerable difficulty was encountered since none of the subjects could read. The use of oral and non-verbal tests, however, finally yielded usable data. It was found that the subjects were uniformly below IQ 71. This highest IQ was attained by the editor of a widely-read psychology journal who was himself the author of one of the intelligence tests used. The pattern of abilities measured by the specialized aptitude tests showed the subjects to be well below the standardization group (which was made up of college sophomores, white rats, and some persons from mental hospitals) on verbal reasoning, numerical reasoning, perceptual speed, spatial reasoning, verbal recall, clerical ability, map-reading ability, needle-threading ability, and the capacity to pronounce words of more than three syllables. Indeed, the only tests on which the subjects performed well were one requiring the use of a spade to pick up and transfer material from one pile to another and the ability to ignore noxious odors.

Interest tests. On the Strong Vocational Interest Blank and the Kuder Preference Record-Vocational, subjects tended to score lower than average on activities

and occupations associated with originality, critical thinking, creativity, scientific research, and literary production and appreciation. They scored relatively high, however, on scales measuring interest in mild manual labor and evading work altogether.

Personality tests. All subjects were found to register insane on Rohrschach Ink-Blot Test, Thematic Apperception Test, Minnesota Multi-Phasic Personality Inventory and the House-Three-Person Test. Exceptions were two subjects, both neurotic, ulcer-ridden, and compulsive shoe-lace cleptomaniacs. All subjects perceived themselves as God, except for one who claimed that he had created God. Another signed his name omitting all vowels.*** Yet another claimed that the ink-blots were actually reprints of old copies of his Journal, and sued the experimenter for plagiarism.

CONCLUSIONS

The reasons for the success of these subjects in editing journals is clear. First, by preventing new ideas from appearing in print, they make it easier to keep up with the literature. Second, by requiring the experimenter to repeat his study dozens of times and re-write his paper hundreds of times, they enforce the consumption of materials and labor, thus stimulating the national economy. Third, if *they* can understand a paper, *anyone* can.

---

*e.g. J. Bacteriol. (the rest of description does not apply).
**censored
***He was from Israel.

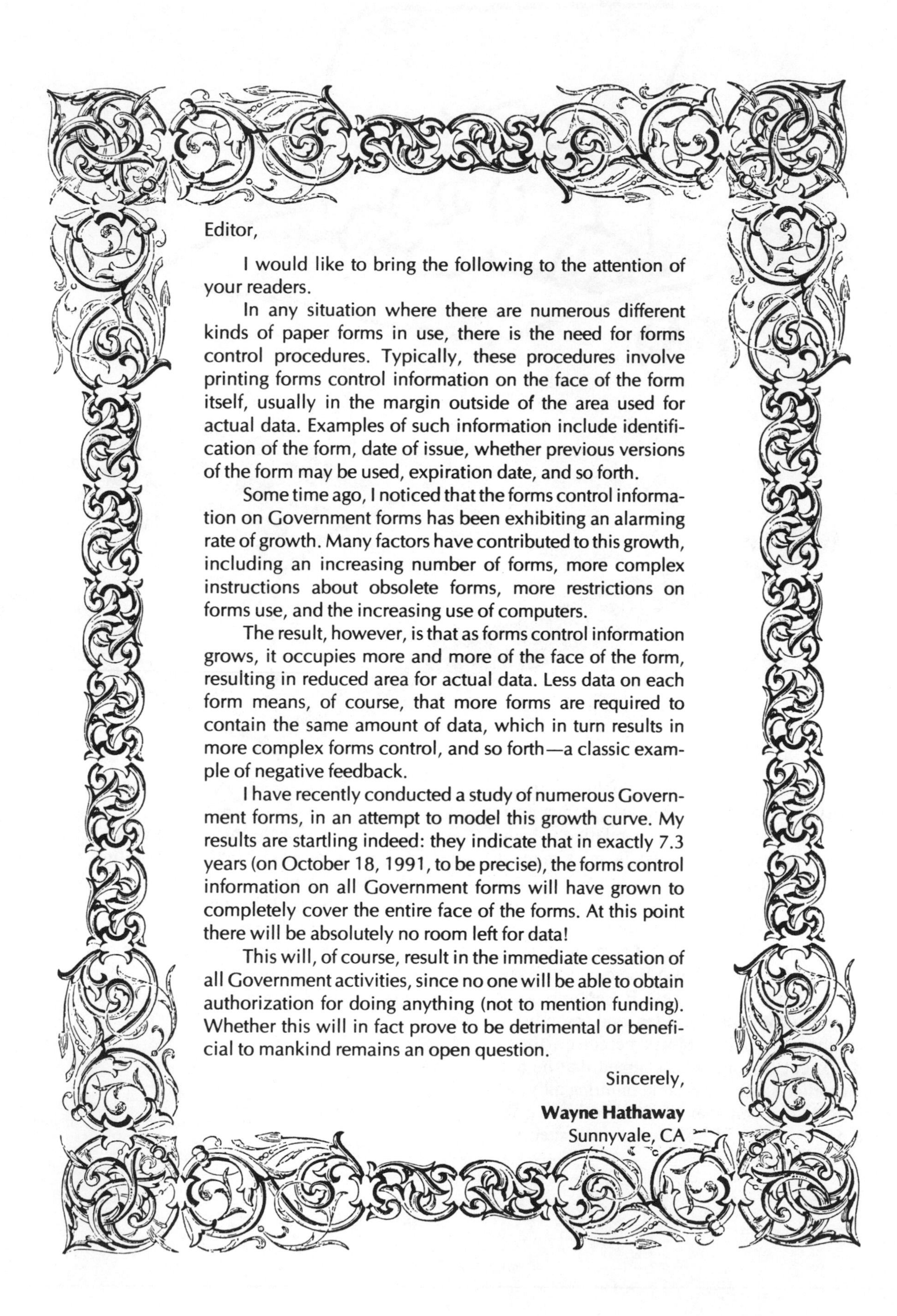

Editor,

I would like to bring the following to the attention of your readers.

In any situation where there are numerous different kinds of paper forms in use, there is the need for forms control procedures. Typically, these procedures involve printing forms control information on the face of the form itself, usually in the margin outside of the area used for actual data. Examples of such information include identification of the form, date of issue, whether previous versions of the form may be used, expiration date, and so forth.

Some time ago, I noticed that the forms control information on Government forms has been exhibiting an alarming rate of growth. Many factors have contributed to this growth, including an increasing number of forms, more complex instructions about obsolete forms, more restrictions on forms use, and the increasing use of computers.

The result, however, is that as forms control information grows, it occupies more and more of the face of the form, resulting in reduced area for actual data. Less data on each form means, of course, that more forms are required to contain the same amount of data, which in turn results in more complex forms control, and so forth—a classic example of negative feedback.

I have recently conducted a study of numerous Government forms, in an attempt to model this growth curve. My results are startling indeed: they indicate that in exactly 7.3 years (on October 18, 1991, to be precise), the forms control information on all Government forms will have grown to completely cover the entire face of the forms. At this point there will be absolutely no room left for data!

This will, of course, result in the immediate cessation of all Government activities, since no one will be able to obtain authorization for doing anything (not to mention funding). Whether this will in fact prove to be detrimental or beneficial to mankind remains an open question.

Sincerely,

Wayne Hathaway
Sunnyvale, CA

Results

that

would be

impossible

without

irreproducible

MANAGEMENT

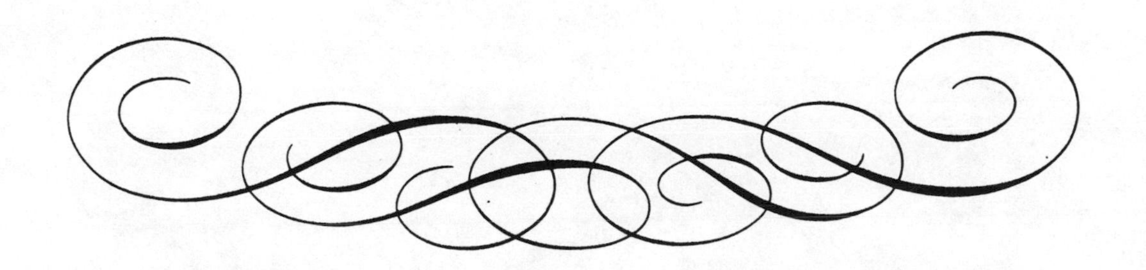

Administration of Research:

A NEW FRONTIER

DON DORRANCE
Milwaukee, Wisconsin

Abstract: The editors of this journal like to have the footnotes at the bottom of each page. I prefer to have the footnotes collected all together at the end. If you take some scissors and cut the footnotes from the bottom of each page and paste them on a blank sheet of paper and then paste this sheet of paper at the end of the article, the editors won't know.

It is said[1] that 90% of all scientists who have ever lived are alive today. This presupposes a great deal of scientific activity, and brings several questions to mind. However, one question which has never been seriously asked is, quite simply, what are we to do with all these people?

Management of research is either a technique or a science, or, neither a technique nor a science. It is the purpose of this paper to clarify some of the management principles which have formed about the business of science.

The Director (D) of a large Research Organization (RO) will have different motivations and goals than that of the Principle Investigator (PI). This difference affects the management of the whole project.

The Director (D) has a unique point-of-view: being in charge of the whole operation there is no way he can gracefully pass the buck. His position exists and it is not necessary, or probable, that he know what is going on:

> "The investigator in charge of the apparatus told me that it provided from 2,000 to 3,500 photographs a week, each photograph including sufficient data to plot the orbit of one meteor. He added that the job of translating one week of records into meteor orbits required nearly a full year of paper work. I asked him how the laboratory managed to keep up with the machine: 'We don't. We shut the machine off. Most of the time it's not operating.' "[2]

This state of affairs leads the Director (D) into what has come to be called the "Chicken Little Complex." This is because he hasn't the slightest idea of what's going on. He just knows that the machine is off. And it's an expensive machine. There has been found[3] a direct correlation between the frivolity of the research being conducted and the seriousness of the Director's (D's) complex. But, as is usually the case, when the research is either useless or redundant, or, as is frequently the case, both, the Director (D) will sense the hum of activity and know that his scientists (the Team) are choked by their work-loads. Something is going on. He will not know what, or, very commonly, be afraid to find out. This causes him to go about muttering, "The sky is falling."[4]

The result of this, from an interactive point-of-view, is that only the Director (D) knows the sky is falling. This is Management information, and, as such, restricted. The Team must be urged to do their utmost, while the exact nature of the disaster facing them is concealed. The Team will never understand the pressures facing Management.

The Principle Investigator (PI) has a different typical function and responsibility. He has to direct the Team towards a suitable goal. Since progress, as all science, is random, he becomes obsessed by the day-to-day triviality of doing something that seems worthwhile.

> "I have been told of a university researcher who, on parking his car each morning, noted down the numbers of adjacent cars and then went to the Library to look at the corresponding Dewey numbers, rarely failing to find something of interest."[5]

One common problem faces the Principle Investigator (PI). Quite frequently[6] he will be ordered by the Director (D) to solve the problem by his method. The PI quickly finds that the D's method violates the Laws of Nature (LN) in several ways. This allows him two courses: he may inform D that it can't work that way, or, the usual course, he will take both the problem and the recommended solution and—in terms of electronics—patch in a by-pass circuit. This will allow him to isolate the "preferred" solution while allowing the project to continue. After the third major change the Director's solution may be dropped. Progress has been achieved while harming no-one.

If there is such a thing as management of science it must be understood as a discipline. Under the conditions outlined above, I would like to propose the First Law of Scientific Management: *Activity is the preferred substitute for intelligence.*

As has been demonstrated, true research cannot be fully understood by managers. And scientists cannot understand that inherently the sky *is* falling. The result can't help but be confusion.

> "No one man knows the entire blueprint of any large-scale computer, and certainly no one man knows whether or not it is wired according to that blueprint."[7]

There is a final hazard in the management of science. The major purpose of science is to publish papers. The quality of scientific prose becomes a function of administrative confusion: if someone actually undesrtands the project, it should be the Director. Since the Director doesn't, no-one should. Competent writers will report accurately what has gone on. Therefore the final report must be written by the most disorganized person on the Team, or, for the same effect, by the Team as a whole. This paper will be signed by the PI and the D, but not in that order. The D will then see that the paper is published in the appropriate journal, and then decide what they're going to do next.

Results reported in this paper were unwittingly supported by several government contracts.

[1] *Readers Digest.* Any issue.
[2] Pfeiffer, John. *The Thinking Machine.* Lippincott 1962 p. 7.
[3] Harvard Business School. Management Paper 127 (Suppressed).
[4] Lang, Andrew. *The Blue Fairy Book.* "Chicken Little" p. 43.
[5] Line, M. B. "On the Design of Information Systems for Human Beings." *Aslib Proceedings,* 22, 7. July 1970 p. 330.
[6] According to Sokol & Meyer, 90% of the time. ■
[7] Pfeiffer. *op cit* p. 51.

PEACEFUL USES OF NUCLEAR EXPLOSIVES
Nuclear News, Vol. 11, No. 3, Page 29 (March, 1968).

"Development of hydro power in the desert of North Africa awaits only the introduction of water . . ."

BLEACH GIANT ACQUIRES STAIN MAKER

In a transaction expected to be finalized, Clorox Co., Oakland, best known as a leading manufacturer of liquid bleach, will acquire Comerco, of Tacoma, known for its Olympic brand wood stains. Clorox, which is paying $123 million for the privately-held Comerco, has been steadily diversifying into household products ranging from kitty litter to frozen onion rings.

From: Alaska Airlines Private Line, Business News. Sept. 1981
Submitted By: Milton Gordon

Important Laws in Science [A Review]

A. KOHN

There are many laws in science. Many are taught in high schools: such as the laws of Newton, Boyle-Mariotte, Thermodynamics, Ohm, etc.

In the last decade some new important Laws were discovered:

1. *Parkinson's Laws.* Most of these laws are well described in Parkinson's book[1]. Recently a new Parkinson's Law for Medical research has been described[2]. It states: SUCCESSFUL RESEARCH ATTRACTS THE BIGGER GRANT WHICH MAKES FURTHER RESEARCH IMPOSSIBLE."
2. *Maier's Law*[3]: IF FACTS DO NOT CONFORM TO THEORY, THEY MUST BE DISPOSED OF.
3. *Murphy's Law*[4]: IF ANYTHING CAN GO WRONG WITH AN EXPERIMENT, IT WILL.
4. *Paradee's Law*[5]: THERE IS AN INVERSE RELATIONSHIP BETWEEN THE UNIQUENESS OF AN OBSERVATION AND THE NUMBER OF INVESTIGATORS WHO REPORT IT SIMULTANEOUSLY.
5. *Hersh's Law*[6]: BIOCHEMISTRY EXPANDS SO AS TO FILL THE SPACE AND TIME AVAILABLE FOR ITS COMPLETION AND PUBLICATION.
6. *Old & Kohn's Law*[7, 8]: THE EFFICIENCY OF A COMMITTEE MEETING IS INVERSELY PROPORTIONAL TO THE NUMBER OF PARTICIPANTS AND THE TIME SPENT ON DELIBERATIONS.
7. *Gordon's First Law*[9]: IF A RESEARCH PROJECT IS NOT WORTH DOING AT ALL, IT IS NOT WORTH DOING WELL.

REFERENCES

1. Parkinson, C. N. Parkinson's Law, Houghton Mifflin Co. 1937.
2. Parkinson, C. N. Parkinson's Law in Medical Research, New Scientist 13:193 (1962).
3. Maier, N. R. F. Maier's Law, The American Psychologist (1960).
4. Michie D., Sciencemanship, Discovery 20:259 (1959)
5. Pardee, A. B., pU, a new quantity in biochemistry, American Scientist 50:130A (1962).
6. Hersh, R. T. Parkinson's Law, the squid and pU, American Scientist 50:274A (1962).
7. Old, B. S. Scientific Monthly, 68:129 (1946).
8. Kohn, A. Boardmanship, J. Irrepr. Res. *10*:24 (1962).
9. ——— ——— J. Irrepr. Res. *9*:43 (1961).

THE MOLECULAR THEORY OF MANAGEMENT

RUSSEL DE WAARD

It is interesting and informative to liken the growth and behavior of a company to that of a synthetized organic molecule. Take as example PVC. In unplasticized form this material is very hard, elastic, and brittle. There are many strong bonds between closely spaced charged atoms. When one adds so-called diluent plasticizers, the bonds are weakened as the inverse square law is invoked by the big putty-like molecules. The material becomes flexible, can absorb shock, and shows promise for commercial purposes; for example, shower curtains, automobile window winders, steering wheels, seat covers, and the like. These big molecules here serve a very useful purpose and furthermore they contribute in just the right way to the modern economy of "planned obsolescence." The molecules slowly lose their footing and leech out of your window winder, leaving behind a cracked handle at just about the same time as the rust eats through your right front fender.

Now, a company also starts with a small group of charged particles. A lot of energy is produced, efficiency is high, and hence business is good. This nucleus rapidly acquires more particles. Because there is no time for careful screening, a large number of the big rubber-like molecules are added to the mix. Many other types of particles are admitted in random fashion. Their quality is inversely proportional to the management stature of the interviewer and their salary directly proportional. Unfortunately, many of the big molecules are completely inelastic diluents whose function appears to be to absorb the energy generated by the charged particles.

It is interesting to contrast these rubber-like additions to the neutrino of atomic physics, whereas the latter has no charge and no mass, the former has no charge and is all mass. If a growing company can avoid gathering too many of these particles it can survive. If the converse holds, all of the energy generated by the charged particles is absorbed by the diluents and no energy or products can escape. Although the evidence pointing this way is sometimes strong, it is hard to believe that a company would consciously plan its own obsolescence. ■

The Innovation Myth

Actually, management is very uncomfortable with creative personnel and, with rare exceptions, has never provided an environment that motivates them to produce. Creative talent always seems to be frustrated, irascible, unpredictable, and unproductive in the corporate world.

Great numbers of corporate executives consciously are not aware of their hostility to new ideas. On a personal basis they believe to a man that new products are essential to the future of the company. But when given an opportunity to implement this conviction within the corporate environment they inevitably demonstrate negative attitudes. And in so doing, they totally reflect the basic negative corporate attitude on this point.

What prudent man is going to put his career on the line for so hazardous an undertaking as a new product venture when he personally has everything to lose and nothing to gain?

In 5 or 10 years, when the product finally begins generating profits, considering the present mobility of top executive talent, it is more than likely that he will no longer be with the company.

The Dept. of Commerce reports that of the 11 major inventions in the steel industry, four were products of European firms, seven came from independent inventors, and *none* came from American steel firms.

And in the petroleum industry, that giant among giants, of the seven major inventions in the refinig and cracking of petroleum *all* were made by small independent inventors! ■

Abstracts from *The Innovation Myth* by Louis Soltanoff, President, Louis Soltanoff & Associates. From *Industrial Research*, August 1971.

A Study of Basic Personality Behavior Patterns Which Characterize the Heirachry of Social Dominance or Rank in Large Corporations.

JAMES A. CUNNINGHAM
Personnel Department
Alabama Instruments, Inc.
Huntsburg, Alabama

ABSTRACT

A quantitative definition of corporate rank is established. Results are presented which show correlations between various types of salutations the audibility of salutations and behavior patterns in meetings.

INTRODUCTION

In any group of higher animals a heirachry of social dominance or rank is quickly established. A classic example is the basic pattern of social organization within a flock of poultry in which each bird pecks another lower in scale without fear of retaliation and submits to pecking by one of higher rank (1). Primates also exhibit similar behavior (2). For example, R. Nasini and A. Scala (3) describe an experiment where five East Indian gibbons were placed in a cage with a small door such that only one could exit at a time. Three males, one castrated, and two females were included. The right to exit first, i.e. the social rank, was quickly and reproducibly established with the largest male first, followed by the second largest male, a female, the castrated male, and finally, the smaller female. These studies clearly show the importance of strength and body weight in establishing social rank in animals, but they are of limited value, of course, in analyzing human behavior which in a modern society is intellectually rather than physically oriented.

Nevertheless, modern zoological and anthropological studies do show interesting correlations between lower primate behavior and man regarding social ranking and status. For example, Desmond Morris (4) lists ten basic laws which govern dominance which he claims are operative both in man and in baboons. The ten laws are:

1. You must clearly display the trappings, postures and gestures of dominance.
2. In moments of active rivalry you must threaten your subordinates aggressively.
3. In moments of physical challenge you (or your delegates) must be able forcibly to overpower your subordinates.
4. If a challenge involves brain rather than brawn you must be able to outwit your subordinates.
5. You must suppress squabbles that break out between your subordinates.
6. You must reward your immediate subordinates by permitting them to enjoy the benefits of their high ranks.
7. You must protect the weaker numbers of the group from undue persecution.
8. You must make decisions concerning the social activities of your group.
9. You must reassure your extreme subordinates from time to time.
10. You must take the initiative in repelling threats or attacks arising from outside your group.

It is not difficult to recognize that many of these behavior patterns are indeed operative in a large corporation.

Although the corporate environment represents a very dynamic system of social rank and dominance as various individuals move up and down and in and out, it is, nevertheless, a well defined system which is relatively easy to describe in quantitative terms. It is the purpose of this paper to present the results of various tests carried out which attempt to relate behavior patterns and characteristics to a precisely defined corporate rank.

RESULTS

I. Definition of Corporate Rank

The corporate rank or CR is defined as

$$CR_i = \frac{S_i}{S_p} + \frac{G_i}{G_p} + \frac{N_i}{N_T} + \frac{R_i}{R_N} + \frac{D_i}{3} + \frac{A_i}{A_T} + \frac{E_i}{E_p} \quad (1).$$

CR_i = corporate rank of ith individual
S_p = President's salary
S_i = individual salary
G_i = individuals job grade as set by personnel department
G_p = President's job grade
N_i = number of people reporting to individual
N_T = total number of people in company
R_i = number of organizational levels from president
R_N = number of organizational levels in company
D_i = number of college degrees up to 3 maximum
A_i = number of unique office accouterments
A_T = number of unique office accouterments in president's office
E_i = number of years of related experience
E_p = number of years of related experience for the company president

Values range from a low of about 0.13, which might represent the CR of a newly employed dock worker, to a value of 7.0 held by the company president. A typical distribution of CR values in a corporation of 50,000 people would look approximately like Figure 1.

FIGURE 1

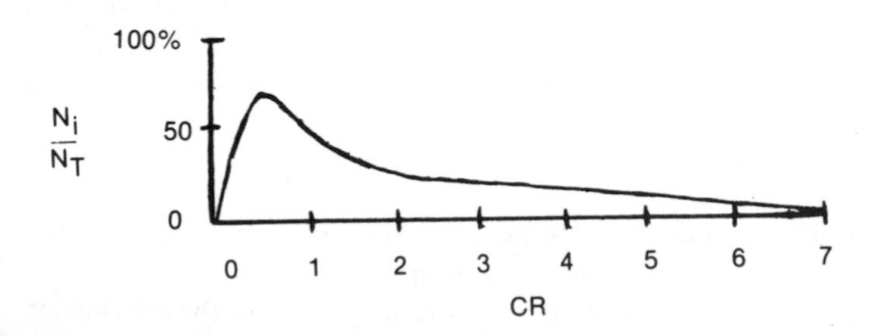

It is believed that the validity of Equation 1 is essentially axiomatic and, therefore, warrants no further discussion.

II. Basic Relationships

First, it may be mentioned that certain intracompany patterns such as (1) placing ones feet on the table, (2) the wearing of only black suits, (3) wearing bow ties, (4) wearing only blue shirts, (5) wearing only white sox, (6) smoking a pipe and others appeared to be related more to conformism and mimicry of a very highly ranked and individualistic company official than to intercompany behavior characteristics of a more general nature. Also, as far as males are concerned, no strong correlations were found between the CR and body weight or height, hair or eye color, or other such physical characteristics except for a rather mild correlation at the very highest levels with the presence of severe hemorrhoids.

Clear and unequivacal correlations were found with sex. No women with CR's above a value of about 3 appear to exist in American industry.

III. Meetings

It became clear in the very early phases of this research that a fruitful area of study would be an examination of the behavior of individuals while attending company meetings where people of various CR's are present. Accordingly, personality traits and patterns were observed and recorded by a team of three trained observers during the course of attending approximately 1600 meetings in four corporations over a period of two years.

The following personality characteristics were found to be typical of the highest ranking member of a given meeting or conference.

1. Asks questions rather than answers questions.
2. Asks questions about timing and schedules rather than about how and why.
3. Sits at the head of the table.
4. Gives commands and advice.
5. Stays on the offensive rather than on the defensive.
6. Usually comes late to a meeting.
7. Exits a meeting by having a secretary deliver a seemingly important note.
8. Rarely gives a presentation — usually is the recipient of a presentation.
9. Shows greater interest in financial matters than in technical matters.
10. Rarely appears to be snowed.
11. Never becomes emotional.
12. Responds quickly and with an authoritative tone.
13. Never lets another member of the group get the upper hand.

14. Rarely reprimands a subordinate of greatly lower CR. Does reprimand subordinates of close CR.
15. Assumes the characteristics of an extrovert rather than an introvert.
16. Looks everyone straight in the eye.
17. Never mumbles, twitches or fidgets.
18. Assigns responsibilities to others rather than to himself.
19. Is in a hurry and tries to speed up the meeting.

Universal use of all these tactics in almost all situations is, of course, reserved for the highest ranking corporate officials that is CR = 6-7 company presidents or chairmen of the board. Much lower ranking individuals, i.e. CR = 0-2, would be expected to behave, in general, according to the antithesis of the above list.

It was observed that such petty tactics as: the use of rough language, asking highly technical memorized questions or obvious attempts to embarrass a competitor were not typical of very high ranked individuals.

Let's take a look at how some of these behavior patterns operate by listening in on a meeting being held by an engineering manager, several of his engineers, three or four corporate staff people and the company president.

Dr. John Galaway, Ph.D. physics, is at the board. He is speaking and directing his remarks in the direction of Markus Hutson, President of Impact Ecotronics, Inc. Mr. Hutson is seated at the head of the table with his corporate staff members positioned immediately to his left and right. John is near the end of his presentation and is saying, "The pyridinium imino complex (related, of course, to ZFQ* activation) will probably react or enable the zeta phase of PFP via the bolonium state. This could lead us to new ortho megatypes. We feel a computer analysis of the quantized spin-orbital Gaűdsmit integrals of the L_{II} L_{III} intermediate may lead us to the answer. (The entire analysis will have to be carried out in vector phase space, of course.) Our main delay, at this time, Mr. Hutson, is corporate funding."

Gerald Hutson, company president, unhesitatingly replies in a forceful and authoritative manner, "John, this is clearly one of our must do's for 1975. When do you think you can have the analysis finished and a full report on my desk?"

Even though the company president had no idea what the engineer was talking about, he still asserted and protected his CR by responding in the classical aggressive style. That is (1) he asked when, not how or why, (2) he answered the implied question by asking another question, (3) he passed the responsibility back to the subordinate, (4) he made it appear that he thoroughly understood the topic under discussion, (5) he postponed any decisions he might eventually have to make without the slightest hint of insipidity and finally, (6) he issued a command. A statement similar to Mr. Hutson's could be made by anyone with the highest CR in a given meeting. Occasionally the 2nd ranked man may resort to similar tactics but at the risk of being put down by the number one man. Such statements and tactics delivered by a 3rd or 4th ranking person would be a faux pas of the first magnitude.

IV. Salutations

A brief study was made of the manner in which two people who know each other greet or acknowledge one another upon passing in the company corridors. Approximately 2500 greetings were recorded in three large corporations. The following table contains the findings.

Type of Greeting	Frequency of Use, Per Cent
1. Say Hi or Hello	33
2. Say hello, followed by name of person greeted	22
3. Raise eyebrows	10
4. Nod head	10
5. No response	10
6. Point index finger and "shoot"	5
7. Raise hand like Indian "How" sign	4
8. Military salute	4
9. Stop at shake hands followed by back slapping and considerable conversation	2

An attempt was made to correlate the various types of greeting response with the CR*. Personnel records were pulled for 10% of the greeter sample and CR's were calculated. The following correlations were found between types of greetings and the approximate difference in CR values (ΔCR).

Category	ΔCR
I. Shooters, brow raisers, howers, no responders, nodders, saluters	0-1
II. Hi'ers, name callers	1-3
III. Name callers, back slappers	3-6

From these results, it is clear that certain greeting techniques, such as shooting or howing, are avoided between greeters of high ΔCR's such as between the Executive Vice President and the Custodian. On the other hand, a response such as howing or shooting could be employed by two assembly line workers of equal CR or even the Controller and the Assistant Vice President without any fear of social recrimination.

We were also able to correlate the audible intensity of greetings as a function of the difference in CR (ΔCR) of the

* Male-female encounters were found to be anomolous and thus are not included in this data.

two individuals who happen to meet. Data was taken by planting a miniature transmitting device on the backs of the picture badges of a representative sample of employees. These persons were then followed with a receiving device and the response noted. The data represents the average intensity of both speakers' responses although it was noted the person of lower CR was usually more audible. Figure 2 is a plot of the response in dB vs ΔCR.

FIGURE 2

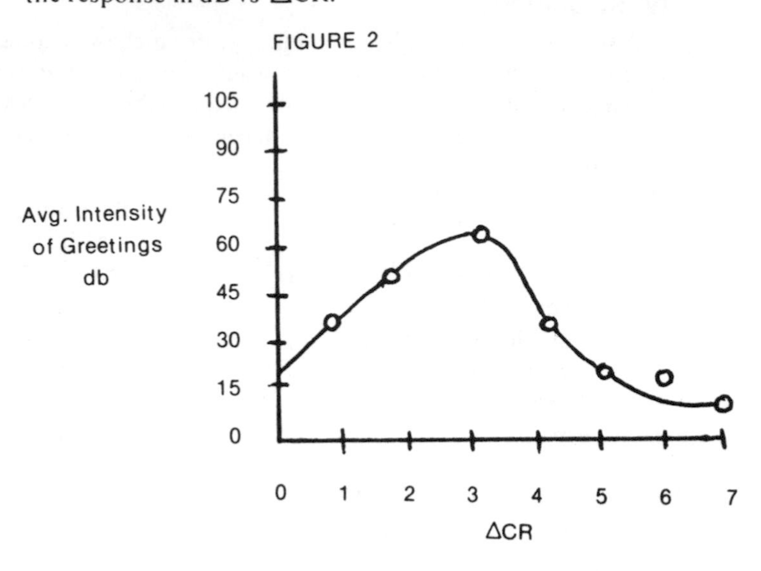

In general, the results indicate that the importance and/or intensity of salutations is proportional to the degree of competition that exists between the greeting parties. Competition is low both at low and at high values of ΔCR and peaks in the range of CR = 1-3.

CONCLUSION

These results clearly establish that strong correlations exist between an individual's corporate rank and his behavior. It is hoped that these findings can be used to further corporate harmony. The use of certain high CR behavior patterns by an individual of low CR would, of course, be highly unethical.

ACKNOWLEDGEMENTS

This research was made possible by a grant from the Mexican Institute de la Gringos (MIDLG-68946-AF126). The assistance of Horst Mackintose and Ellen Grantges who conducted the greeting experiments is gratefully acknowledged. Special thanks are extended to Max Thorbes for participating in the castration experiment (this experiment was, unfortunately, unsuccessful and the results are not reported here); it is hoped his new career in The American Union of Women Libs is rewarding. Finally, the assistance of Henry Tomlinson* who fabricated and planted several of the listening devices is acknowledged. ■

* Present address, Alabama State Prison, Katoola, Alabama

BIBLIOGRAPHY

1. Maximilian Grűtlich, Deut. Einsetzen fűr Umheimlich Vogel, *39*, 864 (1939).
2. R. Nasini and A. Scala, Anal. Psychologie de Pucelage, *14*, 6 (1948).
3. See for example, S. Z. Himalya, "Primate Behavior," p. 86, Venuzian Book Co., Spatoona, Louisanna, 1964.
4. Desmond Morris, "The Human Zoo," pp. 42-53, McGraw-Hill Book Co., New York, 1969.

SOME SOCIOLOGICAL OBSERVATIONS OF REPATRIATE TECHNICIANS

R. W. PAYNE
Quiriquire, Venezuela

(With apologies to Sir Joseph Porter, K.C.B. and the famous operatic pair Goldberg and Solomon).

When I was a lad I was no fool.
I went to a technological school
I learned a little, but not a lot
And a lowly BS was all I got.

That isn't very much, the doctors say,
But I run a little project of my own today.

- - -

That small project was no accident.
I won't tell how the maneuvering went
But will quote the advice that was given to me
"A chief seldom labors in his own country."

That advice was sound and it paved my way
To the fascinating project that I run today.

- - -

As the Dutch "voortrekked" to the old Transvaal
And the Wehrmacht to Cape Canaveral,
So a scientific Gringo will waste his time
If he doen't seek his fortune in a foreign clime.

In a jungly climate, I earn my pay
On the stimulating project that I run today.

- - -

When a man has trouble with the local tongue
And butchers the grammar, like the very young,
The people smile, but still defer
To the "educated, scientific foreigner."

That "foreign complex" can take effect
For the thinking man who wants his own project.

- - -

Now Bachelors all: no matter what they say
You can have a project of your own someday.
Be careful to be guided by this golden rule:
Get a passport on the day you graduate from school.

And prolong your project with the ultimate retort,
"With a little more data, we could publish a report;"

上海机床厂

Discontinuities in Social Research

SAMUEL E. WALLACE, Ph.D.

This is the only report of a study of a cross-cultural research project studying adolescents who show superior effectiveness and creativity in the academic, extra-curricular, and inter-personal spheres and especially to point up contrasts with the features observed in the adaptive behavior of students inside and outside the continental United States. (The real point, however, will be made clear in a moment).

A sample of 20 high school seniors in Spanish-speaking Puerto Rico, half boys and half girls, was selected for participation in this study. The students were interviewed various times during their senior year in order to determine the factors influencing their superior performance.

All twenty students attended an upper-middle class high school, known for its favoritism in admission practices. Although the high school accepted only those students with intelligence quotients above a certain level, the absence of a certified school psychologist permitted the parents to "shop around" the various psychological testing clinics until their child received the correct score. Therefore, students admitted tended to be those who parents had the necessary funds for "I.Q. score shopping." But the high school, interesting though it may be, is not the principal subject of this paper.

In the Fall of 1960, the author was contacted by the Institute which was carrying out (literally) the research described above. The Executive Director, Owner, President and Board of Directors of this Institute explained that a research staff was being organized, and invited me to participate. In additional conversations, the author learned that the staff was to include a psychologist, a social worker, a sociologist, and an anthropologist.[1] A federal agency was to provide the necessary funds, and work was to begin immediately, or rather, as soon as the funds became available.

After observing the Institute and talking with several members of the staff, the idea struck me that here was an excellent opportunity to study a much neglected field of research – "Discontinuities in Social Research."

During World War II, several eminent scholars had come up with the idea of studying "Continuities in Social Research".[2] They had made a lot of yardage with this idea, producing a series of books and several score of articles.

Every time I run across a reference to these "Continuities," I ask myself, what about *Discontinuities*? There are certainly projects where the investigator has a hopeless design for his research problem. There are certainly projects which adopt the worst methodology for their study area. There are certainly projects where staff changes become so frequent that not one inch[3] of continuity is possible. There are certainly projects whose publications[4] are completely unrelated to the data collected. Why is it that no one writes about these projects?

As I observed the Institute and its Executive Director, I had a hunch[5] that Fate had, at last, given me an opportunity to begin my work on "Discontinuities in Social Research." In the course of two of the most memorable years of my life, I watched a nearly complete change-over in janitorial, service, clerical, administrative, medical, clinical, etc., staffs[6]. The procession was simply fantastic. People would report for work in the morning and clean out their desks in the afternoon. The former senior accountant, bookkeeper, and treasurer, for example, seemed as well-entrenched as the Executive Director. He had worked for the Director for nearly two decades, worked hard, seemed to fulfill all the expectations of any employer, and was a personal friend. You can imagine my joy when I was told one morning that he had been summarily dismissed in the regular course of events one day, and given the remaining portion of the day to clean out his desk and go.

The (former) chief of the social workers enjoyed an excellent reputation in the social work community, devoted all her time and energy to her profession, and discovered

[1] An anthropologist was interviewed, but neither he nor the social worker who was supposed to be on the staff was ever seen.

[2] See: R. K. Merton and P. F. Lazarsfield, *Continuities in Social Research:* Studies in the Scope and Method of the American Soldier, Glencoe, The Free Press, 1950.

[3] I suppose that is at least one way of measuring continuity.

[4] Usually few and quite accidentally produced in "shipsaving" operations. Their most distinguishing characteristic is their almost total lack of any relationship to the data collected. This is not to say, however, that they are always worthless. Rather one would ask, why waste money collecting data?

[5] Unkind colleagues have let it be known that perhaps it would not take a genius to have this hunch at this particular Institute.

[6] Unfortunately the record was not perfect. There were one or two notable exceptions. But the reader must keep in mind the short time period of this study.

her desk occupied by another upon returning from vacation one year.

Those who had been around twenty years seemed to go proportionally as fast as those who had just arrived. Those who worked hard departed at about the same rate as those who did nothing. Those who claimed special connections with the "inside office" cleaned out their desk about as fast as those without connections.

In September of 1960 I talked with the Executive Director, and in April of 1961 the funds finally became available and I officially began work. One week later the Research Director, the Director of the project for which the Institute had just received funds, was fired. I confess I was simply ecstatic.

My joy turned to rapture when I filed away a second Discontinuity the second week of my employment. According to the project design[7], field work for this (necessarily unnamed) study was to and did begin in the Spring of 1959. After the sample had been selected, the research staff (two psychologists) and an odd assortment of M.D.s who were studying psychiatry interviewed the students in the Fall (and maybe even the Spring) of the 1959–60 academic year. Interviewing was then suspended until funds became available, that is, until April of 1961. The students were now completing their freshman year in college. With a rather admirable dash of temerity, this federal agency was sponsoring a study of the transition from high school to college of a sample of twenty students who were completing their first year of college at the same moment the funds were approved. Talk about Discontinuities!

Naturally, I began to more closely observe the man who had helped make all this possible—the Executive Director. Here was what Max Weber would have called a Perfect Ideal Type for the study of Discontinuities[8]. The Executive Director never; (1) wrote or read a research publication, (2) interviewed an interviewer or a research subject, (3) had the least idea what data was, how it was collected, what had been collected, or what was to be collected, (4) involved himself in administration, and (5) maintained a professional relationship.

During the summer a Social Psychologist put in an appearance. He had a Ph.D. from Harvard, had published several articles, and came highly recommended. This new Research Director, Project Director, and Co-principal investigator not only knew research and was determined to get at it, but was given *carte blanche* to do anything he wanted.

The new Research Director was first told to write up the data on another major project, which had already had its own Discontinuities. The other psychologist was asked to "secure some funds" to do major study. One research assistant and a part-time sociologist were left to carry on the study of the transition from high school to college.

The next nine months were fairly dull, that is, within the research staff. However, in the late spring the Research Director announced his resignation. Actually, he had resigned himself to the situation several months earlier. It was announced that the other psychologist was to be co-principal investigator and principal writer along with the announcement that he was leaving Puerto Rico! In February the fourth Discontinuity materialized when the nonresident psychologist's salary was stopped without notice.

The two secretaries in the Research Department and the only research assistant had departed sometime earlier, providing interesting, though minor Discontinuities. The research assistant was the only staff member (ever) who spoke fluent Spanish. Consequently, when others translated certain material, Spanish words like *molestar* (to bother, annoy, make or take trouble) were translated as *to molest*. This led a later investigator to wonder about the teacher who was said to molest his female students.

In the summer of 1962, the sociologist was told the project had "shifted away" from his specialization. When this did not produce the expected resignation, his salary was simply stopped. I now had a perfect case of Complete and Continuous Discontinuity. There was now no one who knew anything about the original design. No one who knew what changes in methodology had been made. No one who knew the blind alleys that had been explored. No one who knew how the subjects had been selected. No one who knew how the subjects had been interviewed. No one who knew who interviewed the students. No one who could separate fact, fiction, and interpretation. No one who would even recognize one of the research subjects. Every possible type of Discontinuity had taken place!

Rarely does an investigator have such good fortune as I have had. With this record there is really no point in continuing. I must also confess research on Discontinuities is a bit fatiguing. Now I've taken up stunt flying, sky-diving, and shark hunting; they're so wonderfully relaxing, by comparison, you know.[9] ■

---

[7] A curious document which I finally discovered behind a file cabinet in a deserted office. For some strange reason it had a habit of returning to such places and was located when strongly requested only with the greatest difficulty.

[8] Max Weber, From Max Weber: *Essays in Sociology*, trans. by H. Gerth and C. W. Mills, New York: Oxford University Press. 1946.

[9] The author must add that all references to persons and places have purposely been removed and an attempt made to conceal the identity of the true location of study. I must also add that any resemblance to persons living or dead is purely coincidental, and, of course, not intended.

VETERANS ADMINISTRATION HOSPITAL

Sam Jackson Park
Portland, Oregon

EMPLOYEE RECOGNITION COMMITTEE

Minutes for the March 12, 1969, meeting which convened at 2:00 p.m.

Present: Assistant Hospital Director; Chief, Dietetic Service; Chief, Medical Administration Division; Chief, Social Work Service; Nursing Supervisor; Chief, Buildings and Grounds; Budget Analyst; Section Chief, Processing and Records; Acting Chief, Personnel Division and Personnel Specialist

COMMENDATION MEMORANDA

Dietetic Service: Albert R. Kenney, Sr. commended for reporting to Hospital due to fire alarm on 11-11-68 at 3:30 a.m.

Bldg. Mgmt. Div.: Ronald W. Agee commended for work performance as related to additional work assignments and frequent changes of tours.

Mrs. Lee L. De Berry commended for work performance on ward subjected to constant construction and completion of additional duties.

SUGGESTIONS PROCESSED UNDER DELEGATED AUTHORITY:

S-775: Installation of hinged grills in main surgery area; disapproved.
S-876: Specimen identification stickers; disapproved.
S-892: Installation of acoustical tile in dishwashing area; disapproved.
S-905: Drainage ditch near pedestrian walkway: approved with modification; certificate.
S-936: Ward designation stickers on arm bands; disapproved.
S-952: Extra special foot care; disapproved.
S-955: Automatic rinser tank; approved with $25 award.
S-959: Installation of sign giving instructions to visitors for use of phone for information; approved with certificate.
S-962: Improved lighting to metered parking lot; disapproved.
S-972: Safety hazard – automatic doors; approved with certificate.
S-976: Combining prostate study patients with tumor registry; disapproval.
S-980: Elimination of parentheses; disapproval by C.O.
S-981: Pilot program for use of radiology technologists; disapproved by C.O.
S-982: Standard evaluation sheet; approval with $15 award ($7.50 each)
S-985: Improved lighting in halls, Bldg. 25; disapproved.
S-996: Eliminating hazardous conditions in crosswalks; approved with $15.
S-998: Increase availability to dressing room #335, Bldg. 25; disapproved.
S-1004: Microphone for checker; disapproved.
S-1005: Modification of suggestion to re-use patient's toilet tissue; approved with $15 award and certificate.
S-1009: Form "Request for Volunteer Service"; approved with certificate
S-1011: Drains for cart holding green trays used on tray line; disapproved.

Publishers: We are moving heaven and earth to find out how S-1005 was implemented.

Project Management

B. SPARKS*

From time to time, every professional journal publishes articles to aid their members in a particular field. Much has been written on Project Management; however, certain aspects are often overlooked. These are the practical, everyday, down-to-earth aids that make a project move. In an attempt to fill this void, the following three aids are offered to assist in obtaining irreproducible results.

I. PROJECT VOCABULARY. Knowing the terminology is the first key step in project management.

A PROJECT — an assignment that can't be completed by either walking across the hall or by making one telephone call.
TO ACTIVATE A PROJECT — to make additional copies and adding to the distribution list.
TO IMPLEMENT A PROJECT — acquiring all the physical space available and assigning responsibilities to anyone in sight.
CONSULTANT (or Expert) — anyone more than 50 miles from home.
COORDINATOR — the guy who really doesn't know what's going on.
CHANNELS — the people you wouldn't see or write if your life depended on it.
EXPEDITE — to contribute to the present chaos.
CONFERENCE (or Meeting) — that activity that brings all work and progress to a standstill.
NEGOTIATE — shouting demands interspersed with gnashing of teeth.
RE-ORIENTATION — starting to work again.
MAKING A SURVEY — most of the personnel are on a boon-doggle.
UNDER CONSIDERATION — never heard of it.
UNDER ACTIVE CONSIDERATION — the memo is lost and is being looked for.
WILL BE LOOKED INTO — maybe the whole thing will be forgotten by the next meeting.
RELIABLE SOURCE — the guy you just met.
INFORMED SOURCE — the guy who introduced you.
UNIMPEACHABLE SOURCE — the guy who started the rumor.
READ AND INITIAL — to spread the responsibility in case everything goes wrong.
THE OTHER VIEWPOINT — let them get it off their chest so they'll shut up.
CLARIFICATION — muddy the water so they can't see bottom.
SEE ME LATER ON THIS — I am as confused as you are.
WILL ADVISE YOU IN DUE COURSE — when we figure it out, we'll tell you.
IN PROCESS — trying to get through the paper mill.
MODIFICATION — a complete redesign.
ORIENTATION — confusing a new member of the project.
REORGANIZATION — assigning someone new to save the project.

II. TEN COMMANDMENTS FOR THE PROJECT MANAGER. These are the rules by which the Project Manager must run his project.

1. Strive to look tremendously important.
2. Attempt to be seen with important people.
3. Speak with authority, however, only expound on the obvious and proven facts.
4. Don't engage in arguments, but if cornered, ask an irrelevant question and lean back with a satisfied grin while your opponent tries to figure out what's going on — then quickly change the subject.
5. Listen intently while others are arguing the problem. Pounce on a trite statement and bury them with it.
6. If a subordinate asks you a pertinent question, look at him as if he had lost his senses. When he looks down, paraphrase the question back at him.
7. Obtain a brilliant assistant, but keep him out of sight and out of the limelight.

*Consulting Engineer

8. Walk at a fast pace when out of the office — this keeps questions from subordinates and superiors at a minimum.
9. Always keep the office door closed — this puts visitors on the defensive and also makes it look as if you are always in an important conference.
10. Give all orders verbally. Never write anything down that might go into a "Pearl Harbor File".

III. PROJECT PROGRESS REPORT. Below is a standard report that can be used by just about any project that has no progress to report.

The report period which ended .... has seen considerable progress in directing a large portion of the effort in meeting the initial objectives established.[1] Additional background information and relative data have been acquired to assist in problem resolution.[2] As a result, some realignment has been made to enhance the position of the project.[3]

One deterent that has caused considerable difficulty in this reporting period was the selection of optimum methods and techniques; however, this problem is being vigorously attacked and we expect the development phase will proceed at a satisfactory rate.[4] In order to prevent unnecessary duplication of previous efforts in the same field, it was deemed necessary to establish a special team to conduct a survey of facilities engaged in similar activities.[5]

The Project Control Group held its regular meeting and considered the broad functional aspects of all levels of coordination and cross fertilization of relevant ideas associated with the general specifications of the evolving system.[6] At the present rate of progress, it is believed that most project milestones will be met.[7] During the next quarter a major breakthrough is anticipated and will be fully covered in progress report No. ....[8] ■

[1] The project has long ago forgotten what the objective was.
[2] The one page of data from the last quarter was found in the incinerator.
[3] We now have a new lead-man for the data group.
[4] We finally found some information that is relevant to the project.
[5] We had a great time in Los Angeles, Denver, and New York.
[6] Fertilizer.
[7] Would you be happy with one or two?
[8] We think we have stumbled onto someone who knows what's going on.

Management by Partial Reinforcement

William Wokoun
B.W. Research Laboratories Inc.
P.O. Box 413
Aberdeen, Maryland 21001

Just how should a top executive manage his employees? How can he motivate them to do more work, and better work? Over the years, a parade of psychologists, personnel men, psychiatrists, and would-be philosophers have grappled with this problem, often eloquently but never effectively. These experts have not led us to the millennium they promised. What is more, they have exhorted us to set out in several different directions in pursuit of it. This sorry state of so-called scientific management has forced the harried executive to operate by blind, seat-of-the-pants intuition.

As so often happens, someone had already stumbled onto the real solution — partial reinforcement — but it lay buried and virtually unknown in the dusty annals of the psychological laboratory. Only now is it beginning to take its rightful place in the everyday worlds, as a dramatically effective management tool in offices and industry. Before we examine what partial reinforcement is, and how it works its magic, let us first ask why more conventional methods have failed so dismally.

According to the psychological brain trust, there are three ways to motivate an employee. You can praise him (reward, or positive reinforcement); you can reprove him (punishment, or negative reinforcement); or you can say nothing at all (no reinforcement). Even a moment's reflection is enough to show why none of these methods can be effective.

Suppose you are trying to motivate a group of salesmen. Should you praise them? Of course not. Only the best salesman deserves praise and, if you give it to him, he is only too likely to demand a raise or some other form of coddling. On the other hand, praising an unsatisfactory salesman will only encourage him to rest on his laurels. It is difficult to praise an unsatisfactory worker anyway, at least with a straight face; if he deserved praise, he would not be unsatisfactory. Praising everyone, regardless of what they do, suggests that you are not paying attention. Clearly, praise is not a viable approach.

At first glance, punishment seems to be a better solution, but it is not. Experiments have established that reproof has widely varying and unpredictable effects. There is little point in punishing your best employees, and little benefit from punishing your worst ones. When you do, you appear to be either a hopeless, impossible perfectionist or a merciless tyrant who cannot resist beating dead horses. Some employees reply with their own angry recriminations, a few brood quietly, and most of them merely tune you out and ignore you. Apart from the personal satisfaction that criticizing people may give you — an area, incidentally, that traditional management theory usually overlooks — fault-finding is mostly counterproductive. It does produce changes in behavior, but not the ones you want.

The third possibility — zero reinforcement, or saying nothing at all — is so difficult for most executives to follow, that it has never enjoyed much popularity. Nevertheless, it is interesting from a theoretical point of view. What would happen if you never praised or found fault with your employees, but merely maintained a noncommittal silence? According to proverb, they should continue to improve anyway. After all, practice makes perfect.

Unfortunately, that is not what happens. As the military once discovered, sheer practice by itself guarantees nothing but consistency. When people keep firing at a target, without knowing whether or not they hit it, they do become less variable — but, sadly, no more accurate. Stony silence, then, is the poorest bet.

The telling weakness that all these methods share is that they are rigidly predictable. They have no flexibility. The time was ripe for an eminent animal-learning theorist, Dr. Burrhus F. Skinner, to demonstrate the superiority of a more versatile approach to reinforcement. Skinner dusted off a procedure that had been discovered many years earlier by the dean of Russian psychologists, Ivan P. Pavlov himself.

Testing rats in his justly famed "Skinner box," Skinner trained them to press a bar for a reward of food pellets. But while traditional psychologists had always clung to the same dreary predictability — a pellet of food for each and every response — Skinner cleverly introduced an ingenious innovation. His rats did *not* get food every time they pressed the bar. Sometimes they were rewarded, but other times they were not.

Skinner called this procedure *partial reinforcement,* and his results revealed a dramatic improvement over ordinary learning. His rats learned almost as fast as if they got food every time they pressed the bar — and then the rats kept on pressing and pressing it, even when their efforts no longer produced any food at all! Ordinarily, removing the reward would end the responses — without food, animals would stop pressing the bar. Yet after partial reinforcement, the reward may have been taken away, but the bar-pressing response lingered on and on. It is recorded that one especially determined rat made nearly 200 responses to get a single pellet of food.

Here, obviously, was a technique worthy of further investigation. With just a fraction of the usual reward, it could evidently conjure up a sort of perpetual motion that kept rats responding indefinitely. Nevertheless, precious years trickled through the hourglass of time before anyone had the vision to grasp this breakthrough's true significance: that what works with rats can also be applied to people.

Today, this insight is ushering in a completely new philosophy of management that obsoletes previous approaches.

Using partial reinforcement effectively demands *random* praise and punishment. Admittedly, behaving randomly is difficult at first. It is not enough to do what you *think* is unexpected, or the opposite of what you did last time. These are uniformities in themselves, which can defeat your aims by robbing the technique of its freshness. Thus the group that has given you a progress report today will probably relax for the rest of the week, confident that you have no further questions for them yet. You must break through their apathy by suddenly announcing a reappraisal conference tomorrow, preferably just before or just after lunch.

Long random intervals are just as effective as short ones. If the group has not been called in for a briefing in a week or so, you may be sure they are making careful, nervous preparations for one. Since they are convinced they will have to account for themselves at any moment, they are already working feverishly anyway, and there would be little point in actually having a briefing now. It is best to wait. With every passing day, they feel the briefing grow more imminent, and they will work harder and harder. Appropriately vague reminders — "We've got to have another briefing soon" — can prolong this productive phase into an Indian summer that might last for weeks.

Meetings of your key executives are ideally suited to verifying that no one has come in late. It goes without saying that you should never announce these meetings in advance. five or ten minutes' notice allows more than enough time for most people to get to the conference room, thus preserving spontaneity and training them to think on their feet.

It probably will not jeopardize randomness if you usually convene such meetings without warning half an hour or so before work begins — provided the days themselves are chosen unpredictably. These meetings need not be long to serve their function; you will probably want to adjourn them shortly after the last lost sheep reaches the fold. You will soon begin noticing that all of your people get to work early, to avoid embarrassing you by entering after a meeting has begun. If you comment on this, frame your remarks so they convey partial reinforcement. Temper your praise with a mild admonition about overextended lunch hours.

You have now succeeded in infecting your subordinates with extra motivation to begin and end the day. How can you give them similar boosts in between? Aside from early morning and late afternoon, the times when they need your most help are just before and just after lunch. Once again, the key to solving the problem is judicious application of partial reinforcement. Fortunately, there are simple, yet very effective, ways you can help: by making courtesy telephone calls and personal visits.

At intervals during the day, select some individual randomly for your personal attention, and place a "courtesy" telephone call to him. He should answer immediately: "Yes, Mr. Smith?" The employee develops an uncanny way of sensing that you are genuinely interested in him. No matter how many other people *could* call him, somehow he will know that the one who *is* calling him is *you*.

At this point in the conversation, you have achieved your objective. You have demonstrated your concern through a simple but thoughtful bit of concrete behavior, and the employee knows you care. Rather than wasting the call by hanging up immediately, take advantage of this chance to kill two birds with one stone, so to speak. As for someone else. Say something like, "That you, Hank? How's it going? Listen, is Joe there?"

Whether or not Joe is actually there, you may be sure Hank will tell him you asked about him. In this way, a single telephone call serves to stimulate at least two people, and sometimes even more.

Let us analyze the roots of employee dissatisfaction. Nowadays most employees have too little work to do. Their monotonous jobs offer no real challenge or excitement. Small wonder that so many workers consider their jobs meaningless and trivial.

Your inspirational value will be greatest when subordinates see you as a stern taskmaster, basically fair but demanding. The reason is obvious: they find this image gratifying because it helps them understand how much they are accomplishing.

The simplest approach is assigning far more work than your employees can hope to finish. This tactic has merit, as far as it goes, but it is limited and unimaginative.

A better solution is enriching jobs by adding assignments in new areas. If you need a management analysis of your engineering department, do not assign an engineer to do it. Instead, give the job to someone who can contribute a fresh and unbiased point of view, such as your controller or your sales manager.

Teach your employees how to schedule their time so it is used to best advantage. Even after you enrich the controller's job by assigning him the engineering survey, you will find he clings to his old ways.

When it is his turn to receive one of your courtesy calls, ask how the engineer survey is progressing. A typical narrow-gauge controller will be forced to equivocate because, as your friendly discussion will reveal, he has been fiddling with income tax forms or something of the sort and has not begun the survey. You may rest assured that he *will* begin it, however, once you have focused his attention on it. So far, so good.

The next time he receives a courtesy call — this is very important! — remember the principles of partial reinforcement. *Do not mention the engineering survey!* Refuse to discuss it, if necessary. Instead, ask him how he is handling the depreciation allowance on your computer facility.

By now, the astute reader has grasped the basic point: help employees grow by making sure their initially narrow attention is focused sharply on every conceivable aspect of your business.

How effective is partial reinforcement? The technique is based on solid experimental findings, but applying it to personnel management is still a new frontier. Thus far, there has not been time to marshall a full-blown scientific validation.

Wherever objective, quantitative measures are available, the results have been uniformly positive. Statistical tabulations prove that management by partial reinforcement all but eliminates lateness, early departures, and absenteeism. Employees take shorter breaks and lunch periods. They make fewer visits to the water cooler and the rest room. Indeed, the well-motivated employee seldom leaves his desk at all, except in case of direst crisis.

There is likewise a remarkable improvement in office deportment. Telephones are answered after (or even during) the very first ring. Idle chatter virtually ceases. Time-sampling observations disclose that practically all employees are busy constantly, or seem to be, which is much the same thing. These are not the tired, sluggish employees one typically sees in an ordinary office. They are alive, vibrant and alert. Anyone who doubts it has only to clap his hands or drop a pencil, then observe the immediate and vigorous response.

In fairness one must admit there have been a few isolated reports that partial reinforcement may produce negative effects under certain circumstances. These criticisms are largely subjective, unsupported by any meaningful data. They probably reflect only a predictable, tradition-bound reluctance to accept forward-looking management methods.

Some critics have charged that partial reinforcement increases turnover. But there is a semantic problem in defining turnover: turnover is not merely the crude number of people who leave their jobs. Properly speaking, turnover refers to the loss of capable, well-motivated employees. It does not encompass such things as weeding out the weak links.

Remember that the employee who leaves is really acting out his psychological admission that he is not a good or desirable employee. Nine times out of ten, you will be forced to agree with him if you reevaluate him retrospectively after he has departed.

Finally, it has been claimed that partial reinforcement causes nervous breakdowns, ulcers, heart attacks, and other vaguely defined physiological ills. These statements are so nebulous as to be practically meaningless, from the scientific point of view.

Throughout recorded time, people have always had coronaries, ulcers, and the like. Since sound bodies and sound minds go together, these physical weaknesses all too often betray the weak mentality of an unmotivated individual — the incapable underachiever who is not doing his share.

Such unfortunates are to be pitied, but they should not be allowed to blame partial reinforcement for their troubles. They will, naturally, be happier and generally better off under enlightened personnel management. Nevertheless, even partial reinforcement can scarcely hope to rescue them from the ravages of basically weak constitutions and, in all probability, their lifelong exposure to an increasingly polluted environment.

Overall, the preliminary results have been very encouraging. As a modern technique for tailoring management methods to human abilities and limitations, there is bright promise for management by partial reinforcement. ■

Corporate Biorythms[1]

Robert A. Steiner
El Cerrito, CA

Many people now totally understand and accept the profound influence that biorhythms have upon their personal lives.

By way of introducing you to the concept of corporate biorhythms, it seems reasonable to present a review of the generally accepted concepts of biorhythms for people, or people-biorhythms, or, as we refer to them in our tiny scientific community, PB's.

The reason that our scientific community on this topic is so tiny is that, while we were discovering and testing the ideas, it seemed to be a good idea to keep our research a secret. Since this might appear to be a somewhat selfish idea, an explanation is in order.

All of the marvelous benefits you derive from consulting your PB before making any move in this life inure solely to your benefit, and, of course, to the benefit of those with whom you deal. There is no competition on that score.

However, with CB's (that's corporate biorhythms) the case was different. We were making a study of the earnings of corporations, as well as fluctuation of the prices of the stocks on the open market, and correlating them with our ever-increasing knowledge of CB's.

This enabled us to predict with considerable accuracy the future earnings of companies, as well as the future prices of the stocks.[2]

Armed with this knowledge, we were able to make shrewd investments in the stock market, realizing literally tens of millions of dollars from our study.

Unlike PB's, had the information about CB's gotten out[3], everybody would have jumped upon the corporate biorhythm bandwagon, and our precise investments would not have been unique. Thus, in order to protect our investments, secrecy was the watchword.

Now, having amassed reserves in the high eight figures, we feel that it is our altruistic duty to let the investing public in on this latest scientific-financial CON.

As with any new scientific or financial discipline, there is unavoidably a new lexicon of terms, abbreviations, and acronyms which must be learned by the student. The term "CON," as used in the previous paragraph, stands for "centralization of knowledge." The final letter "n" was used, rather than "k," for the following two reasons:

1. The "k" in "knowledge" is silent, so that the acronym "con" corresponds to the phonetic sound of the words for which it stands.

2. Initially we used the term "COK"; however, we found that periodically prurient levity interrupted the serious study at hand.

Your PB commences on the day of your birth. Actually, you have three PB's which continue in repeating cycles throughout your life: a physical PB of 23 days, an emotional PB of 28 days, and an intellectual PB of 33 days.

The CON on PB's is that these cycles are immutable throughout your entire life, regardless of traumas, illnesses, accidents, or statistics which conclusively prove the opposite conclusions.

Well, it occurred to us that corporations must also have biorhythmic cycles. We tested, and sure enough, they do!

It was no simple task to ascertain the components of CB's. CB's are far more complex than PB's. PB's rely upon birth date only, and nothing more.

CB's, on the other hand, are based upon a blending of final formulas (that's "BOFF") of the PB's of the Board of Directors, top management, and the assistant shop foreperson of the corporation. It also takes into account the "birthdate" of the corporation, which is, of course, the date of first filing (that's "DOFF") with the appropriate state official.

In brief summary and review, we see that after we do our DOFF with a state official, we can have a BOFF. And from the BOFF comes the CB. And with the CB, we can make a fortune, and avoid disaster.[4] What a CON!

Had the general investing public been cognizant of our CON prior to the stock market crash of 1929, there would have been no crash!

Using CB information for that period, which we have, with the benefit of hindsight, reconstructed, the exact day of the impending calamity would have been luminously evident *in advance*. All the signs pointed to it!

Armed with the foreknowledge of foreseeing the crash, everybody in the country could have sold all of their stock precisely eight days *before* the crash, thereby avoiding the crash.[5]

Space and prudent business judgment dictate that we disclose no more in this article.

For further information, including step-by-step methods for computing the BOFF for any corporation, please send us a certified check for $1,000, and we will disclose to you our complete CON. ■

1. All of the supporting data for this article are considered so important that they are included in the body of the article. Consequently, there are no footnotes.
2. Refer to Chart I—Chart of Perfect Financial Mesh.
3. The grammatical construction is adequate. "Gotten—past participle of 'get', as in 'had the information about CB's gotten out'."
4. Refer to Chart I—Chart of Perfect Financial Mesh.
5. Refer to Chart I—Chart of Perfect Financial Mesh.

CHART I. Chart of Perfect Financial Mesh

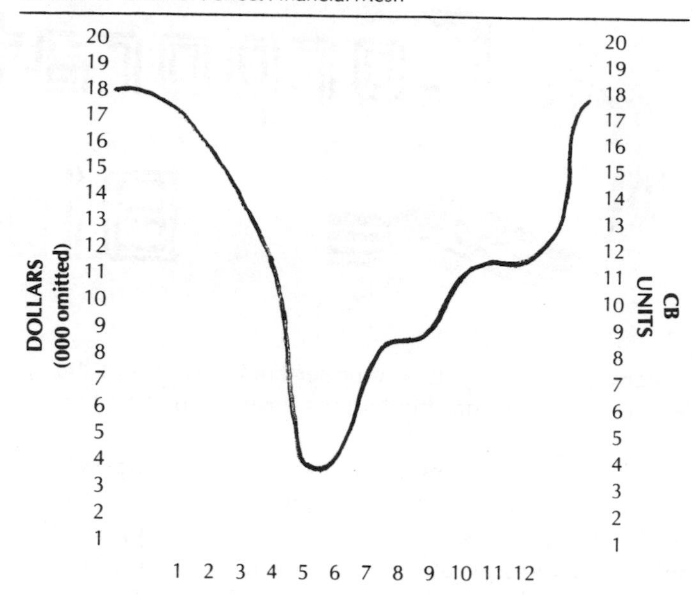

KEY:
---Market Price of a Randomly Selected Stock.
---Corporate Biorhythm for the Same Company.

You will notice the appearance of only one solid line. This is actually the perfect meshing of the two dotted lines: one for the Market Price of a Randomly Selected Stock, and the other for the Corporate Biorhythm for the Same Company, which, as stated above mesh perfectly.

The application of the Corporate Biorhythm CON invariably leads to a perfect financial mesh.

D. V. Nalivkin
THE GEOLOGY OF THE USSR (Translated by S. I. Tomkeieff)
Oxford Univ. Press, 1960, p. 153.

". . . Great progress has been made in the study of the bowels of our country . . . Great and prolific is our motherland and much wealth is hidden in her bosom".

Chernyshev, V. I.
ADAPTATION OF MONKEYS TO THE CONDITIONS OF MOSCOW SUBURBS
Int. Congress Anthropol. Ethnol. Sci. 1968, 7(3), 216–219.

(In Central Moscow they are probably already adapted.)

* * *

LUMPS FROM A LUMP SUM SETTLEMENT

Sylvan D. Schwartzman
Albuquerque, NM

THE PATRICK HENRY INSURANCE COMPANY
"Your Lifelong Friend with a Heart"

File No. 839-750-299-9173AL
In correspondance please
refer to this file number

January 2, 1980

Prof. Clarence Already
Custer University
Sky Valley, Idaho

Dear Prof. Already:

Congratulations! After 40 years of participation with Patrick Henry, your retirement annuity is now payable. We note your desire for a lump-sum settlement.

Kindly forward the following promptly so we may process your claim speedily:

(1) Form 27385 (Request for Lump-Sum Settlement) to be filled out and notarized. When you have completed all 18 pages in quadruplicate, please also have your wife, children and grandchildren (if any) sign each copy.
(2) Three (3) certified copies of your birth certificate and marriage license, together with two (2) recent photographs of you and your wife.
(3) One (1) set of fingerprints, obtainable from the FBI or local police department.
(4) Two (2) validated copies of your driver's license and Social Security card.
(5) Forms 1382 (Physical Examination Report) and 1383 (Dental Examination Report) to be completed in full by your physician and dentist. All recent X-rays, cardiograms, CAT-scans and test results should accompany same.
(6) If you have ever been involved in litigation, provide all details, including the names, addresses, phone numbers of your attorneys.

We look forward as always to serving you.

Sincerely yours,

Harold Skinner, President

Koerner's MY PARENTS, Mr. and Mrs. Charles Brisk

File No. 839-750-299-9173AL

May 15, 1980

Dear Valued Client:

In response to your inquiry, please know that your claim is being processed. However, at this time we require a list of all your home addresses since Jan. 1, 1940, including zip codes.

Kindly send same by Registered Mail.

CLAIMS DEPARTMENT

By: Eloise Skinner, Head

File No. 839-750-299-9173AL

Mar. 22, 1981

Dear Prof. Already:

Your reminders of Dec. 8, 1980 and Feb. 13, 1981 were really not necessary since we are moving right ahead with your settlement.

Unfortunately, in transferring our offices to the new Patrick Henry Tower, your file was misplaced, but we have since located same.

Thank you for your understanding.

ACTUARY DEPARTMENT

By: Herman Skinner, Chief

File No. 839-750-299-9173AL

July 6, 1981

Prof. Clarence Already
Custer University
Sky Valley, Idaho

My Dear Professor:

I am informed of your two long-distance calls, plus your telegram. Our company has been working overtime on your settlement. Unfortunately, we have had much trouble with the operation of the elevators in our new building, and many of our employees are exhausted by the time they reach our offices on the 104th floor.

Nevertheless, the final calculations of your lump-sum settlement should be ready within a very short time. Because of prevailing interest rates the amount may vary somewhat, but that is nothing to be concerned over.

I regret any inconvenience the delay may have caused.

Sincerely yours,

Harold Skinner,
President

PS. Please be advised that our company does not accept collect calls.

File No. 839-750-299-9173AL

October 22, 1981

Professor Clarence Already
Custer University
Sky Valley, Idaho

Dear Professor Already:

Responding to your numerous communications to date, I am pleased to inform you that your settlement check will shortly be in the mail. We are sure you will be more than satisfied with the amount.

Meanwhile, should you be anywhere in our vicinity, I do hope you will take the time to visit the new Patrick Henry Shopping Complex surrounding the Tower. be sure to inform the clerks that you are one of our valued clients.

I know you appreciate all our efforts to be of help.

Sincerely yours,

Harold Skinner, President

File No. 839-750-299-9173AL

Feb. 3, 1982

Dear Professor:

We could not understand why you had not received your settlement check, but we have finally located same, together with our letter of Dec. 18, 1981. It was stuck in the mail chute at the 73rd floor.

Your understanding is greatly appreciated.

ACCOUNTING DEPARTMENT

By: Harold Skinner, Jr., Supervisor

File No. 839-750-299-9173AL

March 22, 1982

Professor Clarence Already
Custer University
Sky Valley, Idaho

Dear Professor Already:

I was pleased to learn that you received your lump-sum settlement check, but I am disturbed by the tone of your remarks to our Chief Executive Officer.

A thorough review of your account shows the amount of the check to be absolutely correct. You must certainly realize that all you paid in was $1,000 a year for merely 40 years, and in our industry the settlement you received is considered far above average.

Please know that I will be personally happy to answer any questions you might still have.

Sincerely yours,

Harold Skinner, President

File No. 839-750-299-9173AL

August 17, 1982

Professor Clarence Already
Custer University
Sky Valley, Idaho

Dear Professor:

It is really not necessary to phone us weekly. As our Chief Executive Officer explained, the high altitude of our offices interferes with the operation of our computers and has delayed the preparation of the accounting you requested.

However, we are now in a position to supply you with the information:

LUMP-SUM ACCOUNTING
Professor clarence Already
File No. 839-750-299-9173AL

40 annual annuity payments $1,000 per year . . $ 40,000
Less front-end load investment charge 4% 1,600
Total Investment . . . . $ 38,400
Plus accumulated interest at annual average of 3% . . 1,152
Account Balance . . . . $ 39,552
Less 2-year inflation factor since annuity was due:
(14% for 1980, 11% for 1981) = Total of 25% 9,888
Purchase Cost of Annuity . . . . $ 29,664
Less 40.8% discount for Lump-sum option 12,103
Lump-Sum Principal . . . . $ 17,561

Less annuity conversion rate 18.3% (short-term interest rate, Aug., 1981) 3,214
Lump-Sum Value $14,347

Less cash withdrawal charge 4% 574
Total Settlement Amount . . . . $ 13,773
(Check #8269173)

I trust you are now satisfied that your settlement check was correct.

Sincerely yours,

Harold Skinner, president

PS. Unfortunately the company cannot reimburse you for the expenses you incurred in connection with your claim.

(telegram)

Oct. 13, 1982

Prof. C. Already
Custer U.
Sky Valley, ID

Responding to your latest call. STOP. Further review has uncovered error in your account. STOP. Patrick Henry neglected to apply additional 3% present-value discount rate over 40 years. STOP. Your $13,773 now reduced by .3066 or $4,223. STOP. Your remittance of this amount due within ten days. STOP.

Harold Skinner, Pres.,
Patrick Henry

■

Edward H. Katz
Barrington, R.I.

The Central Intelligence Agency (CIA) has recently declassified a number of intercepted[1] memoranda between the offices of various Soviet heads of state and their KGB counterparts. These notes shed some light on aspects of Soviet intelligence activity within the Military-Industrial Complex[2] of the United States over a 20 year period. Five of these translated[3] memoranda are presented here.

OFFICE OF THE CHAIRMAN CCCP

29 October 1962

TO: Ivan Stroganoff (KGB)
FROM: Boris Metinicki
SUBJECT: Disruption of U.S. Technology

My Dearest Ivanovitch,

The Chairman has indicated to me that the USSR has just suffered a humiliating defeat at the hands of the Americans over missiles in Cuba. He does not intend to let it happen again! So he is ordering a massive military building program (Army, Navy, Air Force). this will necessitate that our country develop a technological capability inferior to none. However, he also would like to destroy the technological capability of the United States.He gets so mad he wants to pound their technology into the ground with his shoe. These Americans are tremendous innovators who can react very quickly with new ideas for any military situation imaginable. Ivan, do you have any suggestions?

Boris Metinicki

OFFICE OF THE COMMISSAR KGB

27 November 1962

TO: Boris Metinicki (CCCP)
FROM: Ivan Stroganoff
SUBJECT: Operation Beany

Dear Niki,

Received your note of 29 October 1962. Not to worry; with the approval of my superiors, I have the situation well in hand. I have initiated "Operation Beany" which will paralyze the U.S. Military-Industrial Complex. It gives me goosebumps just to even think about it. One day they will wake up and find that they are so constipated with red-tape (that's a joke, Niki), accounting procedures, and excess paperwork, that nothing gets done productively. I believe they call it bean-counting. I have a sleeper in Washington, D.C. who will be our operative at the Pentagon. He very successively executed Project EDSEL which is a forerunner of what is to come. Talk to you at the next meeting of the Presidium.

Ivan Stroganoff

OFFICE OF THE CHAIRMAN CCCP

10 July 1972

TO: Uri Borsht (KGB)
FROM: Nicolas Lensky
SUBJECT: Disruption of U.S. Productivity

Comrade Borsht,

My predecessor, Boris Metinicki, enlisted your office to institute "Operation Beany," which is being carried out successfully in the United States. The results are even more astounding than even your predecessor, Ivan Stroganoff, envisioned. It takes years to get any new technology into their military inventory because of mountains of paperwork. The staff at the Pentagon keeps increasing in size and tries to invent new accounting methods to hamstring and harass their industry. They have budget reviews, cost reviews, program reviews, configuration management reviews, logistic support reviews, reliability reviews, and even a few technical design reviews. However, our current Chairman has conveyed to me his dismay in that the Americans can still outproduce us with technological inventions. Uri, do you have any suggestions?

Nicolas Lensky

By the way, Uri, regards to your wife, Helga, on her birthday.

OFFICE OF THE COMMISSAR KGB

12 August 1972

TO: Nicolas Lensky (CCCP)
FROM: Uri Borsht
SUBJECT: Operation Bells and Whistles

Comrade Lensky,

The KGB is aware of your concern, as expressed in your memorandum of 10 July 1972. I have recently initiated "Operation Bells and Whistles" which will work directly on negating the United States' ability to provide their military with good (my friend Admiral Gorshkov's philosophy) technological weapons instead of the best. This new Operation will complement "Operation Beany" to provide, I hope, synergistic results. Since the Americans are enamored with

technology, gadgets, and the like, we will try to convince them that all their weapons systems must be the best. The best meaning that it must have everything that money can buy. This operation is so sensitive that I cannot now divulge the name of our operatives in Washington, D.C. to you, Nicolas, or even to the Chairman. (My wife says thank you.)

Uri Borsht

OFFICE OF THE CHAIRMAN CCCP

21 December 1982

TO: Dmitri Zil (KGB)
FROM: Lev Petrovich
SUBJECT: Continuing U.S. Counterproductivity Efforts

Comrade Zil,

Our lately deceased Chairman (may he rest in peace) was greatly interested in "Operation Beany and Operation Bells and Whistles." Our new Chairman maintains similar interests. You, of course, must by now be aware of these counterproductivity efforts in the United States. They have been remarkably successful, as can be seen from the U.S. ships sailing without suitable armament (only missiles and pop guns), rapid-fire guns that shoot at islands and expensive tanks that break down. The Americans themselves keep inventing new ways to increase their procurement cycle, the latest being an accounting system called Department of Defense (DOD) Instruction 7000.2. Besides, they can't afford to buy in quantity any of the systems that do manage to get out of design to production.

The Americans have also developed what they call consultants. Sometimes these consultants are referred to as "Beltway Bandits." (The beltway is a circular road around Washington, D.C., Dmitri). I don't know what these Beltway Bandits do, but I think they are like beanies! In fact, most American defense establishments and laboratories are surrounded by these highway helpers. Dmitri, my friend, the bean-counters and the consultants now outnumber their working scientists and engineers by ten to one. But, alas, our new Chairman has confided to me that the Americans can still produce technological marvels like the F-14, F15 and F16 fighter planes. Dmitri do you have any suggestions?

Lev Petrovich

■

[1]The agent was a 65 year old cleaning lady who emptied wastepaper baskets in the Kremlin.

[2]As defined by President Dwight David Eisenhower

[3]Several other intercepted memoranda have not been published, since they were only reminders for stopping off on the way home for a quart of milk or having the front end of the car aligned.

A photomicrograph of a "question mark" photographed in the electron microscope and enlarged here to 38,000x. The subject is in fact a platinum-carbon replica of a polishing scratch in the epoxy resin embedding some grains of moon rock from Apollo 12 (sample number 12057). When studying these replicas this clearly fortuitous and irreproducible punctuation mark was quite appropriate as I was pondering the structures in the mineral grains themselves.

Kenneth M. Towe, Geologist
Nat. Museum of Natural History.

The

sometimes

fickle

handmaiden

of the

sciences:

MATHEMATICS

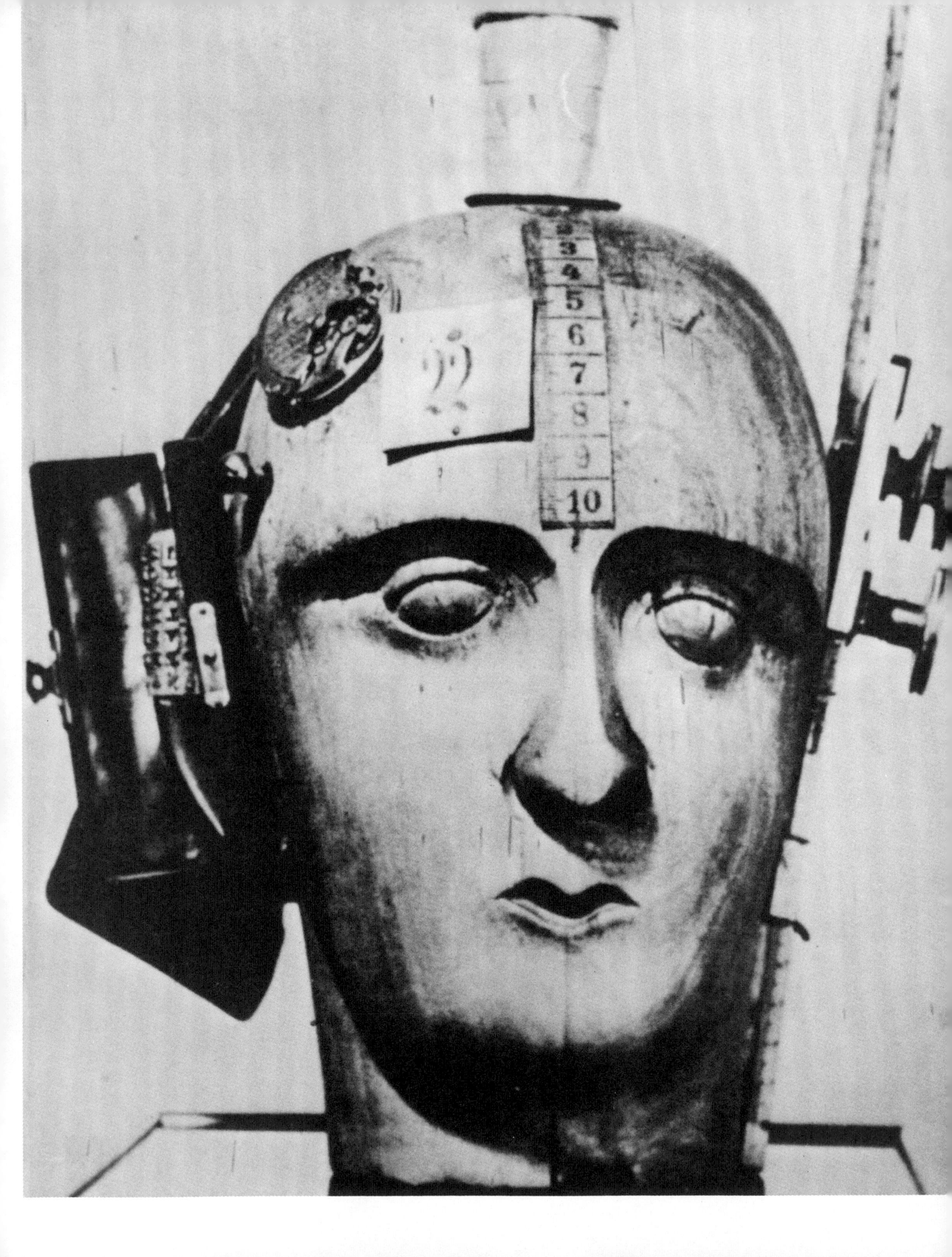
3
4
5
6
7
8
9
10

THE DATA ENRICHMENT METHOD*

HENRY R. LEWIS

The following remarks are intended as a nontechnical exposition of an interesting method which has been proposed (not by the present author) to improve the quality of inference drawn from a set of experimentally obtained data. The power of the method lies in its breadth of applicability and in the promise it holds of obtaining more reliable results *without recourse to the expense and trouble of increasing the size of the sample of data.* The method is best illustrated by example. Two such examples are outlined below; the first is somewhat routine, but the second is a striking illustration of what "data enrichment" can achieve.

Consider an experiment performed to test the ability of a specific sound receiver to detect an audio signal. The experiment is performed in such a way that in each of a series of trials one learns either that detection was accomplished or that it was not accomplished. Suppose, moreover, that the sound source and the receiver are fixed in space and trials are made with the source intensity set at six different levels. At each of the six source intensity levels a number of tests are made and the result, detection or no detection, is recorded. The data from such an experiment are summarized in Table 1.

sity. Using these simple facts, the data collected at one source level can be used to add to the data available for other levels. For example, looking at Table 1 we see that three of the trials made at a source level of 77 db resulted in no detection. These trials would also have led to no detection had the source level been at 62 db. Consequently, we can add the results of these experiments to our body of knowlege about 62 db *since we know how these experiments would have come out had we performed them.* Similarly the five trials made at 62 db and resulting in detection would certainly have resulted in detection had the signal been as high as 77 db at the source. Thus five more trials resulting in detection can be added to those actually made at 77 db. Treating all the data in this fashion, we can compile Table 2.

Two things are apparent at once: the probabilities of detection given in Table 2 are quite different from those which might have been deduced crudely and directly from Table 1; in addition the number of "virtual" trials at each level of source intensity is much larger than the actual number of trials. Hence one may be more confident of the results of Table 2 than of any results one might get directly from Table 1.

TABLE 1.

Raw data

Source level (db)	Number of detections	Number of failures to detect
62	5	40
65	10	30
68	15	20
71	20	10
74	25	5
77	30	3

It is desirable, of course, to increase the amount of data available at each source level. It is reasonable to assume that detectability is a function of source level and that, if all other parameters are held constant, a loud sound is easier to detect than one of smaller intensity. Thus it is safe to assume that if a signal was detected at a given level, it would have been detected at all higher source intensity levels. (The electronics are not such that overloading of the receiver would prevent detection.) Moreover, if a signal was not detected at a given level, it would not have been detected at any lower level of source inten-

A second example, even simpler than the first, should make the advantages of this method of analysis quite clear now that the details are fixed in the reader's mind. It has been known to those interested in psycho-physical phenomena that a man's tendency to flip a coin in such a way that when it lands he will be faced by a head rather than by a tail increases with the altitude at which the experiment is performed. The effect is small but a vast

*Reprinted with permission of the Editor from Operations Research, 1957, *5*, 551.

TABLE 2

ENRICHED DATA

Source level (db)	Number of virtual detections	Number of virtual failures	Probability of detection
62	5	108	5/113
65	15	68	15/83
68	30	38	30/68
71	50	18	50/68
74	75	8	75/83
77	105	3	105/108

number of trials conducted on Mount Everest, from base to summit, have shown that the effect indeed exists. With due respect to the hardy band of men who invested so many years and Sherpas in this effort, it is of interest to show how the same result can be obtained by one man with no more athletic ability than that required to climb a flight of stairs and no more equipment than an unbiased nickel. Our advantage over the pioneers in this field lies, of course, in our knowledge of the "enriched-data" method.

Consider a set of stairs with ten levels and number them in the order of their increasing altitude. The experimenter climbs the stairs, slowly, and at each level flips a coin ten times and records a head as a success and a tail as a failure. The results of an actual test are recorded in Table 3.

The results of Table 3 are not conclusive. The altitude effect may be present but is not evident, at least to a naive observer. Suppose we now attempt to increase the data available by recourse to logic in the manner already illustrated in the first example. The altitude principle tells us that if a trial on the first step resulted in a head, then it would certainly have resulted in a head if the trial had been made at the loftier tenth step. Similarly, if despite the height of the tenth step a trial made there resulted in a failure to throw a head, then the same trial would surely have been a failure on the lower steps. Using this added insight, the data can be enriched by a large number of virtual trials as is shown in Table 4.

A glance at Table 4 shows that the altitude principle, which was skulking almost unnoticed in the raw data of Table 3, has been fully brought forth by the data enrichment method. The probabilities in Table 4 are shown in Fig 1 to further emphasize the point. It might be mentioned in passing that the altitude effect in the Pentagon

TABLE 3

RAW DATA : COIN EXPERIMENT

Step number	Number of successes	Number of failures
1	4	6
2	5	5
3	7	3
4	4	6
5	6	4
6	5	5
7	6	4
8	6	4
9	3	7
10	4	6

TABLE 4

ENRICHED DATA : COIN EXPERIMENT

Step number	Number of virtual successes	Number of virtual failures	Probability of throwing a "head"
1	4	50	4/54
2	9	44	9/53
3	16	39	16/55
4	20	36	20/56
5	26	30	26/56
6	31	26	31/57
7	37	21	37/58
8	43	17	43/60
9	46	13	46/59
10	50	6	50/56

FIGURE 1 . ALTITUDE EFFECT IN THE PENTAGON

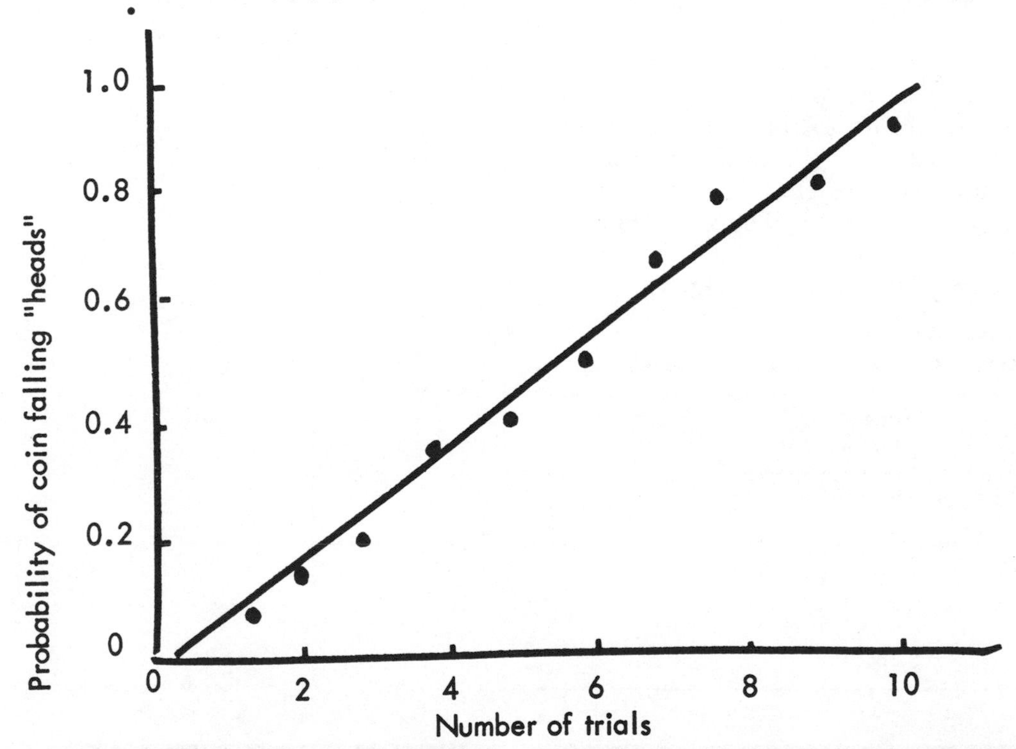

appears to be 10^5 times as large as that found in the Himalayas. Whether this is a temperature effect, a geographical effect, or the result of psychical factors as yet unknown should be the object of further study.

A final remark on the strength and weakness of the method is in order. As mentioned earlier, its strength lies in its breadth of applicability, and the method is as pertinent to experiments in classical physics as it is to experiments in psychical phenomena. In short, the method will give new meaning to data quite without regard to the status of the hypothesis used to increase the sample size.

Despite its evident power, however, the method requires further study. Its principal shortcoming is that before the enrichment process can be started, some data must be collected. It is quite true that a great deal is done with very little information, but this should not blind one to the fact that the method still embodies the "raw-data flaw." The ultimate objective, complete freedom from the inconvenience and embarrassment of experimental results, still lies unattained before us. ■

========

Measuring The Primadona Factor For Odd Numbers . . .

Y. RONEN et al.
Department of Experimental Mathematics
Beer Sheva, Israel. P.O.B. 9001

Recently Arbinka[1] proposed a revolutionary theorem in the field of falstional analysis. According to this theorem (Appendix A) all odd numbers are primary numbers. Due to the importance of this discovery our group at the department of experimental mathematics has proposed an experiment to verify this idea.

In our experiment we measured the amount in which odd numbers differ from being primary. The Primadona factor P is defined as,

$$P = \frac{\text{primary number}}{\text{odd number}} \qquad (1)$$

In other words, P is the factor by which you have to multiply an odd number in order to get the closest primary number. According to the Arbinka theorem this factor has to equal 1. In the experiment a large domain of odd numbers were chosen, namely all the odd numbers between 1 and 13. The temperature at each measurement was carefully checked. The results obtained are given in Table 1 and in Fig. 1.

TABLE 1

THE PRIMADONA FACTOR FOR ODD NUMBERS

odd number	Primadona factor	Temp °C
1	0.99±0.1	20.4
3	0.98±0.1	19.8
5	1.03±0.1	19.0
7	1.05±0.2	22.1
9	0.1 ±0.1	40.0
11	0.97±0.1	21.5
13	0.88±0.2	25.0

In Fig. 1 we see an excellent agreement between our experiment and Arbinka's theory. The only difference occurs for the odd number 9. A possible explanation for this difference is the effect of temperature. Table 1 shows that all the measurements, besides the number 9, were taken at about the same temperature, whereas the measurement of 9 was taken at the high temperature of 40°C, which might explain the difference. We are now designing equipment for measuring the dependence of the Primadona factor on temperature in order to get a conclusive explanation of the anomality of the number 9. ■

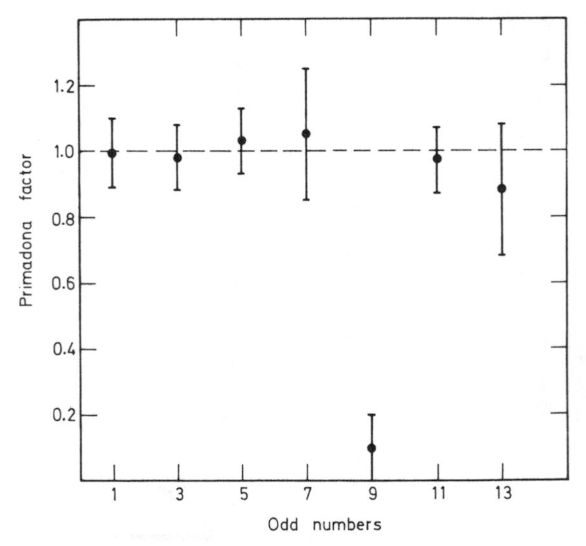

Fig 1. The dependence of the primadona factor for odd numbers

REFERENCES

[1]ARBINKA: Private Communication (1972)
[2]ibid.

ACKNOWLEDGEMENTS

The authors are indebted to many people for useful comments and smiles.

APPENDIX — A

Theorem: All the odd numbers are primary numbers.

Proof: One is primary number
Three is primary number
Five is primary number
Seven is primary number.

Thus using the induction technique every odd number is primary.

Advanced Applications Of The Theory Of The Bar Chart

PICROCOLE RASHCALF, et al.

In a report prepared for the U.S. Department of Labor[1] the following quotation is found:

> "Fiscal Year 1975 was not included in Figure 2 because the graph was not long enough to show the projected state and local funding for the four projects combined."

Fundamental to the performance of basic irreproducible research is an understanding of the basic theory of the bar chart. We find, however, dangerous lacunae among our fellow scientists in their apprehension of this basic statistical skill. It is regretable that the authors found themselves in such a pickle for FY 1975 was a vintage year for local, state and Federal statistics on vocational education. Even simple means and standard deviations based on FY '75 data have been known to throw the casual statistical analyst of these data into paroxysms that border on the orgasmic.

Had the authors been truly interdisciplinary they would have known, for instance, that econometricians had long since cracked the statistical problem they were confronted with. The technical name for the solution of Miller and Miller's problem is the "Procrustean Schtick." One either lengthens or widens the page (the procedure is perfectly general) or one changes the scale on the bar. Corollaries to the Procrustean Schtick involve several dimensions. If one also has a large number of bars as in the case of M-M, where they were confronted with a fixed page size, fixed scale, but an almost infinite historical record to display, one can narrow the width of the bars and reduce the type size. Or, since six years of data were reported, FY 1969 could have been dropped, thus creating room for FY 1975. The only problem remaining could then have been solved by dropping the analysis of State and local funding and reporting only on Federal funds, which would have easily fit the page.

One expedient, but not a very professional response, would have simply been to omit the sentence quoted above, whereupon the reader would have likely assumed that FY '75 data were not available. On the chance that FY '75 data are known by one or more additional colleagues to be available, the authors could have pleaded lack of resources necessary to publish and discuss FY '75 data or to buy the required larger size paper. In fact, the authors had almost hit on this solution on page 1:

> "The writers of this report did not intend to present a complete exposition of all the major projects . . . It would not be possible within any reasonable time and cost allowance."[2]

But, apparently, this train of thought was lost in the shuffle between pages 1 and 3, a common problem in jointly authored papers.

Let us assume, however, that none of these alternatives were acceptable to M-M. The Procrustean Schtick proves inoperative. The solution here, then, is a fail-safe known as the "Phallic Schtick." This simple but elegant technique merely involves a bending of the bar up along the vertical right hand margin (See Figure 2). Often, depending on the dimensions of the page and the placement of the figure on it, one can double or quadruple the amount of data to be displayed. One can either have a smooth, twice differentiable curve at the bend or a simple right angle. Either is correct.

Finally, if one is truly constrained by the scale on the horizontal axis it is possible to generalize the Phallic Schtick to take advantage of the fact that every page has four margins. This is known as "Coil-Relational Analysis"[3] (See Figure 3). Depending on the thickness of the bar and the size of type (given the margins, of course) one can wind the bar around the page margins literally millions of times, constrained only by the fact that the real number set ends at plus infinity. This, of course, represents the very apex of bar chart theoretical development. Quod erat demonstrandum.

---

[1]Robert Miller and LaRue W. Miller, *Impact of Vocational Education Research at the State and Federal Levels: Project Baseline Supplemental Report*, Project Baseline, Northern Arizona University, Flagstaff, Arizona, October 25, 1974, p. 3, paragraph 2, lines 1, 2 and 3.

[2]*Ibid*, p. 1, paragraph 3, lines 1, 2, 3 and 4.

[3]Coil-Relational Analysis is derived from "Coillon Analytics," which, in turn was developed by Chaucer in his memorable *Canterbury Tales*. The particular quote on which the theory is roughly based is "Your Coillons Shal be enshrinéd in a Hoggés Tordé."

Figure 1: Basic and Deviant Bars

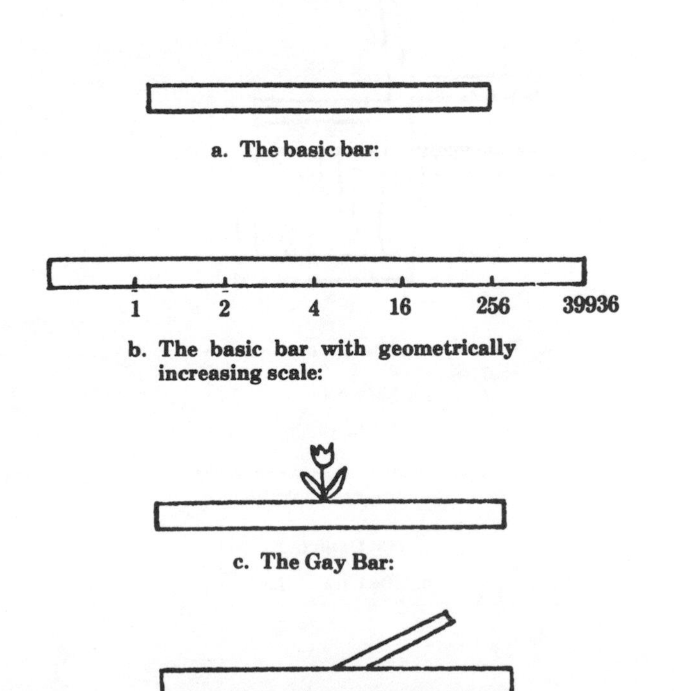

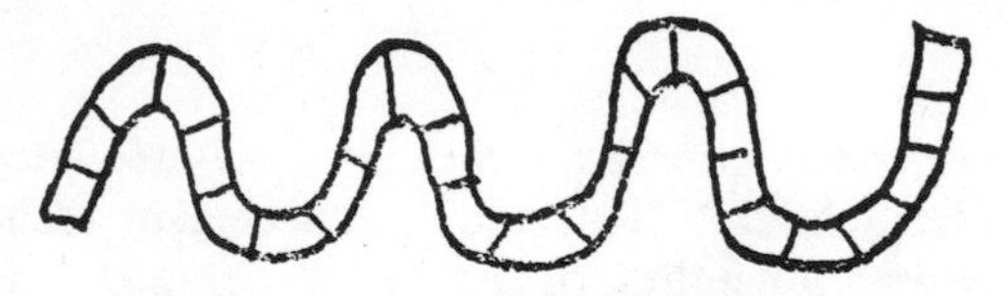

"Don't tread on me"

e. **The sine wave bar for short or narrow pages:**

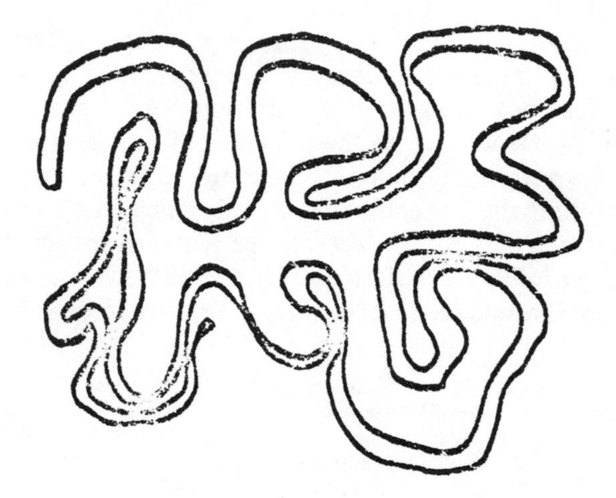

f. **The convoluted bar for small pages and tightly enclosed spaces:**

g. **The polygonal bar for multivariate analysis:**

Not Depicted

h. **The Grizzly Bar:**

Figure 2: The Phallic Schtick

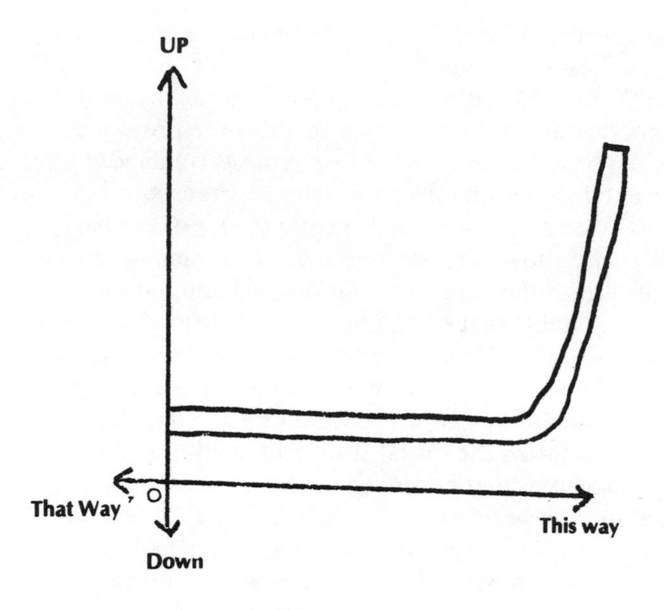

Figure 3: Coil-Relational Analysis

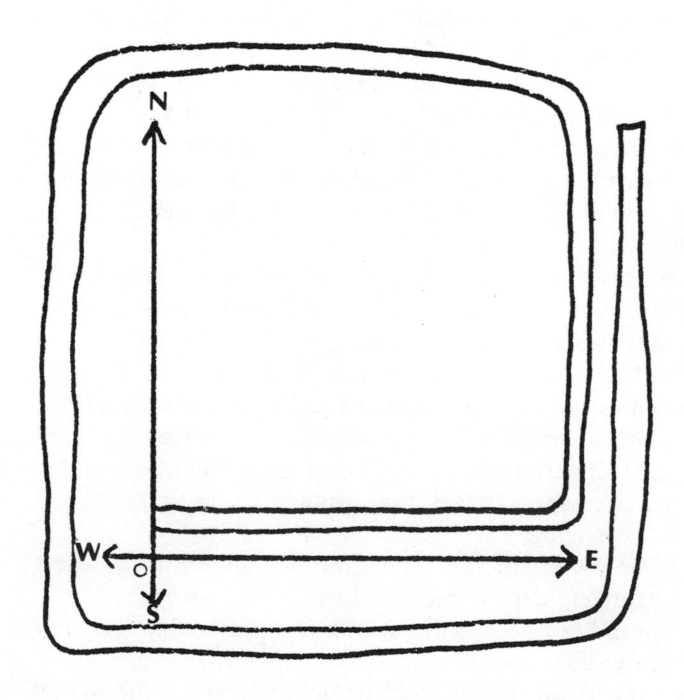

SWINGER FUNCTION (S.F.)^{4}*

RICHARD J. SASSETTI
H. HUGH FUDENBERG
UNIVERSITY OF CALIFORNIA Medical Center
San Francisco

The "information explosion," and its attendant increase in the number of scientific publications, has forced scientific journals to employ more rigid editorial policies. The enforcement of these policies requires that work submitted for publication be evaluated by people sufficiently expert in the field to provide critical evaluation. This burden has fallen on those people who are already preoccupied in producing the "information explosion."

In an effort to alleviate the burden and speed the laborious process of refereeing scientific articles we have devised a method for rapid evaluation of articles submitted for publication. This method can be summarized in the following formula:

$$N = \frac{P}{f} \cdot \frac{R(r/a)\, r!(Oo - Os)}{2\ Oo}$$

where N is the Numerical Score; P is the number of pages in the article; f is the number of figures; R is the total number of citations; r is the number of citations of works authored by the referee; a is the number of citations of works by the author of the submitted manuscript; Os is the number of citations of work by others whose data support the hypotheses of the referee; Oo is the number of citations of works by others whose data oppose the referee's hypotheses.

From this, one can see that the size of N is dependent on a number of parameters. For example, an increase in P, the number of pages (which is often proportional to the vagueness of the subject), increases N directly, while f, the number of figures, roughly proportional to the amount of hard data obtained in the study, tends to decrease N directly.

Since R, which represents the total number of citations, could reflect incorporation of much meaningless information, a factor to moderate this effect is necessary. To do this we have incorporated the factor r/a, where r, the number of works by the referee cited is divided by the number of citations of works by the author, a. In this way, any inadvertent exclusion of significant information, will tend to moderate the effect of R. Therefore, to insure proper weighting of N, the exponent r factorial has been added to overcome small differences between r and a.

Finally, to minimize the adverse effects, should the data be controversial, we have added the normalizing factor (Os − Oo). By this means, bias introduced by the citation of too large a number of papers not in accord with accepted theories is minimized in favor of the more accepted hypothesis.

The interpretation of the numerical score is relatively simple and falls into four categories.

1. *N is greater than 0 but equal to or less than 1.*
   This could arise if P and f are nearly equal of if f (number of figures) is very large but r is large relative to a. If r is small or zero, N will be zero. Also a large (Oo − Os) (which could arise if Os is small or Oo inordinately large), would result in a small N.
2. *N is greater than 1 but equal to or less than 10.*
   This could arise where either P, a, or Os are relatively large, or f and r relatively small.
3. *N is greater than 10 but equal to or less than 100.*
   This could arise when P, r, or Oo are relatively large.
4. *N is greater than 100.*
   This can occur where r is significantly large or Oo is negligible.

The interpretation of N, then, is as follows:

N = 0 − 1:	REJECT	(Author is not familiar with pertinent literature)
N = 1 − 10:	ACCEPT	(It is difficult to determine how well the author has mastered the subject material but his proposals certainly do not cause any controversy. To be on the safe side, demand revision).
N = 10 − 100:	ACCEPT	(The article should be accepted since it has obvious merits).
N = 100:	REJECT	(Excellent work. It should be rejected because experts in the field, namely, the referee (or his friends) are about to publish the same material).

This investigation was supported by grants-in-aid from NSF (Non-sufficient Funds) and NIH (Not Immediately Helpful.) ■

* Scientific Facilitation, Special Formulation, Sassetti-Fudenberg, San Francisco.

The Art of Finding the Right Graph Paper to get a Straight Line

S. A. RUDIN

As any fool can plainly see, a straight line is the shortest distance between two points. If, as is frequently the case Point A is where you are and Point B is research money, it is important to see to it that the line is as straight as possible. Besides, it looks more scientific. That is why graph paper was invented.

The first invention was simple graph paper, which popularized the straight line (Fig. 1). But people who had been working the constantly accelerating or decelerating paper had to switch to log paper (Fig. 2). If both coordinates were logarithmic, log:log paper was necessary (Fig. 3).

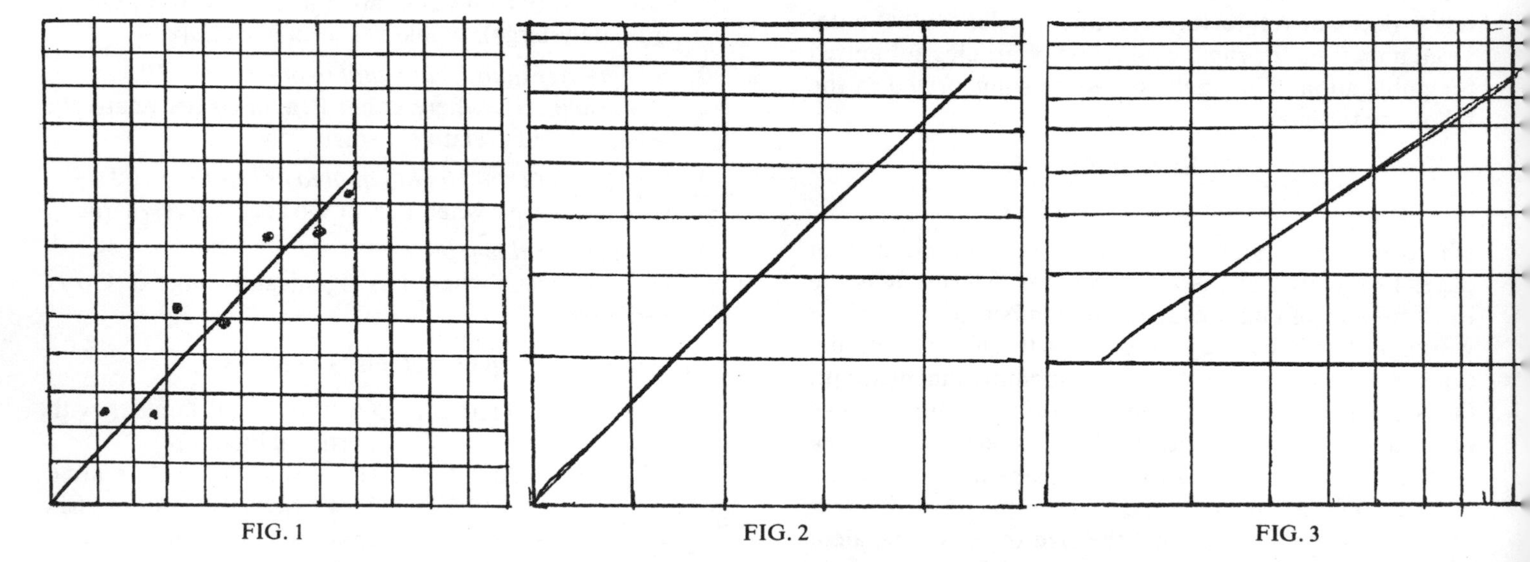

FIG. 1 FIG. 2 FIG. 3

Or, if you had a really galloping variable on your hands, double log-log paper was the thing. And so on for all combinations and permutations of the above (Fig. 4).

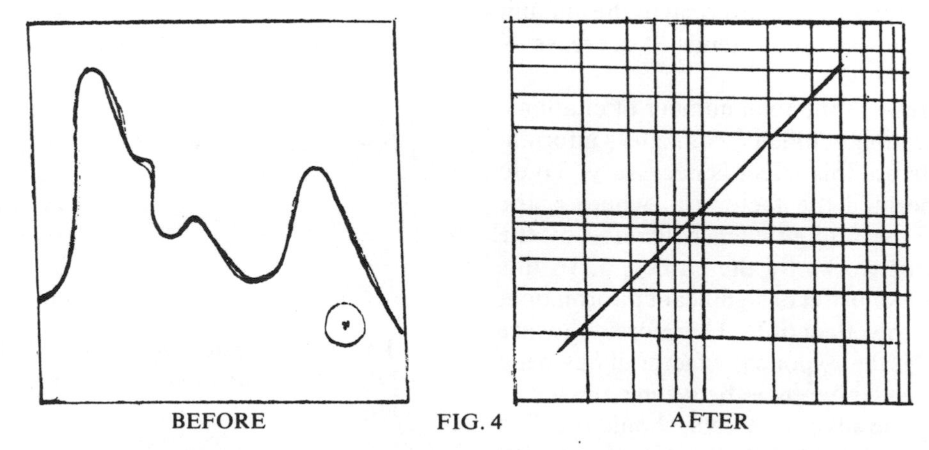

BEFORE FIG. 4 AFTER

For the statistician, there is always probability paper, which will turn a normal ogive into a straight line or a normal curve into a tent. It is especially popular with statisticians, since it makes their work look precise (Fig. 5).

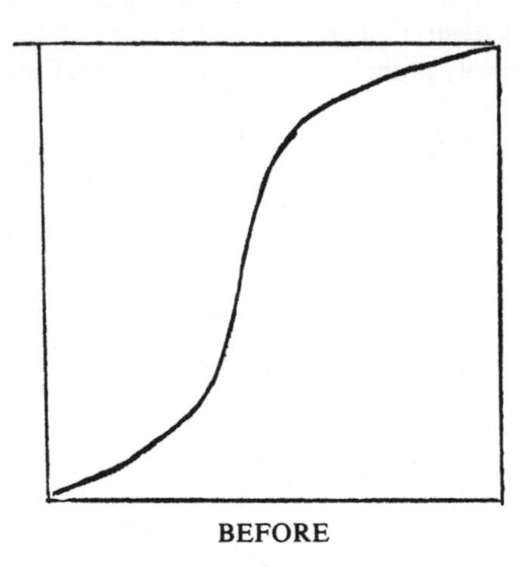

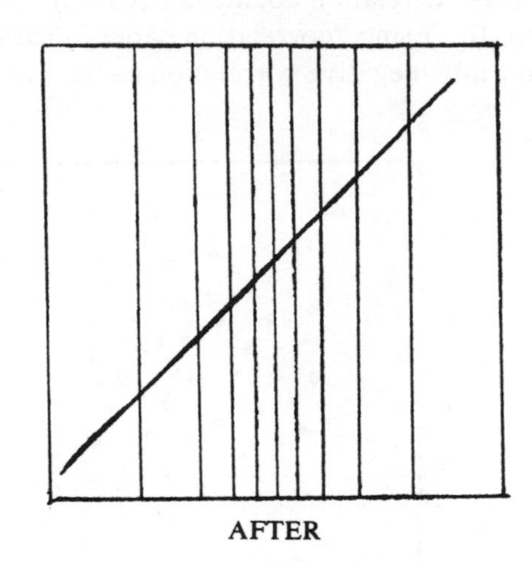

BEFORE FIG. 5 AFTER

For sailing submarines under the North Polar Cap, there is polar coordinate paper (Fig. 6).

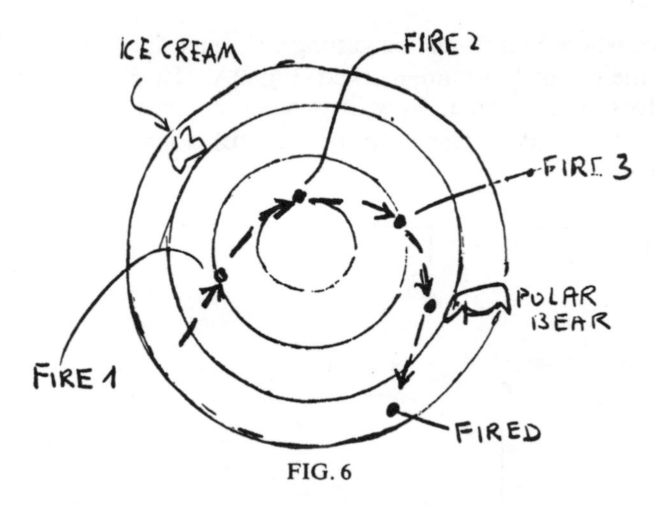

FIG. 6

But one does not really get into the swing of things until one is familiar with the latest line of graph papers, turned out by the great-great-great-great-great-grandson of the inventor of the graph paper, Harry Graph. Psychologists, in particular, are plagued by curves that increase, decrease, and do other unnerving things. For example, if your lab assistant is a lazy slob, his work output is zero. If you yell and threaten him, it goes up. The more you yell at him, the higher it goes, up to the point where he becomes a nervous wreck and starts dropping things. From there on, the more you scream, the worse he gets. For this situation, curvilinear paper will show the increase, but shrivel up the decrease so much it looks like a mere error (Fig. 7A—ordinary paper, 7B—curvilinear paper).

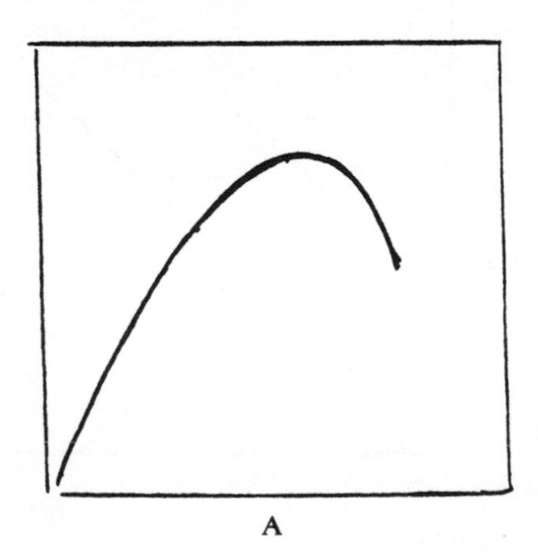

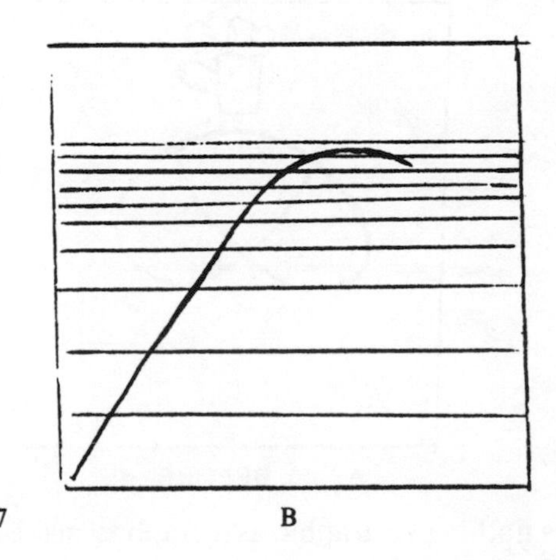

A FIG. 7 B

Sometimes correlation coefficient scattergrams come out at .00, with a distribution shaped like a matzo ball (Fig. 8A). But using "correlation paper" Pearson r's of any desired degree of magnitude can be obtained (Fig. 8B). Naturally, negative correlation paper is available; it simply points the diagram the other way.

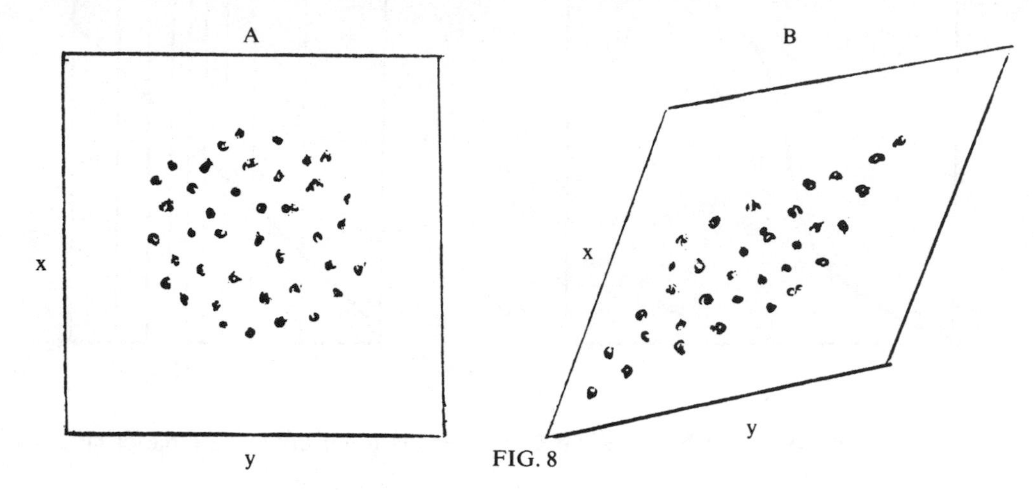

FIG. 8

When you get a curve where you should be getting a straight line, you use the following method. First, the peaks and troughs of the original plot are marked (Fig. 9A). Then, an overlay of transparent plastic sheet is put over it, and the dots alone copied. Now, it is obvious that these points are simply departures from a straight line, which is presented in dashed form (Fig. 9B). Finally, the straight line alone is recopied onto another graph paper (Fig. 9C).

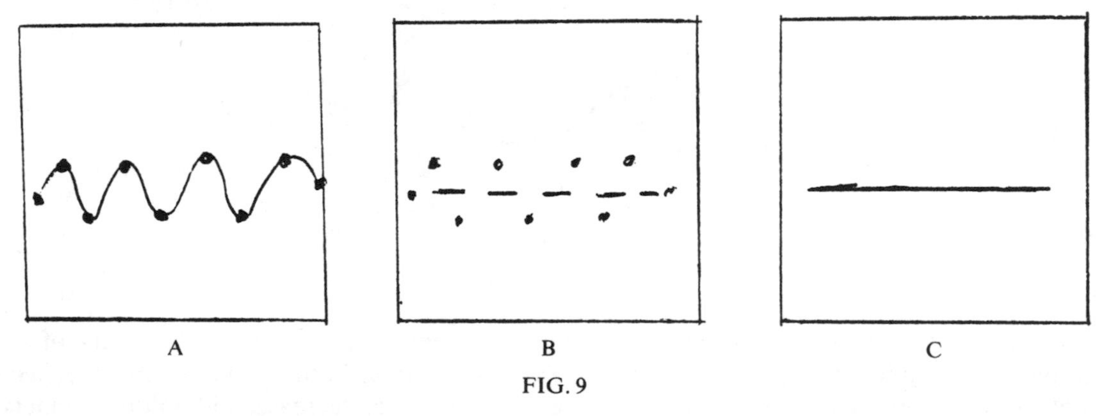

FIG. 9

But by far the latest wrinkle in graph paper (ha!) is the do-it-yourself kit. This consists of an ordinary graph of equally spaced lines, at right angles to one another, as in Fig. 1, but printed on a large sheet of transparent rubber. The user is then free to make up his own technique by stretching the sheet according to his requirements (Fig. 10A and 10B).

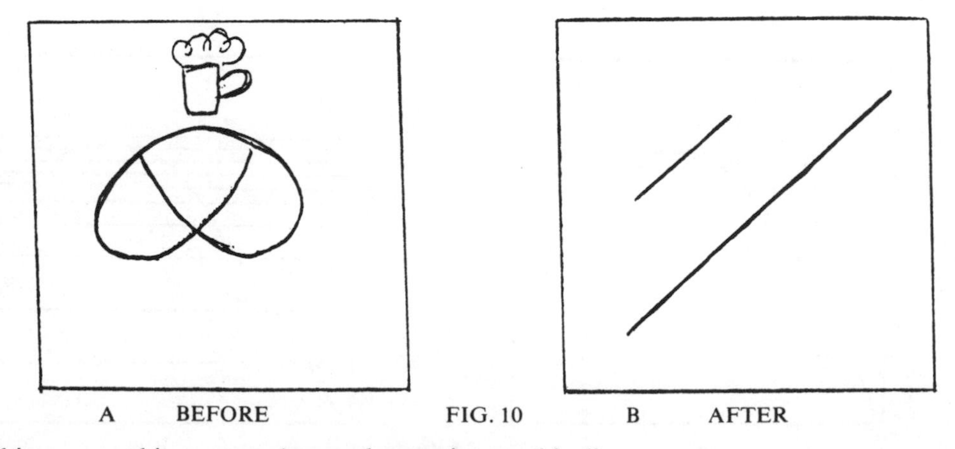

FIG. 10

There is nothing so graphic as a graph to make a point graphically.

THE GLASS AND SPLEEN EXPLOSIONS

SIDNEY L. SALTZSTEIN (M.D.)
San Diego County General Hospital

"Explosions" and their dire consequences occupy much of the current scientific and popular literature. The world's population is growing at such a rate that if it were maintained until the year 6000, a solid mass of humanity would be expanding outward from the earth at the speed of light.[1] The number of authors per scientific paper is increasing so rapidly that by 1980 it will reach infinity.[2] Another pair of phenomena, possibly more to be feared, is the glass and spleen explosions.

MATERIALS AND METHODS

When the surgical pathology laboratory at Barnes Hospital, St. Louis, Missouri, was remodeled in 1959 the number of specimens received annually in the laboratory was tabulated so that storage space for past and future slides could be planned. At the same time it was necessary to count, and review the slides, of all spleens either biopsied or removed and received in the same laboratory since the hospital was opened.[3]

RESULTS

The annual number of surgical specimens received is plotted in Figure 1 on semilogarithmic co-ordinates. It is apparent that the number of specimens is increasing loga-

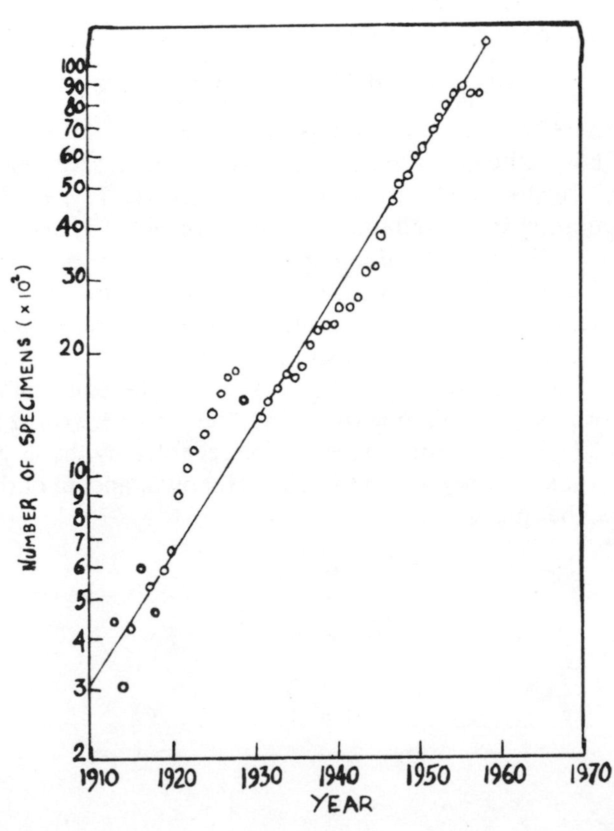

rithmically (exponentially). Close examination of the graph also shows the effect of the Depression of the late 1920's and early 1930's and of World War II. Determining the line of best fit by the least squares method, the following linear regression equation is obtained[4]:

$$\mathrm{Log}_{10}N_1 = 2.175 + 0.0317(T - 1900);$$

where N_1 is the number of specimens received in the year T. The coefficient of determination, correcting for degrees of freedom lost, is 0.916, indicating that 91.6% of the variation in the number of specimens received is explained by time alone. A "doubling time" of 9.49 years can be obtained from this equation. Extrapolating backwards, the year when the first specimen should have been received (the year when $N_1 = 1$) is 1832. While this may be absurd since Barnes Hospital did not open until 1912, Dr. William Beaumont, the first modern surgeon in St. Louis, arrived in the area in 1834 (an error of only 0.1%)[5]. Perhaps the onset of this growth of surgical specimens should be correlated with his arrival.

Extrapolating forward, in the year 2224, 2.854×10^{12} specimens will be received, making a total of 5.7×10^{12} specimens to be stored. The slides from approximately 1119 specimens can be stored per cubic foot[6], so 5.1×10^9 cubic feet of storage space will be needed. As the area of the city is 61 square miles[7] this will be enough slides to bury the city of St. Louis 3 feet deep with glass!

The annual number of spleens received is plotted in Figure 2, again on semilogarithmic co-ordinates*. Again, it is apparent that the number of spleens is increasing logarithmically (exponentially), and the linear regression equation turns out to be:

$$\mathrm{Log}_{10}N_2 = 0.917 + 0.0457(T - 1900);$$

where N_2 is the number of spleens received in the year T. The coefficient of determination, correcting for degrees of freedom lost, is 0.773, indicating that 77.3% of the variation in the number of spleens received is explained by time alone. A "doubling time" of 6.59 years can be obtained from this equation, showing that the number of spleens received is increasing considerably more rapidly than the total number of specimens received.

If one then sets $N_1 = N_2$, and solves the equations for both N and T, one will find that in the year 2121, 8,271,800,000 specimens will be received by the surgical pathology laboratory at Barnes Hospital, and all of them will be spleens! ■

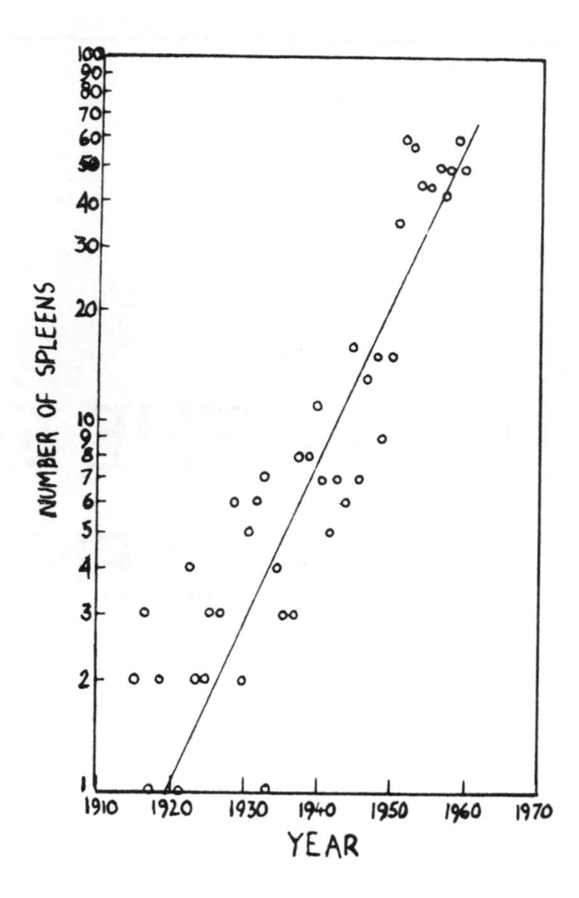

CONCLUSION

Be careful when extrapolating biological data.

REFERENCES

1. Editorial: The Weight of Humanity. St. Louis Post-Dispatch, Sunday, July 22, 1962.
2. Price, D. J. de S.: Little Science, Big Science. Cited in Review: J. Irrepr. Res. *13*: 22, 1964.
3. Saltzstein, S. L.: Phospholipid Accumulation in Histiocytes of Splenic Pulp Associated with Thrombocytopenic Purpura. Blood *18*:73–88, 1961.
4. Hirsch, W. Z.: Introduction to Modern Statistics. N.Y. The MacMillan Co. 1957.
5. Pitcock, C. DeH.: The Involvement of William Beaumont, M.D., in a Medical-Legal Controversy: The Dames-Davis Case, 1840. Missouri Historical Review, 1964, 31–45.
6. Saltzstein, S. L.: Unpublished data.
7. World Almanac, 1960. p. 294.
8. Furst, A.: On the Treatment of Annoying but Incontrovertible and Inexplicable Facts. J. Irrepr. Res. *13*:10, 1964.

* In accordance with Furst's first modification of the scientific method, the years when no spleens were received ($\mathrm{Log}_{10}0$ indeterminant) are ignored.[8]

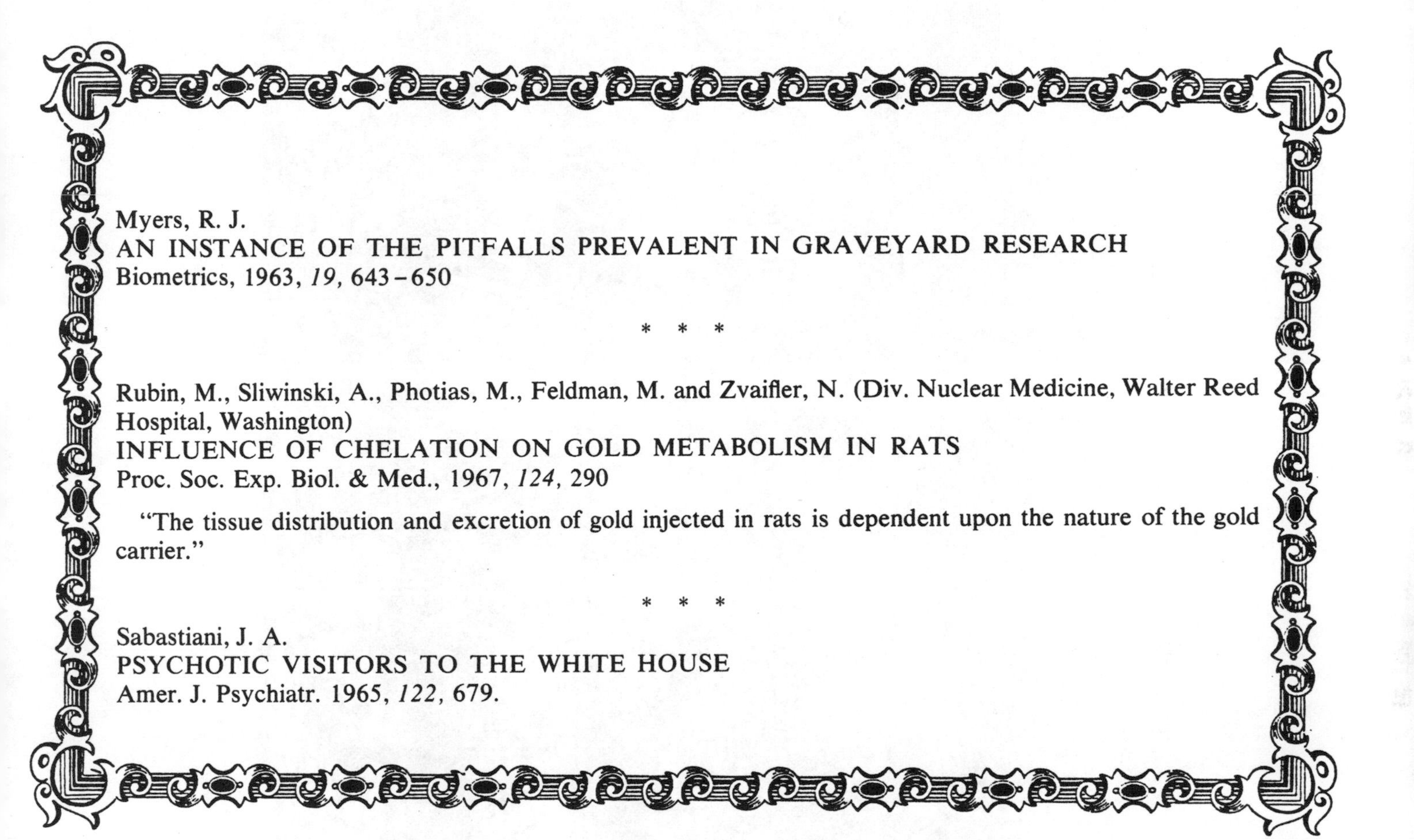

Myers, R. J.
AN INSTANCE OF THE PITFALLS PREVALENT IN GRAVEYARD RESEARCH
Biometrics, 1963, *19,* 643–650

* * *

Rubin, M., Sliwinski, A., Photias, M., Feldman, M. and Zvaifler, N. (Div. Nuclear Medicine, Walter Reed Hospital, Washington)
INFLUENCE OF CHELATION ON GOLD METABOLISM IN RATS
Proc. Soc. Exp. Biol. & Med., 1967, *124,* 290

"The tissue distribution and excretion of gold injected in rats is dependent upon the nature of the gold carrier."

* * *

Sabastiani, J. A.
PSYCHOTIC VISITORS TO THE WHITE HOUSE
Amer. J. Psychiatr. 1965, *122,* 679.

L. JOHNSON & CO.

Some
improbable
investigations
into

FOOD
AND
NUTRITION

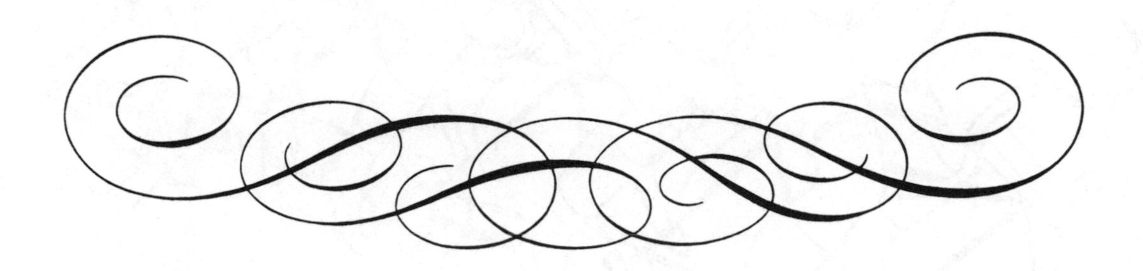

HOW TO EAT AND LOSE WEIGHT

A new solution to an old problem

The problem of energy equilibrium in the human body has received a great deal of attention in the USA[1]. Many methods have been advocated to reduce the caloric value of the food without affecting its taste or the appetite of the consumer. Some pharmaceutical houses have even expanded by wise investment of the monetary equivalent of the weight theoretically lost by their customers.

The various methods advocated involve either a drastic reduction in the quantity of the food ingested, use of drugs that affect the appetite, or increase in the amount of physical exercise. The first method leaves the patient permanently hungry, the second makes him sub-human by dulling his senses, and the third defeats its own purpose by increasing the appetite and leading to a vicious circle.

In this communication we propose a new method of reducing, without changing the quantity of the food ingested. The method is based on the principle that the caloric value of the food also depends on its temperature. For each degree of temperature of the food below the temperature of the human body, the body has to supply heat (energy) to raise the temperature of the food to that of the body.

When food is frozen, heat of thawing has to be supplied to it at the rate of 80 calories/gm of water contained in the food in addition to 1 calorie/gm for each degree centigrade of temperature rise. One can easily see that a glass of frozen milk (200 gms) at the temperature of deep freeze, i.e. $-20°$ C, needs 200×80 calories for thawing and 57×200 calories for heating up to 37° C (98° F), i.e.

$$16{,}000 + 11{,}400 = 27{,}400 \text{ calories or } 27.4 \text{ Cal.}$$

The caloric value of a glass of whole milk is 138 Cal and of skim milk only 74 Cal.[2], thus $(27.4 \times 100)/74 = 34\%$ of caloric milk energy is lost for its heating. By having the milk diluted in the ratio of 1:2, 68% of the energy would be lost on consuming the same amount of milk.

Similarly one can calculate that consumption of a precooked frozen steak calls for an expenditure of at least ⅓ of its caloric value on the thawing and heating up. If one adds to this the amount of energy supplied as heat by the body and the mechanical energy required to crush the food with the jaws, the loss of the caloric value of the food becomes even greater.

For those individuals who do not care to crush their food with their teeth, or those who use valuable and fragile plates, and finally for those who have lost their teeth and did not replace them, one can advise the use of homogenized frozen food in form of popsicles which would be sucked instead of chewed.

The suggestion of some research workers (personal communication) to achieve an additional loss of energy by consuming the food frozen in liquid helium or nitrogen is not considered practical, because of the temporary rarity and high cost of these two elements in their liquid form. With the increased use of liquid oxygen, however (rocket fuels, etc.) one may hope to have it supplied in a domesticated and easily handled form in the near future. ■

==============

[1] E. Haveman: The wasteful, phony crash dieting craze, Life, 1959, *46*, 102–114.

[2] H. C. Sherman: *Essentials of Nutrition*, Macmillan, N.Y. 1945, page 377.

On the Purchase of Meat and Fruit

By PIROCOLE RASHCALF

These are troubled times. A conspiracy exists in this nation between the paper converter and meat packing industries. Through the most unstinting casual empiricism this author was able to discover that enormous costs are being imposed upon the American housewife. Did you know that the housewife who shops for meat pays for a heavy cardboard tray and cellopane wrapper at meat prices? In the green grocer department, the trays that hold fruits and vegetables are paid for at fruit and vegetable prices! The paper on the meat is the same as that used on the vegetables, and so the housewife is being cheated by paying meat paper prices when she buys vegetables instead of vegetable paper prices.

We have contrived a clever test to expose this conspiracy. We went out and bought some minute steaks, ham hocks, best chanks, tomatoes, lemons, and such and made observations on cardboard tray size, tray weight, price of the item per pound and color of tray. In fact, the trays differ as between meats and vegetables only in terms of the tray color. Meat trays are white. Non-meat trays are green and purple. In doing our data search and collection, we noticed this difference in color almost immediately.

We cast about for a method to test our hypothesis and decided to fit the following linear econometric model:

$$Y = a_0 + a_1X_1 + a_2X_2 + a_3X_3 + a_4X_4 + a_5X_5 + e,$$

where,

Y = weight of cardboard container in ounces;
X_1 = size identification of tray;
X_2 = total weight of the purchase in lbs.;
X_3 = a dummy variable where 1 = a fruit tray, 0 = a meat tray;
X_4 = total cost of item in cents;
X_5 = price per pound of item in cents;
e = an error term.

The results were

$$Y = \underset{(.24)}{.35} + \underset{(.0009)}{.0016X_1} + \underset{(.00123)}{.00023X_2} + \underset{(.10)}{.19X_3} + \underset{(.0020)}{.0026X_4} - \underset{(.0036)}{.0017X_5}$$

$R^2 = .79$ F- Ratio = 11.5 SSE = 11.01 N = 15

Two things stand out from these results. First, the model is hopelessly misspecified. Second, further research needs to be done. But we are not dismayed.

We find, for instance, that a fruit tray weighs on net 1/5 oz. more than a meat tray, yet for each increase of one cent in price per pound of item purchased, the weight of the tray increases .0026 ozs. There is no doubt in our mind that the average weight of the meat item is greater than that of the fruit item. Thus, starting with the fact that the fruit tray already weighs more than the meat tray, and recognizing that as the price of the item increases, people pay more for fruit trays than meat trays. This can be verified for it is a fact that while we bought two pounds of apples, we bought only .85 lbs of cube steak.

Give nthis incontrovertable proof, how can such a scandal be perpetuated? Simply by the old Madison Avenue trick of differentiating the product by selling meat in plain white trays while vending fruit and vegetables in trays of mind-boggling green and purple. What do you think of that?

Scientific research isn't hard, you know, if you put your mind to it. ■

SCHOENFELD'S NOSTRUM: *THE UNIVERSAL THERAPEUTIC COCKTAIL* A Proposal for World Salvation

Myron R. Schoenfeld,*
Scarsdale, New York

Enough! The economy of the western world is collapsing, Mickey Mouse countries are calling the shots, mental midgets are piloting the ships of state, small men are casting long shadows, and, if anything, the future looks even bleaker. The time has come for bold action, not further debilitating equivocation; and for creative new approaches to the novel problems of today, not the continued use of the same old clumsy methods that have been tried and found wanting for years. Serious diseases require strong remedies, and justify risk-taking. Since our leaders obviously lack the guts or the imagination to grapple with today's desperate situation, I am taking up the gauntlet. Here is what I propose:

The government should add to our drinking water...just as floride is added.. a cocktail of drugs which alleviate most of the common ills of mankind, and which have a proven record of wide usage, great therapeutic efficacy, and minimal side effects. All of them, in fact, today are used with such increasingly wide medical indications that their prescription often seems to be near indiscriminate. If so, why should the physician have to be the middle-man? I say that society should dispense these wonder drugs on demand via slot machines, or, better yet, in the drinking water, so that all of mankind, and organized society as a whole, should get their benefits. These drugs, and their wondrous expected effects, are:

...*Valium:* This is the near-perfect panacea. It will calm our restive youth, prevent urban riots, and tranquilize away the everyday tensions associated with job and home.

...*Tagamet:* Should anxiety manage to break through the protective shield afforded by Valium and cause heartburn and ulcers, Tagamet will come to the rescue.

...*Mylanta:* Just in case a few cases of dyspepsia get by Valium and Tagamet, Mylanta will turn the trick. Much more important, its gentle laxative effect will cure that most dreaded scourge of humanity—I speak, of course, of bowel irregularity—and will ensure, each and every one of us a delightfully refreshing movement every day.

...*Aspirin:* Aspirin will help prevent heart attacks and strokes, and will cure headaches and rheumatism too. Best of all, some serious scientific studies indicate that even a smidgen is effective...a baby aspirin every two days...so that some of us are quietly wondering if just holding an aspirin in our hand every now and then might not be all that really is needed.

...*Inderal:* Should aspirin fail and angina occur, Inderal should take care of the situation. As a bonus, it will cure people's migraine, lower their blood pressure, and prevent any paralyzing stage-fright that commonly besieges all of us on crucial occasions...very considerable advantages indeed.

...*Lidocaine:* If, somehow, occasionally, a heart attack should manage to occur anyway despite Aspirin and Inderal, Lidocaine will prevent the arrythmias that cause so many of the deaths due to this terrible disease. No problem.

...*Hygroton:* this should adequately lower the blood pressure of the milder hypertensives, and, when laced with Inderal, take care of most of the severer cases too. Inasmuch as the mortality rate of even mild hypertensives now has been unequivocally proven to be cut almost 25% by Hygroton, it logically follows that chronic shock is good for your health, and that, therefore, the time has arrived to share this extended life expectancy with normotensives too. After all, simple decency demands it.

...*Vitamin C and apricot pit extract:* Until recently, cancer had been the one important disease that seemed to defy such a simple and safe kind of solution, and this was the reason that I had held off making my revolutionary announcement to the world until now. Fortunately, this important gap now has been filled. If testimonials from Nobel Laureates and dying cancer patients are to be believed (and if you can't believe them, who can you believe?), a liberal daily intake of vitamin C will help prevent cancer, and, should cancer inexplicably occur despite this, apricot pit juice will handle the problem.

...*Vitamin E:* Vitamin E deficiency in chickens causes muscle weakness and infertility. It therefore follows to the perceptive mind that dietary supplements of this chemical in man should act as a tonic for the heart; brace tired muscles and hence relieve fatigue; and make the sexual juices flow again, and thereby shore up faultering sexual potency. Truly, no household should be without it.

...*Chicken Soup:* Chicken soup will tidy up the formula, curing anything not already cured by the other drugs. Four thousand years of Jewish history can't be wrong!

There you have it! I put it to you that widespread implementation of my program will substantially blunt social unrest; it will make rebellious youth more tractable; it will go a long way to preventing most of the important diseases ravaging mankind-anxiety, heart disease, stroke, cancer, sexual impotence, and chronic fatigue; it will cure or ameliorate most of those symptoms that cannot be prevented; and, most important of all, it will allow all of us to revel and rejoice in the mellow glow of a daily morning bowel movement - all at a trivial expense, and at minimal risk. How can you beat such an idea?

The only problem that remains to be solved is dosage of medication, but such details easily can be worked out by technocrats. I will leave such trivia to lesser minds...while I pack my bags to go to Stockholm. ■

*While there is every reason to believe that the proposal herein described truly will be the long sought-after elixir of life and happiness, Dr. Schoenfeld, with typical modesty, has declined the suggestion that he declare himself the new Messiah. Still for the good of us all, and with The Nostrum as his platform, he is considering throwing his hat into the ring for the next Presidential election. The reader's support in this regard is humbly solicited.

MORAL, NUTRITIONAL AND GUSTATORY ASPECTS OF CANNIBALISM

J.B. Garble
Highgrove, CA

The Piltdown Newsletter Vol. II No. 4, 1972

The recent flurry of anthropological interest in the nutritional and other aspects of cannibalism (Brown 1971, Garn and Block 1970, Hay 1971, Posinsky 1971, Randall 1971, Rohrl 1970, Vayda 1970, Walens and Wagner 1971) has prompted me to contribute an account of one more case regarding the practice under discussion. Where Garn and Block concern themselves with "regular people-eating," their calculations (indicating that "one man...serves 60, skimpily") lead them to conclude that "the nutritional value of cannibalism may...be viewed as questionable" (Garn and Block 1970: 106), a conclusion found questionable in turn by Randall (1971). Walens and Wagner review demographic and other circumstances under which the food value of human flesh could be very high indeed (Walens and Wagner 1971: 269-270), and Vayda directs his attention to the therapeutically nutritional effects of anthropophagy where the consumer is suffering the dietary deprivations of stressful circumstances (Vayda 1970: 1462). Rohrl's article is addressed to the culturally determined but socially condemned craving for human flesh manifested in the disease called windigo psychosis that is instigated, it is argued, by conditions of famine and cold (Rohrl 1970: 99-100; but see also reply by Brown 1971). As the name of windigo psychosis indicates, the symptoms of the disease are largely in the form of mental aberrations ("the affected individual may see the people around him turning into beavers or other edible animals," cited by Rohrl 1970: 98, from Barnouw 1963: 366), and at least one author has indeed approached the etiology of windigo in psychoanalytic terms of repressed desires (Parker 1960). Hay cautiously asserts that one doesn't have to be psychotic to indulge, and suggests that cannibalism's wide distribution "testifies to the commonness of the desire to eat human flesh" (Hay 1971: 1).

I came across the first person reflections on cannibalistic experience, including its nutritional, gustatory, psychological and even moral considerations, in the autobiography of Diego Rivera, the famous Mexican muralist.

In a curious chapter, called "An Experiment in Cannibalism" (Rivera 1960: 44-46), Rivera relates how his venture in unusual eating was first stimulated by reading about a French furrier in Mexico who decided to economize by feeding his cats (he was a cat furrier) on the internal remains of their predecessors who had progressed to the fur coat stage. "On this diet," he relates, "the cats grew bigger, and their fur became firmer and glossier. Soon he was able to outsell his competitors, and he profited additionally from the fact that he was using the flesh of the animals he skinned." Rivera, at the time, was studying anatomy at the medical school in Mexico City, and he and his fellow students decided "to extend the experiment and see if it involved a general principle for other animals, specifically human beings, by ourselves living on a diet of human meat." They therefore bought cadavers from the local morgue, choosing only those that had died by violence and were not victims of disease, and followed the diet for two months. "Everyone's health improved," he claims.

Far from regarding such a diet as macabre, Rivera lauds it as both socially virtuous and rational: "Cannibalism does not necessarily involve murder," he points out, "and human flesh is probably the most assimilable food available to man. I believe that when man evolves a civilization higher than the mechanized but still primitive one he has now, the eating of human flesh will be sanctioned. For then man will have thrown off his superstitions and irrational taboos."

Despite Rivera's somewhat naive and optimistic assumptions regarding the evolutionary progress of mankind, and even apart from the manifest ecological significance of *Homo sapiens* recycled (see Posinsky's remarks relevant to this point 1971: 269), we, as students of human proclivity, are surely in debt to this dauntless investigator. ■

REFERENCES

Barnouw, V. 1963 *Culture and Personality*, Homewood, Ill.: The Dorsey Press.

Brown, J. 1971 "The Cure and Feeding of Windigos: A Critique," *American Anthropologist* 73: 20-22.

Garn, S.M. and Block W.D. 1970 "The Limited Nutritional Value of Cannibalism," *American Anthropologist* 72: 106.

Hay, T.H. 1971 "The Windigo Psychosis: Psychodynamic, Cultural, and Social Factors in Aberrant Behavior," *American Anthropologist* 73: 1-9.

Parker, S. 1960 "The Windigo Psychosis in the Context of Ojibwa Personality and Culture," *American Anthropologist* 62: 603-623.

Posinsky, S.H. 1971 "Cannibalism," *American Anthropologist* 73: 269.

Randall, M.E. 1971 "Comment on the 'Nutritional Value of Cannibalism'," *American Anthropologist* 73: 269.

Rivera, D. 1960 *My art, my life*. New York: Citadel Press.

Rohrl, V.J. 1970 "A Nutritional Factor in Windigo Psychosis," *American Anthropologist* 72: 97-101.

Vayda, A.P. 1970 "On the Nutritional Value of Cannibalism," *American Anthropologist* 72: 1462-1463.

Walens, S. and Wagner R. 1971 "Pigs, Proteins, and People-Eaters," *American Anthropologist* 73: 269-270.

THE ANSWER TO THE PROBLEM OF BULIMIA

EUROPA & THE BULL, by Jost Amman

GORGE (Goals for Overcoming Regurgitive Gluttonous Eating)

Kathleen M. Donald
Livington, MT

A recently-identified problem for many of today's young college women is the gorge-purge syndrome, also called bulimia.[1] The *Diagnostic and Statistical Manual of Mental Disorders (DSM-III)* specifies certain criteria for the diagnosis of bulimia: Recurrent episodes of binge eating; consumption of high caloric, easily-ingested food; inconspicuous eating during binges; repeated attempts to lose weight by self-induced vomiting or use of cathartics or diuretics; depressed mood and self-deprecating ideation following binges; and, awareness that the pattern is abnormal, and a fear of not being able to stop.[2] Boskind-Lodahl and White coined the term "bulimarexia" to describe those young women who use gorge/purge behavior in an attempt to lose weight and conform to society's standards of svelteness.[3]

At a recent conference on eating disorders,[4] nine out of ten doctors recommended the term bulimarexic in labeling the problematic behavior. However, a certain few preferred the term bulimarectic. Dr. Gunnar Ralph brought up that each diagnostic label referred to a specific aspect of the binge/purge syndrome. According to Ralph, most binge/purgers choose self-induced vomiting as their preferred purgative modus operandi (P.P.M.O). The most familiar of all the recognized P.P.M.O.'s, the vomit-purge has resulted in two colloquialized, tongue-in-cheek labels: The vernacular, Anglo-Saxon version, "Scarf and Barf," and the more erudite, Latinate term, "Masticate/Regurgitate."[5] The overwhelming consensus of the conferees was to use the most predominant diagnosis, bulimarexic, to identify the most prevalent P.P.M.O., vomiting.

The other, less-renowned P.P.M.O. involves the utilization of laxatives and enemas. Due to the rectal nature of this method, it emerged that bulimarectic was the more useful and descriptive term. Unfortunately, despite the fact that the laxative/enema technique has received less coverage, apparently the modality has not escaped notice. People being what they are,[6] they have also derived descriptive derogations for the bulimarectic type of purge. The vulgar version, "Scoop and Poop," and the Latinate label, "Masticate/Defecate,"[7] are currently being bandied about by the ill-informed.

Although many therapists have written *ad nauseum* about the weighty issue of bulimia, we wish to tip the scale still further in the direction of diarrheic dicta. What follows is some food for thought for those bulimarexics and bulimarectics who cannot stomach the condition any longer and who are tired of feeling down in the mouth.

We believe that it is possible for you to lick the problem of bulimia. In no way are we wishing to heave an impossible challenge at you. Please don't throw up your hands in despair. However, we acknowledge it will take guts to eliminate this habit. But we are not just mouthing generalities. Nor do we wish to cram our ideas down your throat. We're offering no pie-in-the-sky solutions. However, we do suggest that you stop your bellyaching, take in our advice, let it settle, and digest the ideas thoroughly. At first it may seem hard to swallow our suggestions but as they become more palatable and start sinking in, you'll be back for more. All it will take is a stiff upper lip. As Sherlock Holmes used to say, "It's alimentary, my dear Watson." And alimentary it is. The problem of course cannot be expelled in one sitting. It is not something to be cured in one gulp. As the saying goes, "One swallow does not make a spring." In fact, when attempting to set goals for change, it is important not to bite off more than you can chew. Sometimes people tend to be perfectionistic and set their goals too high. Their eyes are bigger than their stomachs. You must avoid such a position. Instead, set smaller goals and take them one at a time in bite-size pieces. This advice is straight from the horse's mouth. We hope it will feed your imagination and nurture your desire to get well. Of course we realize our limitations. We cannot spoonfeed you. Spit it out. Only then will you discover the stuff of which you are made. Only then will you get a taste of the power you've kept hidden deep within the lower regions of your personality. All you need to do is relax and let it surface. Once you let your hidden strength rise up, chuck away those self-defeating habits. Stop feasting on those negative put-downs and swallowing your own self-deprecating statements. Those pessimistic putdowns are self-fullfilling. As the expression goes, garbage in, garbage out. You must forget about girth control. The issue is "worth" control. Stop starving your self-esteem. Feed your self-worth. Fatten your self-confidence. You deserve to have your cake and eat it too.

We have found that the advice given in the previous paragraph has been well-received whenever we have dished it out in group therapy with bulimics. Although they have not gone hog-wild over our suggestions, they have found some morsels of information to be appetizing. It appears that it takes some time for bulimics to ruminate on the problem and hash it over in their minds before they are ready to let it go for good. ■

[1]So prevalent is bulimia, as well as anorexia nervosa and other eating disturbances, that a special professional publication, *The International Journal of Eating Disorders,* has recently begun circulation. This publication, informally dubbed the "Vomit Comet," brings up the heavy issues concerning professionals who treat bulimics.

[2]American Psychiatric Association. *Diagnostic and Statistical Manual of Mental Disorders, Third edition (DSM-III).* Washington, D.C.: APA, 1980.

[3]Boskind-Lodahl, Marlene, and White, William C., Jr. "The definition and treatment of bulimarexia in college women—A pilot study." J.A.C.H.A. 1978: 84-86, 97.

[4]An informal meeting of Dr. Lotta Pugh and Dr. Gunnar Ralph on their way to a local meeting of the Eager Eaters Society.

[5]Many thanks to Dr. Lotta Pugh and Dr. Gunnar Ralph on their way to a local meeting of the Eager Eaters Society.

[6]Disgusting, illiterate, lewd, vile, disparaging, unwholesome, unruly, unholy, messy, filthy, and outright crass.

[7]Chuck Upham

FAKE

Thou shalt not steal, an empty feat
When it's so lucrative to cheat.—

Arthur Hugh Clough
The Latest Decalogue

T.J. Hamblin
Royal Victoria Hospital
Bournemouth, England

In the year that Popeye became once again a major movie star it is salutary to recall that his claims for spinach are spurious. Popeye's superhuman strength for deeds of derring-do comes from consuming a can of the stuff. The discovery that spinach was as valuable a source of iron as red meat was made in the 1890's, and it proved a useful propaganda weapon for the meatless days of the second world war. A statue of Popeye in Crystal City, Texas, commemorates the fact that single-handedly he raised the consumption of spinach by 33%. America was "strong to finish 'cos they ate their spinach" and duly defeated the Hun. Unfortunately, the propaganda was fraudulent; German chemists reinvestigating the iron content of spinach had shown in the 1930's that the original workers had put the decimal point in the wrong place and made a tenfold overestimate of its value. Spinach is no better for you than cabbage, Brussels sprouts, or broccoli. For a source of iron Popeye would have been better off chewing the cans. *British Medical Journal Vol. 28, Dec. 1981*

Submitted by:
L.X. Finegold
Philadelphia, PA

Hunger in America:[1] A Research Proposal

CHARLES L. McGEHEE
Central Washington State College

In recent years much attention has been given to the phenomenon of lack of food among certain segments of the population in the United States. It has been observed that certain[2] immature members of the population, as well as some mature, have occasion to be compelled to pursue their routine tasks in a state of food deprivation, a state which is systematically different from that of certain other members of the population. Moreover, correlative to this state of food deprivation, varying degrees of learning ability[3] as well as motivational level[4] have been observed.

To the everlasting discredit to social science this situation has been seen as a "social problem" and social scientists are to be found becoming increasingly more active in, of all things, trying to change this, the natural order of things.[5] In doing so, they are ignoring the fact that this situation constitutes a *natural laboratory* for the study of learning processes which here-tofore could only be carried out in a laboratory[6] with animals or voluntary human subjects — at best undesirable and unreliable situations. They are, in other words, missing a magnificent opportunity to further social psychological knowledge of human behavior.

It is, of course, well known that food is a very basic reinforcer, one which does not have to be learned in the usual sense. Subject behavior is quite easily modified through the systematic control of reinforcement in the form of food. Moreover, subjects learn best when they are not satiated.[7] Pigeons, for example, learn best when they are at 80% of normal body weight. Although M and M[8] research with human subjects in the laboratory indicates that the same may hold true for humans, we still do not know if humans will respond in the same way when placed in their natural habitat with the aversiveness of genuine food deprivation.

Large urban areas seem, without exception, to have succeeded for us in isolating prime food-deprived subjects into certain educational institutions,[9] and, as mentioned these subjects have been observed to have a lower response rate, experience more rapid extinction of learned patterns, and exhibit greater avoidance behavior by escaping the environment than do non-food-deprived subjects in similarly isolated educational institutions.

We wish, then, to propose the following: since the subjects report to the educational environment in a state of food deprivation which normally runs from 15-24 hours without food (in some cases longer) at best, or at the worst having had a supplemental feeding of gravy or syrup immediately preceding their training schedule, we as social psychologists can use this state to induce and experimentally study the modification of behavior. That is to say, we can make reduction of the state of food deprivation contingent upon the successful performance of programmed tasks in a given learning situation.[10] The extent to which social psychological knowledge will be incremented defies the imagination!

Funding such a project should be no problem. The C.I.A., which has had considerable experience in operant conditioning utilizing food as a reinforcer in foreign countries, has expressed enthusiasm[11] and is already establishing a private philanthropic foundation for the purpose. The Department of Defense also will likely respond positively, as it has consistently furthered the advancement of pure science.[12]

To those critics from the "humanitarian" school who would question the ethics of pure science, may we say that we did not create the situation of food deprivation.[13] It behooves us as scientists, however, to take advantage of every opportunity to further knowledge, permitting the "chips" to fall where they may. History, after all, will be our judge. But, as a matter of fact, we are, in our own small way helping to alleviate the situation by providing food to those who would not otherwise get it[14] — provided they do as we tell them.[15]

[1] From the film of the same name. No pun intended.

[2] The referent should be clear to the reader. Cf. Wallace's article "Bussing and Law and Order: Concepts of Infinite Value." *Rgt. Wng. Rev.*, V.1, 1968.

[3] For discussion of this and related concepts see "Scientific Concepts and What Any WASP Knows Anyway. Part I: The Just Plain Dumb." *Annals KKK*, undated.

[4] *Ibid*, "Part II: Shiftless Bums".

[5] Adherents to this so-called "humanitarian" school of thought would do well to read Ptolemy's epic pronouncement *Things Are The Way They Are: Love It Or Leave It*, Alexandria: Geocentric Publishers, 2nd century, A.D.

[6] The value of the natural laboratory has not been given its proper due. A few dedicated persons have recognized its value, however. See H. Himmler, *Der Konzentrationslager als Natürlabor für wissenschaftlische Forschung*, Auschwitz und Dachau: Verlag der Gefangenschaft, 1944.

[7] This perhaps accounts for the fact that leaders in the 1968 Chicago tests of Daley's theory of social control did *not* learn any more from the Russian Revolution than they did.

[8] Melts in mouth, not in hand. Makes for very neat research.

[9] Certain cognitive theories of learning may have screwed up things for us here. See Wallace, *op. cit.*

[10] No read, no eat! Since lack of food is often accompanied by lack of clothing, the project could also be expanded to use shoes and overcoats (particularly when it is snowing) as reinforcers.

[11] The comment was "Let's get to them before the Reds do!" Source asked to remain anonymous.

[12] For example, see the following DOD reports: "Botulism: Boon to Mankind," "Poverty and its Uses," and "Project Camelot: Milestone in the Unification of Science and Ideology."

[13] "For the poor will never cease out of the land. . . ." Deut. 15:11 "For you always have the poor with you." Matt. 26:11

[14] This is true humanitarianism in the tradition of the Romans who found the plebians could not get along without the Roman elite.

[15] The results of this proposed study will be in every sense irreproducible. It is predicted that subject (and subject parent) interaction will increase in frequency and intensity over time to the point the experimenter will be subjected to extinction along the lines proposed by Marx.

M. A. Schneiderman
THE PROPER SIZE OF A CLINICAL TRIAL. "GRANDMA'S STRUDEL" METHOD.
Jour. New Drugs, 1964, *4,* 3.

"Although clinical trials are not apple strudel, too often they leave a bad taste."

* * *

A FRIGHTFUL DREAM

FERENC MOLNAR

I invented ordinary breakfast coffee, coffee with milk. I was the only person in the whole world who breakfasted on that beverage. I was convinced that if mankind should come to know my discovery, it would become the world's most popular breakfast drink, and hundreds of millions of people would drink it, even several times a day. So I rushed to a great bank that financed various industrial enterprises, and after many difficulties was admitted to see the top boss. When I told him I had discovered a drink for which I prophesied universal popularity, the bank president asked me to explain my discovery. I had to be brief, so I confined myself to saying:

"You hire some people, and ship them to the other side of the globe, where there is a certain kind of bush in each of whose berries are two bean-like seeds. When the berries are ripe, your people gather the beans, put them into an iron vessel, and light a fire under the vessel; they heat the vessel slowly, but not hot enough to burn the beans, only to a degree that will turn them black and make them spread a pungent burning stench."

The president was already eyeing me suspiciously.

"Then," I went on, "you grind these half scorched seeds to powder. But we don't eat the powder, nor a decoction of it, but we construct a vessel in two parts, the lower of which contains boiling water. The steam from this water rises through the black, granular powder, which rests on a sort of sieve above the water; this causes the powder to exude a blackish liquid, which is collected in a separate vessel, and the bitter sour taste of which is unpalatable to most people."

By now the president was looking at me with very wide-open eyes.

"Then," I went on, "we set out to find a certain mammal; for our purposes we require the female. From this female we remove in an artificial manner, by a sort of torture, the white liquid with which it ordinarily feeds its newborn young. This liquid we put on the fire, warm it to the boiling point, then cool it off, but not entirely – only to the point where it will not burn the human mouth. The liquid obtained from the animal and thus prepared is mixed with the black liquid from the plants."

"Ugh," said the president.

"Then," I continued, "in order to make this mixture palatable we go out into a field and grow a certain plant called a beet, which has a very fat root. For our purpose, however, we take not the leaves, flowers or fruit, but strangely enough the root. When the root has reached the desirable size, we pull it out of the ground, slice it and soak it in big kettles of water until this water has turned them into a sweet pulp. Then we throw away the root. The dirty juice thus obtained is then distilled until all the water is driven off, and the evaporating mixture leaves only dirt colored crystals. These crystals we crush, then by a special process bleach them white, and make a solid mass out of them. The solid we cut into little cubes, of which we drop two or three into the previously mentioned vegetable-animal mixture, wait until they dissolve, and then we drink the whole business."

"Dreadful," said the president. He rang. His secretary entered. "Call up the lunatic asylum at once," he said, pointing to me.

"I know," I said, as they were putting me into straight-jacket, "that an inventor must suffer and struggle much in discovering the world's most popular drink, in getting it known and accepted, and in trying to convince bank presidents that this preposterous concoction will one day be popular, nay perfectly commonplace."

Reprinted from *Companion in Exile* (Allen & Co).

PICKLES and HUMBUG

(A bit of comparative logic)

Anonymous

Pickles will kill you! Every pickle you eat brings you nearer to death. Amazingly, "the thinking man" has failed to grasp the terrifying significance of the term "in a pickle." Although leading horticulturists have long known that *Cucumis sativus* possesses indehiscent pepo, the pickle industry continues to expand.

Pickles are associated with all the major diseases of the body. Eating them breeds wars and Communism. They can be related to most airline tragedies. Auto accidents are caused by pickles. There exists a positive relationship between crime waves and consumption of this fruit of the curcubit family. For example:

1. Nearly all sick people have eaten pickles. The effects are obviously cumulative.
2. 99.9% of all people who die from cancer have eaten pickles.
3. 100% of all soldiers have eaten pickles.
4. 96.8% of all Red sympathizers have eaten pickles.
5. 99.7% of the people involved in air and auto accidents ate pickles within 14 days preceding the accident.
6. 93.1% of juvenile delinquents come from homes where pickles are served frequently.

Evidence points to the long term effects of pickle-eating:

Of the people born in 1839 who later dined on pickles, there has been a 100% mortality.

All pickle eaters born between 1869 and 1879 have wrinkled skin, have lost most of their teeth, have brittle bones and failing eyesight—if the ills of eating pickles have not already caused their death.

Even more convincing is the report of a noted team of medical specialists: rats force-fed with 20 pounds of pickles per day for 30 days developed bulging abdomens. The appetites for *wholesome food* were destroyed.

The only way to avoid the deleterious effects of pickle eating is to change the eating habits. Eat orchid petal soup. Practically no one has any problems from eating orchid petal soup.

Scientists

are

always

affected

by

irreproducible

FACULTY CONCERNS

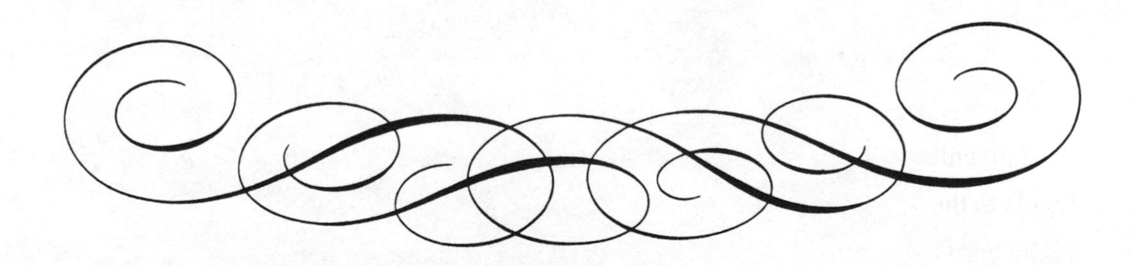

UNIVERSITY OF MONTEREY

Monterey, Mexico

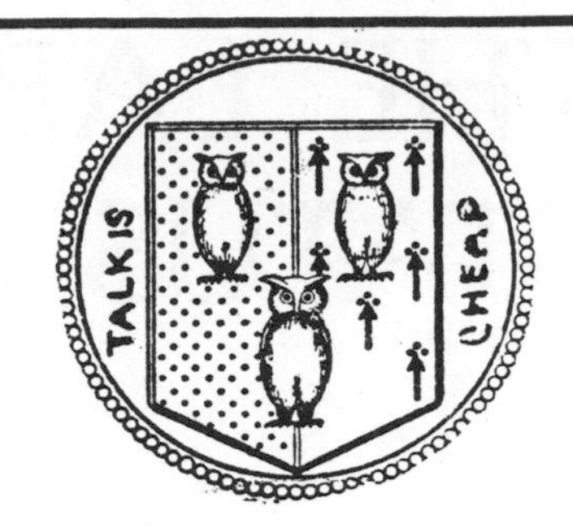

OFFICE OF THE PROCURATOR GENERAL

AGUSTINO BERNARDINO LEGUIA Y SALCEDO
President

Professor A. Lindsay Lewis
Director of Graduate Studies
University of M . . . . .
U. S. A.

Dear Mr. Director:

For six months now our agents, unbeknownst to you, have been carefully reviewing your activities and career, past and prospective. Thus, I act with great confidence when I offer you an appointment as Director of the Graduate School of the University of Monterrey.*

Despite the fact that the University of Monterrey is known as "the oldest University in the Western Hemisphere," I find that few Americans are familiar with our institution. Let me tell you about it.

You have probably not heard of the diabolically clever scheme proposed fifteen years ago by Professor Morris D. Morris (a real person!) of the University of Washington, for raising academic salary levels in American colleges and universities. Morris observed that in the absence of quantitative measures of performance, universities rely on job offers from competing academic institutions to determine whose compensation (promotion) should be raised, and by how much. Almost inevitably an offer from a competing institution results in a matching counter-offer from the "parent" institution. And, over a longer period "equity" considerations operate to generalize these salary increases to those not receiving outside offers.

On the basis of this insight Morris suggested — first to the Ford Foundation — that a "pseudo-University" be established whose sole function would be to raise academic salary levels. Corresponding to this purpose, the University would engage in two activities only: (1) inviting distinguished scholars to give pseudo-papers at the pseudo-University, and (2) making offers and countering counter-offers. Morris estimated that on an annual budget of $500,000 ($100,000 for me, $100,000 for you, $100,000 for administrative expenses and $200,000 for the hotel to house the pseudo-University), faculty salaries in the U. S. could be raised 5% per annum above the previous long-run trend.

The Ford Foundation rejected his proposal out of hand: "You will do too much with too little," they said, "and make us look bad!" A more farsighted institution (and now I must bind you to secrecy), the CIA, saw the merit in the proposal and has supported us since 1958. (I will leave it to you to infer what these initials signify.)

The task of the Dean of the Graduate School requires a delicacy of touch which will challenge you: to make the maximum offer which will *not* be accepted. As you consider this position, I need hardly call the attention to a *labor* economist to the intrinsic attractiveness of being Dean of the Graduate School in a non-University which, if he functions effectively, has neither faculty nor students!

With respect to compensation, it is our view that academic executives, like others, should earn their worth. Therefore, in addition to the basic "support" salary of $60,000 — that should be a nice round target for other deans to aim at — we propose a sliding scale supplemental compensation. Our scale slides both ways. For each pay raise your efforts secure, you receive 2% of the *increase*. For each Monterrey offer *accepted*, you are docked 1% of the acceptee's *annual salary*. If more than 10 offers are accepted in any year, you will be appointed ex-Dean. (Unhappily I must tell you that it cost us some $500,000 to send our former Dean and some 40 academics for whom counter-offers were not received to some place whose name I disremember.)

The next task of our venerable institution — and I hope you will accept it as your challenge — will be to do for graduate students during the next decade what we have done for faculty members in this.

In urging you to accept our position — and let me assure you this is a real offer and not a pseudo-offer — I am sure that I would have the support of our Board Chairman, L. Trotsky (himself a labor economist of sorts), were he alive.

Sincerely,
Agustino Bernardino Leguia y Salcedo
el Presidente de la Universidad

*American printers, somewhat geographicocentric, rendered "Monte*rr*ey" as "Monterey" after some American place of that name. My Scotch soul — you might say — prevents me from discarding the supply of 50,000 letterheads (!) they prepared for us.

THE LAB COAT AS A STATUS SYMBOL*

F. E. WARBURTON

A neat, white, knee-length coat is universally recognized as the uniform of the scientist. The lab coat's primitive function as a utilitarian garment, protective against the dermolytic and vestidemolitive hazards of the laboratory, has bit by bit been replaced by its function as a status symbol. Just as we recognize a bishop by his mitre, or a burglar by his mask, we recognize a scientist by his lab coat. The soldier peels potatoes, cleans his rifle, and even fights his battles in his uniform; the modern scientist rarely works in his lab coat. When work is unavoidable, he will be found in his shirtsleeves, in a coarse brown smock, or in plastic. His lab coat, clean, pressed possibly even starched, hangs safely behind the door, to be worn only when lecturing or greeting official visitors. Like spurs and shakos, the lab coat has been promoted to a new role; it is rapidly becoming, not merely the uniform, but indeed the *dress uniform* of the scientist.

Dress uniforms are worn solely for symbolic and ceremonial reasons, not for practical purposes. Nevertheless, their once-useful features are conscientiously preserved; an infantry-man's sleeve buttons, or the spiked helmet of an Uhlan, are examples. The lab coat is fraught with potentialities for such symbolic survivals. Detachable buttons were highly functional on garments subject to the vicissitudes of frequent vigorous laundering. The modern lab coat should of course be safely drycleaned, but the Chinese puzzles formerly used to hold the buttons in place might well be retained and even elaborated into

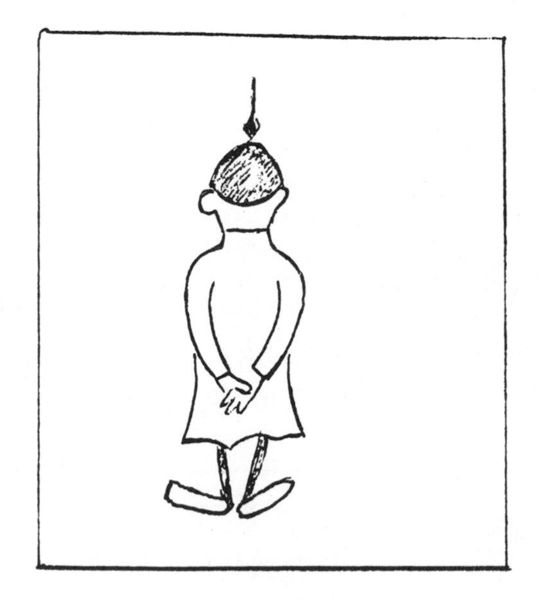

conspicuous ornaments—no longer detachable, of course. The utilitarian lab coat always bore stains characteristic of the work of the wearer. These could be symbolized by chevrons or flashes of suitable colour; purple and red (hematoxylin and eosin) for the histologist; black and orange (sulphuric acid and bichromate) for the chemist; greenish yellow and scarlet (pus and blood) for the pathologist; blue and brown (ball point and coffee) for the statistician. Compact patterns of small holes or a bit of fringe on the cuff, might be other symbols reminiscent of the days when lab coats were worn in the lab. Vertical as well as horizontal status could be shown by such insignia; undergraduates would wear unadorned white; graduate students might claim the right to a single, grey grime-coloured insigne; Ph.D's would wear the colors of the specialties; and Nobel-prize winners, like Admirals-of-the-fleet and Field Marshals, would be privileged to blossom out into creations of their own tasteful design.

These developments cannot be pressed; they must evolve slowly, guided by tradition and respect for the past. But they should be taken seriously. Scientists have momentarily achieved a position of high prestige, but in a democratic society (as in any other) prestige without symbols is but fleeting, while symbols without prestige may endure forever. ■

* Reprinted from SCIENCE by permission.

Umbrella Disappearance, Exchange, and Loss Rates in American Academic Libraries

NORMAN D. STEVENS, Ph.D.
Director, The Molesworth Institute
Storrs, Ct.

"Of course if you leave your umbrella at home, it's sure to rain."[1]

****

"The rain it raineth on the just
And also on the unjust fella:
But chiefly on the just, because
The unjust steals the just's umbrella."[2]

****

As the quotations cited above indicate there are two major natural phenomena associated with umbrellas. These have been the subject of frequent comment but mainly of the folklore variety. Despite Stevenson's brilliant identification of these phenomena in 1894,[3] there has been, as far as an exhaustive literature search can determine, virtually no scientific investigation of these matters.

The relationship of umbrellas to precipitation is, as Stevenson demonstrated, most likely an example of Murphy's Law.[4] Not the least important, and by far the most curious property of the umbrella, is the energy which it displays in affecting the atmospheric strata. There is no fact in meteorology better established – indeed, it is almost the only one on which meteorologists are agreed – than that the carriage of an umbrella produces desiccation of the air; while if it be left at home, aqueous vapor is largely produced, and is soon deposited in the form of rain. No theory competent to explain this hygrometric law has yet been given (as far as I am aware) by Herschel, Dove, Glaisher, Tait, Buchan, or any other writer; nor do I pretend to supply any such theory. I venture, however, to throw out the conjecture that it ultimately will be found to belong to the same class of natural laws as that agreeable to which a slice of toast always descends with the buttered surface downwards."[5] Unfortunately that phenomenon lies largely outside the realm of libraries and information science; the mission of The Molesworth Institute,[6] and its further investigation must be left to others more qualified to treat of it.

Since, as library annual reports have indicated for some time,[7] the disappearance, exchange, and loss of umbrellas is a phenomenon closely associated with libraries, The Molesworth Institute has undertaken a detailed investigation of that problem under the direction of Mr. Adelard Took, whose special qualifications require no further amplification here.[8]

The essence of that natural phenomenon, which curiously does not quite seem to be associated with Murphy's Law, was also stated succinctly by Stevenson. "Except in a very few cases of hypocrisy joined to a powerful intellect, men, not by nature umbrellarians, have tried again and again to become so by art, and yet have failed – have expended their patrimony in the purchase of umbrella after umbrella, and yet have systematically lost them, and have finally, with contrite spirits and shrunken purses, given up their vain struggle, and relied on theft and borrowing for the remainder of their lives."[9]

Based on that initial observation and library reports, The Molesworth Institute undertook an investigation to find some means of measuring the disappearance, exchange, and loss rates of umbrellas in American libraries. Our initial efforts were centered around the work of our CALP (Computer Analysis of Library Postcards) project.[10] Those analyses produced a significant amount of information about the number of people shown carrying umbrellas in exterior views of libraries (17.3%), the number of people shown carrying umbrellas in interior views of libraries (2.3%), and the number of unattended umbrellas (329 in 403 interior views). It soon became evident, however, that this information could shed little light on the basic problem primarily because it represented information gathered at one point in time.

Next we drafted, and tested with a sample of 500 users in 5 libraries, a simple questionary concerning past umbrella experiences and current umbrella behavior. That approach was rejected when the results indicated that people were unwilling to respond honestly to questions on such a delicate subject and when it was apparent that participation in such a survey directly affected the current umbrella behavior which was the subject of the survey.

Direct observation was then attempted but it was too difficult to accurately retain and record the voluminous observations involved. Finally Mr. Took, with the assistance of The Institute's technical staff, designed and constructed a special device designed to photograph automatically umbrellas entering and leaving buildings, whether alone or accompanied by a person, as well as to record, analyze, and tabulate the information contained in the photographs. A number of these devices were built and were ultimately installed in 5 of the largest academic libraries in the United States for the period from January 1, 1974 through December 31, 1974. At the start and conclusion of the survey all umbrellas in the libraries were also photographed and analyzed. The results of this survey are given in Table 1.

TABLE 1
Disappearance, Exchange, and Loss Rates of Umbrellas.

Category	*Library A*	*Library B*	*Library C*	*Library D*	*Library E*	*Totals*
(1) Total Number of People Entering with Umbrellas	63,602	48,718	32,384	27,119	200	172,023
(2) Total Number of People Leaving with Umbrellas	32,301	22,602	15,192	12,041	100	82,236
(3) Total Number of People with Same Umbrella Entering and Leaving	9,363	8,765	6,501	5,853	25	30,507
(4) Total Number of People with Different Umbrella Entering and Leaving	12,622	7,031	5,288	3,206	37	28,184
(5) Total Number of People with Umbrella Leaving but not Entering with Umbrella	10,316	6,806	3,403	2,982	38	23,545
(6) Number of Umbrellas in Library (Start)	529	471	306	602	51	1,959
(7) Number of Umbrellas in Library (End)	832	723	603	935	295	3,388
(8) Umbrellas Unaccounted For	30,998	25,864	16,895	14,745	(+144)	88,358
(9) Umbrella Disappearance Rate (#8 as % of [#1 + #6])	48.3%	52.6%	51.7%	53.2%	(+57.4%)	50.8%
(10) Umbrella Exchange Rate ([#4 + #5] as % of [#1 + #6])	35.8%	28.1%	26.6%	22.3%	29.9%	29.7%
(11) Umbrella Loss Rate ([#1 − #3] as % of #1)	85.3%	82.0%	79.9%	78.4%	87.5%	82.3%

CONCLUSION

This study has determined, for the first time, accurate figures for the disappearance, exchange, and loss rates of umbrellas in American academic libraries. These figures confirm the popular impression that these rates are significantly high. Figures from a number of other American academic libraries are now being collected in the hope of determining standard rates. Figures from other libraries, other environments, and other countries—which are not currently available—would be helpful.

The unexplained appearance and disappearance of umbrellas is a natural phenomenon calling for further investigation. It would appear to be related to the scarf, single glove, and single overshoe/rubber problem in American academic libraries. Investigation should be focused on attempting to confirm the hypothesis that there is somewhere an "umbrella graveyard" similar to the well-known elephant graveyards. ■

FOOTNOTES

1. Gelett Burgess *Are You A Bromide?* N.Y., B. W. Huebsch, 1906. p. 24.
2. Attributed to Lord Bowen (1835–1894). See Walter Sichel *The Sands of Time* London, Hutchinson, 1923. p. 82.
3. Robert Louis Stevenson "The Philosophy of Umbrellas" in *The Mind of Robert Louis Stevenson* N.Y., Thomas Yoseloff, 1963. pp. 110–115.
4. See numerous articles in *The Journal of Irreproducible Results.*
5. Stevenson *op. cit.* p. 114.
6. Norman D. Stevens "The Molesworth Institute" *ALA Bulletin* 57: 75–76, 1963; Norman D. Stevens "The Molesworth Institute Revisited" *ALA Bulletin* 63:1275–1277, 1969.
7. See *Annual Report of the Library at Alexandria for 250 B.C.*, p. 103.
8. Those who wish further information on Mr. Took are referred to J. R. Tolkein's *The Lord of the Rings.*
9. Stevenson *op. cit.* p. 113.
10. Norman D. Stevens "A Computer Analysis of Library Postcards (CALP)" *Journal of the American Society for Information Science* 25:332–335, 1974.

PREPARATION OF PURE CRAP

M. KAYE

(Proposal for research contract in connection with the National Bureau of Standards program for the Preparation of Pure Materials.)

It has been estimated that about 90% of the scientific complement of almost all scientific institutions is engaged in the processing of crap, and some scientists have placed this value at even higher levels[1]. Correspondingly, at least 90% of the world scientific literature can be regarded as describing systems or discussing problems generally within this subject area.

Crap is known to exist in great natural abundance[2, 3]. Some edible forms of the material are known[2, 3]. The metallic form has been reported to possess rather recalcitrant properties and to be of high rigidity[4]. Other forms appear to have a wide range of properties, of which the prominent features seem to be a lack of suitability or adaptability of the material as regards to any possible relevance or practical application[5].

Little work has been done on the purification of this material. This may be due to a lack of demand[6]. However, small amounts of crap have been purified in some laboratories[7]. Lately, there has been evidence of an increased demand for crap, particularly for distribution in high-ranking Government circles[8]. In view of this, it is felt that preparation of the pure material in large quantities might be worthwhile.

The preparation of pure crap requires rather large-scale planning and it is felt that organization of the project as follows will give the best results.

The preparation of pure crap can be carried out at any Research Institute, particularly at Institutes which can provide the necessary large scale facilities.

Stage 1. A Pure Crap Advisory Committee should set up, consisting of 24 members. Other committees should be set up to deal with particular aspects of crap as they arise.

Stage 2. Re-appointment of the committees and arbitration work should then occupy the second year. (Auxiliary psychiatrically trained consultants and industrial arbitration experts should be available, possibly a senior surgeon and one or two first aid men). During the third year, efforts should be made to recruit the technical staff and to set up the laboratory organization.

New appointees should be processed via the various official regulations and procedures for recruitment of scientists which assists in automatically selecting suitable candidates for work on crap (e.g. blood pressure is more important than academic qualifications *per-se*).

Appointment of the technical staff, and re-arrangement of the appointments according to the requests of uncles, old school pals, and ex-army buddies of the department heads should take up the third year of the project.

The fourth year of the project will be occupied by preparation of reports and their discussion in the various committees.

It is hoped that by the end of this period, kilogram quantities of pure crap will be ready for stock piling.

Budget. The budget required is astronomical. The production costs per Kg. of pure crap have been estimated as high as $500,000. However, as most of the facilities are already available, it is felt that with the contribution of $2,000 (to cover the cost of publication of reports) from the sponsor, the project can be realized within existing frameworks. ■

---

[1] Personal communication: figures within the range 90–95% have been quoted by members of 8 scientific institutes, mainly referring to the work of their colleagues.

[2] A Roumanian restaurant on the Petah Tiqua Road, Tel-Aviv, displays a sign listing: "boiled crap, fried crap, stewed crap, baked crap"* (Try the fried).

[3] A. M. Kaye, personal communication, e.g. "I'm starving – serve out the crap." However, see also: "You surely don't expect me to eat that crap?"

[4] As, in the laying of pipes: "I can't get this crappy pipe to bend."

[5] Personal communications, e.g. "You expect me to do anything with that crap?" However, it has been reported that crap output has been used as a basis for promoting of scientists to senior rank and a position with a peak at 0.3 cu. m. crap/year. Non-purified crap, of course.

[6] James Cagney in "World of Gangsters (1949)." "Don't give me none of that crap. I don't take no crap from anybody."

[7] Several sources reporting on the reaction of Research Directors to annual reports: "This is pure, unadulterated crap."

[8] Anonymous Ministry Official: "Don't bother to work out the figure, just give them some crap."

* They mean "carp."

UNIVERSAL SCIENTISTS' LETTERS FOR ATTENDANCE AT CONFERENCES CAN BE ADAPTED ACCORDING TO TASTE AND INGENUITY FOR INDIVIDUAL USE. CAN ALSO BE USED, WITH SUITABLE CHANGES, AS A MODEL IN INITIATING A BI—NATIONAL FELLOWSHIP OR EXCHANGE PROGRAM

Hans Holbein the Younger, ERASMUS, WRITING AT HIS DESK, Art Gallery

Example 1 Initiating letter

Tel-Aviv
June ....

Dear Charles,

I feel very guilty about not having written to you in the last six months or so. Things have been pretty hectic in the lab. We have been winding up the DNA -estrogen story and are fairly satisfied that the peak of stimulation we observe in the newborn animals is the result of a secondary mechanism involving the estrogen cycle. Shlomo has been working on the methylation experiments but is unable to repeat your results using the cross-fertilization technique. I would love to talk this one over with you and also see what you folks are doing.

Which brings me, incidentally, to what is no news for you, that is, the Febs meeting in New York in September. I would have liked to bring Tamar and the kids but we do not want to take the boy out of school. Accordingly, I will be arriving alone in N.Y. on 12th September (flight 342 arriving 2 A.M. at Kennedy) and will probably stay on for about a week or so after the meeting to pay some visits and give a couple of seminars. Are you people interested? — the estrogen story has revealed some quite fascinating features.

I wonder whether it would not be too much of a bother for you and Sheila to put me up? I am not very keen on hotels and relish the family atmosphere. I will be staying approximately 2½ weeks. Also, if Sheila still keeps the old Chevy tucked away in the garage for occasional shopping, would I be able to borrow it over the week-end to go down to Boston — I want to visit Walter's group — and so on?

We were puzzled to know what to bring you from here that you haven't got already, but Tamar suggested a pot of her home-made tomato chutney you were so fond of — it should go down nicely with those terrific T-bone steaks you used to make. Do you still go in for those barbecues in the yard I remember so well?

With much affection from all of us here,

Benjamin

Example 2 Response letter

New York
July ....

Dear Benjamin

We were delighted to hear from you. I had an idea you would be turning up for the Febs meeting. I look forward to a chin-wag Shlomo's work and of course, on what is going on generally in your lab.

As for putting you up, naturally we would have loved it, but there is a complication. Sheila's sister just got divorced and is spending the summer with us. What with her six kids and our four, it is a pretty tight fit, and frankly, absolute pandemonium. I think you will be glad to have a nice quiet hotel room — in fact, I might even join you! Do let me know if I can help with booking arrangements and perhaps Sheila and I could join you for dinner at your hotel, so that we can hear all the gossip, I suggest on the 14th, as you will probably want to catch up on sleep on the 13th, after arriving so late? I also suggest you take a taxi straight to the hotel as the airport bus service is pretty poor during the night hours, even in New York. Even then, you probably won't get in till 5 A.M. or so.

How thoughtful of Tamar to send the chutney. She really shouldn't have gone to such trouble. Barbecuing is difficult, with such a large household as we have at present, but what we usually do is pick up frozen hamburgers at the Supermarket and the chutney will do down very well with these. I hope you'll join us.

As for Sheila's Chevy, in normal circumstances of course, we'd be glad to let you have it. As it is, it's pretty much in use ferrying the kids to Sunday school, as you can imagine. To be on the safe side, I suggest you rent a Hertz car.

Do you remember our second boy Philip? He is now 17, believe it or not, and being the adventurous type wants to spend his summer vacation in Israel, with a couple of pals. They are basically good kids, so pay no attention to the torn jeans and long hair! Sheila and I are convinced it's just a phase. Anyway do you think you and Tamar could generally keep an eye on them for us when they are over there? They'll probably need just the occasional bath and meal and maybe what they call a pad for the odd night or two. You might find their music quite interesting. Phil is taking his electric guitar and one of the other boys plays the drum (or any odd cooking pot as a substitute).

Looking forward so much to seeing you again,

Charles

Example 3 Final Response Letter

Tel-Aviv
August ....

Dear Charles,

Acknowledging your letter — which I just received — I see you really have a houseful! Anyway, I am now all fixed up. I finally remembered that I have an aged uncle who lives in the Palisades, New Jersey, just over the bridge and as I am really not at all keen on hotels I wrote to him. Turned out his room-mate just died and he got permission for me to bunk in with him. Please do therefore come out for supper on the 14th which I understand is a light meal served at 6:30 P.M. followed by community singing.

The address is:

Home for the Aged and Indigent
Fort Lee, 1417
Palisades, New Jersey

I reckon it is not more than 1½ hours drive from your place, so looking forward to seeing you.

Shlomo is pretty convinced the response you people found may be due to impurities in the enzyme system — he does not find it when he works with the doubly purified hydrogen, but it appears when he uses the commercial grade directly. Anyway, it will be interesting to talk this over. We are now preparing a paper for BBA on this work.

As for Philip and his pals, they are more than welcome. This is just the place for adventurous lads. There are several trips for youngsters to border settlements where the firing is. They go with an army escort so I think it is fairly safe! The boys are also welcome to camp out on our lawn (as you know, we have no spare room) but we do have an excellent tent and there have been very few snakes this year as it has been a cool summer. Anyway, as you know the Kaplan Hospital is only minutes away. I really only mention this as I suppose Sheila is the usual anxious mother.

Looking forward so much to seeing you.

Benjamin

P.S. You remember that pretty lab girl Mary you worked with when you were over here? She is now married and has just had a baby (premature). She sends you her love. ■

Lab Productivity I. Correlation to Snack Shop Visits

L. BOONE, J. COLLINS, S. WINSLOW,
C. WHITLEY, W. RUTALA,* D. CLANTON,
G. STEWART, J. OAKES, G. PETERMAN,
D. WENNERSTROM, S. HOLCOMB, S. BROWN,
J. MARTIN, D. ROOP
Department of Microbiology
University of Tennessee,
Knoxville, Tennessee 37916

(Lab Productivity/daily/groups/the)

Communicated by W. S. Riggsby, December 1, 1972

ABSTRACT Single daily visits to the Snack Shop, both alone and in groups, enhance Lab Productivity. Daily visits greater than one correlate with decreases in Lab Productivity. As the number of participants in the visit increases, the Lab Productivity falls drastically with each additional visit until it plateaus at three Lab Productivity Index units in the hole.

[1]Lab Productivity Index:

9 = Hours active lab work on day you thought was a site visit.
6 = Hours in lab on normal day. (may or may not be active)
3 = Hours in lab on day director and/or major professor was away.
0 = Hours in lab on day when funding agency paid unexpected visit.
−3 = (−) Hours ÷ 2 you said you were working in lab last night.

This research was supported in part by Grant No. BLT-7UP-IOU-1.25. L.B. is a Snack Development Award recipient, S.W. is a UT Lunch Fellow, and C.W. borrowed money from Ray. The authors wish to acknowledge Richard's Snack Shop for excellent technical assistance.

For reprints send $1 to 1st author, add your name to the list and send list with instructions to 10 friends. ■

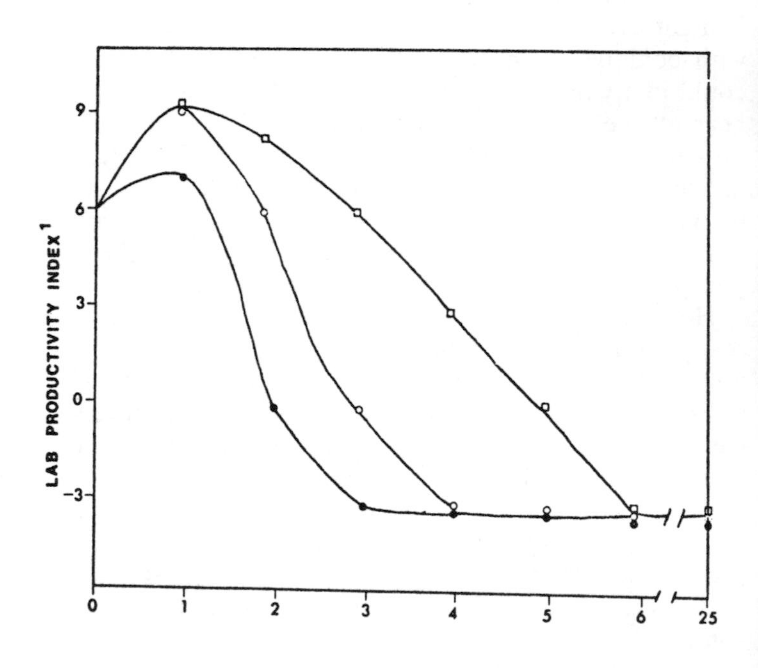

FIGURE 1A.

Correlation between visits to the Snack Shop(SS) and Lab Productivity(LP). (—□—) alone; (—○—) groups of two; (—●—) groups larger than two.

*Present Address:
5th U.S. Army Medical Lab, Ft. Sam Houston, San Antonio, Texas

Toward Greater Efficiency of Information Transfer

HARLEY L. SACHS
Assistant Professor, Humanities Department

Action and discussion in the Michigan Tech. Senate have established a premise which needs to be pursued in the interests of the public and in the interests of this institution. Specifically, a policy was passed stating that our courses are informational in nature and measurable by a single examination in lieu of taking the course. Using this policy as a fulcrum, the Senate has pressured us beyond the point of balance, and we must follow the law of gravity — all the way to the bottom. Let us then proceed.

Since it is only the facts that matter, there is no justification for inefficiency in the transmittal of those facts. Inefficiency is a universally recognized evil, as obvious to a mechanical engineer as it is to a business administrator. What is needed is an elimination of waste and inefficiency in teaching. Some steps already have been taken in that direction here at Michigan Tech. Professor Erbisch has prepared a wonderful series of taped instructions which are used in his biology classes to give individual step-by-step information and training to the students. There can be no doubt about the value and advantage of his program.

What needs to be done is further and broader application of this more efficient method of information transfer. In the case of biology, the tapes will eventually be prepared for distribution to other institutions across the country. This is a pilot effort at an early stage. To pursue our basic premise of efficient transfer of information, there is no logical reason why the same method cannot be applied to other courses.

The programmed courses and tapes are very marketable. They bring in revenue to those who prepare them and revenue to the school administrations that share in the expense of their preparation. This will relieve part of the financial burdens of this university and release progressive educators so they can pursue more basic research. Admittedly, in the cases of some professors who have neither the ability nor the inclination to basic research, but are mere teachers, it may be difficult for them to find employment in a job market saturated by tapes and other more efficient means of information transfer.

This is all to the good. There is a shortage of engineers and skilled technicians, and there is no excuse for tying up this valuable talent in inefficient teaching when there is need for their talents elsewhere. There should be no complaint about releasing talented people who have been saddled with mere teaching when they can make, as one circulated complaint about summer teaching pay scales put it, more money as skilled or semi-skilled laborers. Many of the present faculty could effectively alleviate the shortage of personnel in jobs that cannot be better filled by teaching machines, jobs such as plumber, auto mechanic, or electrician.

Not only may these men find more lucrative jobs, but the budget of the university can be reduced. With adequate contributions from the alumni and adequate income from industry in the form of patent royalties and shared research costs, the university could more closely approach the efficiency of a well-operated business. There might even be hopes of profits.

Ultimately the rental of course tapes and programs will spread nation wide. Instead of being hamstrung by wasteful, inefficient teachers, educational institutions may confine themselves to their real purpose: information transfer. Properly and most efficiently performed, this could be achieved by a single federal agency that supplies facts and information on all subjects.

This leaves one link in the information transfer system that remains a problem. We have taken care of the source of the information. Its recipient must be dealt with in equally efficient manner. Anyone who has seen test scores realizes that, under the present system, the information does not transfer. There is a percentage of mediocrity and a percentage of failure. This is inefficient. Not only that, but experience has shown that the recipient of the information, though improvable, cannot be made efficient. A replacement should be found. We have eliminated the need for the human teacher. Now we must find a replacement for the human student. This may be the subject of a future article. ■

. . . From SINKINGS, Vol. 3(1)

"Dr. Kaiser Emigliano has received $867,483.73 from NSF for a one-year feasibility study to transversely saw Virginia Key from its base and float it. This project will provide valuable clues to the Pleistocene geology of our tight little isostatic island and will yield valuable fossils to be sold at the Seaquarium to replenish the long-suffering overhead account 8800.

Concomitant with Dr. Emigliano's calculated recommendations, the trustees of IMS have embarked upon the remarkable concept of the "Visiting Institution" to replace the Visiting Scientist. This breakthrough will enable our entire Institute to be towed, by LCM, to other areas for periods of 6 to 12 months, depending on tidal flow and neglecting geostrophic forces. This will eliminate the need for sabbatical leaves and vacation periods. Mr. Truly Illman reports this also will greatly reduce paperwork for travel forms, as only a *single* form will be needed. Instead of individual scientists attending meetings, an *entire institute* could be represented! The prestige value alone is staggering to the imagination. Site visitations by NSF committees could be eliminated as the Institute could visit the committee wherever it is. Costly interlibrary loans could be eliminated since the Institute could visit libraries *en masse* to examine a given journal. (Note: This is provided that all faculty members are seeking a given journal at the same time.) Furthermore, the proposed move would eventually eliminate the need for research grants, as floating institutions would be considered research vessels, and thus would be eligible for "block funding" on shiptime programs.

Early reports have reached our desk indicating that other institutions are following our brilliant line of endeavor. Texas A&M, located 200 miles from the coast, has received modest financial assistance from the LBJ ranch and the Mohole Project, while Woods Hole Oceanographic Institution is cross-cutting their peninsula in a gallant, but dastardly attempt to be the first "All-Afloat Institute". Scripps Institution of Oceanography poses a definite threat to IMS progress, as they are located (at the time of this writing) on the San Andreas fault, and need perhaps only a slight push to become insular."

The Seaside Laboratory

COLLEGE OF SEA SCIENCES

University of Southernmost Florida
1.25 Ricketyback Causeway
Miami, Florida 33169

Submitted by
Samuel P. Meyers

Origin of Margarine

To the Editor — I have been interested in the history of bread spreads for some time and I am delighted to see a scholarly and thoroughly researched article on bread spreads (**214**:2312, 1970).

It may be of interest to the readers of THE JOURNAL to receive information on the origin of the use of margarine[1]: In brief, it was initiated as an economy measure by Abbott Francis Geschmiertsparer (1645 to 1712?) of the Order of St. Benedict in Benedictbeuren, Bavaria.

The first, although somewhat illegible, report was found in one of the monastery's food lists, which states in 1701: "Unschlitt and Fischoel, 12 Fass" (lard and fish oil, 12 barrels). On pursuing this matter I discovered the following: The monks were required to attend prayer in 1690 (to about 1712, when this practice was changed) for at least 14 hours a day. Consequently, they developed the habit of supplying themselves with some high caloric nutrient, eg, butter, which for concealment and convenience was often placed on the inner sides, or margins, of their missals. However, the price of butter had risen from 1 Batzen for a Lot (about 20 cents for 1 oz.) in 1692 to 5 Batzen in 1797.[2] Therefore, some means of supplying this food at a lesser cost became mandatory. It is said that Brother Augustus Salbenstirrer was responsible for the invention of the mixture of lard and fish oil, colored with yellow saffron, resembling the high priced spread closely.[3] Since this substance was often attached to the margins of missals (vide supra) for the use during those long vigils the name Marginal Food or Marginanus developed naturally.

Marginanus is first mentioned in the said food lists in 1705 ("Marginanus 6 Fass"). Obviously it spread — and became well known as Margerinanus in all of Southern Germany by 1758.[4] The etymological development of Margerinanus to margarine is obvious.

H. F. CASCORBI, MD, PhD
Cleveland ■

1 Cascorbi HF: Low cholesterol diets in the 18th century. *Allgemein Krankschaft und Gierschift Zeitung* **12**:121, 1962.
2 Von Calb H: *Cycles in Economic Structures*, Urban und Schwarzhaupt, 1905, pp 102-107.
3 Forecastle Hilmar: *The Yellow Rose in Bavaria*, Heimeran, 1936.
4 Crassuss HF: *The Guild System in 18th Century Svavia*, Springel, 1921, p 113.

Delaware State College
Dover, Delaware

Office of the President — **RESOLUTION** — June 15, 1982

TO: All Department Chairpersons

FROM: The Board of Trustees of Delaware State College

WHEREAS, THE BOARD OF TRUSTEES AT ITS MEETING OF MAY 13, 1982, NOTED ITS DISPLEASURE WITH THE USE OF THE PHRASE ABBREVIATED BY THE LETTERS, "M.F.", IN THE INSTRUCTION OF STUDENTS IN THE CLASSROOMS AT DELAWARE STATE COLLEGE,

NOW THEREFORE:

BE IT RESOLVED BY THE BOARD OF TRUSTEES OF DELAWARE STATE COLLEGE THAT THIS RESOLUTION BE SENT TO ALL DEPARTMENT CHAIRPERSONS TO BE PASSED TO ALL TEACHING FACULTY THAT THIS LANGUAGE WILL NOT BE USED IN THE INSTRUCTION OF STUDENTS IN THE CLASSROOMS AT DELAWARE STATE COLLEGE.

Luna I. Mishoe, President
Delaware State College

William G. Dix, President
Board of Trustees of Delaware State College
and Members of the Board of Trustees

THE PRACTICAL APPLICABILITY OF TEST-RETEST RELIABILITY ON A DECISION MAKING ALGORITHM: THE CASE OF THE MUTUALLY EXCLUSIVE AND EXHAUSTIVE ALTERNATIVES

Dan R. Dalton
Graduate School of Business
Indiana University
Bloomington, Indiana

You would think that subjecting oneself to a journal article containing an unconstrained diatribe against a pet theory would be trial enough. Imagine at the same time proctoring an examination in introductory something or other for over 300 students. Add to that a student in the rear left quarter of the lecture hall making distracting gestures and you have the plight of the eminent Professor Coe E. Ficient.

With a stream of consciousness sharing, albeit it disdainfully, a recognition of some 300 number 2 pencils rustling across 150 questions on machine scoreable answer sheets, a completely untenable, not to mention trivial, objection leveled at the pet theory, reflecting on the teaching assistant calling in sick on this of all mornings, and some student flipping coins during the test.

"Flipping coins; flipping coins; I don't believe it!" muttered Professor Ficient under his breath. With his attention now clearly focused and with a combination of amusement, indignation, and curiosity, the good Professor watched intently as a student imperviously flipped a coin into the air; a moment later, again; and, again.

Finally, unable to contain himself further, Professor Ficient managed to catch the eye of the student (no small accomplishment, given the student's rapt attention on the coin) and gestured him to come forward. With relatively little ceremony, considering climbing over and around a dozen or so other students to reach the aisle, the student faced the incredulous Professor.

"Son," asked the bemused Professor, "I have been watching you for some time. It appears that you are flipping a coin at somewhat regular intervals."

The student just nodded his head affirmatively.

"Dare I ask why?" continued the Professor somewhat annoyed at the lack of explanation.

"Frankly, I am not prepared for this examination," answered the student with no hint of embarrassment. "So, to choose the answers, I flip a coin. When I get a 'head,' I mark true; a 'tail' is false."

Rather than reacting strongly as one might expect, the Professor merely asked, "How do you handle the multiple choice questions?"

With no hesitation, even confidently, the student countered, "All of the multiple choice questions have four alternatives. Therefore, I flip the coin twice. If I get two 'tails' in a row, I pick the first alternative; two 'heads' back to back indicates the second. If the first toss is a 'head' and the second a 'tail', I select the third choice; a 'tail' then a 'head' is the fourth."

By this time, several of the other students had finished their examinations and were, with considerable ceremony, bringing them to the front of the class. Aware that the fewer of the other students who knew what was going on the better and inasmuch as further interaction with this coin-tossing student was not likely to be very productive in any case, Professor Ficient sent the student back to complete the examination.

True to form, while the remaining students filed past the front desk depositing their examinations, the coin tosser continued to cast his coin fervently into the air, consult the result, and hover over the examination.

Time passed and fewer and fewer students remained in the lecture hall. Still, the coin rose and fell. There were only three, then two students. Finally, only the student and the ever-active coin remained. Professor Ficient continued to watch the exhibition as the large clock at the back of the lecture hall marked the passing of five, ten, fifteen more minutes. The coin continued to cascade through the air; the student feverishly pressed on alternately entering and erasing responses on the machine sheet.

Finally, in exasperation, Professor Ficient called out to the student, still flipping the coin in the back of the now empty lecture hall, "Aren't you finished yet?" It had been well over fifteen minutes since the last of the other students left.

"Yes, sir", replied the student unabatedly flipping the coin. "I was finished some time ago. I was just checking my answers." And again, the coin tumbled through the air. ■

MR. EVAN ESCENT
QUANTUM
MECHANIC
IN
(BUT NOT
NECESSARILY
HERE)
BARNETTSCHARF

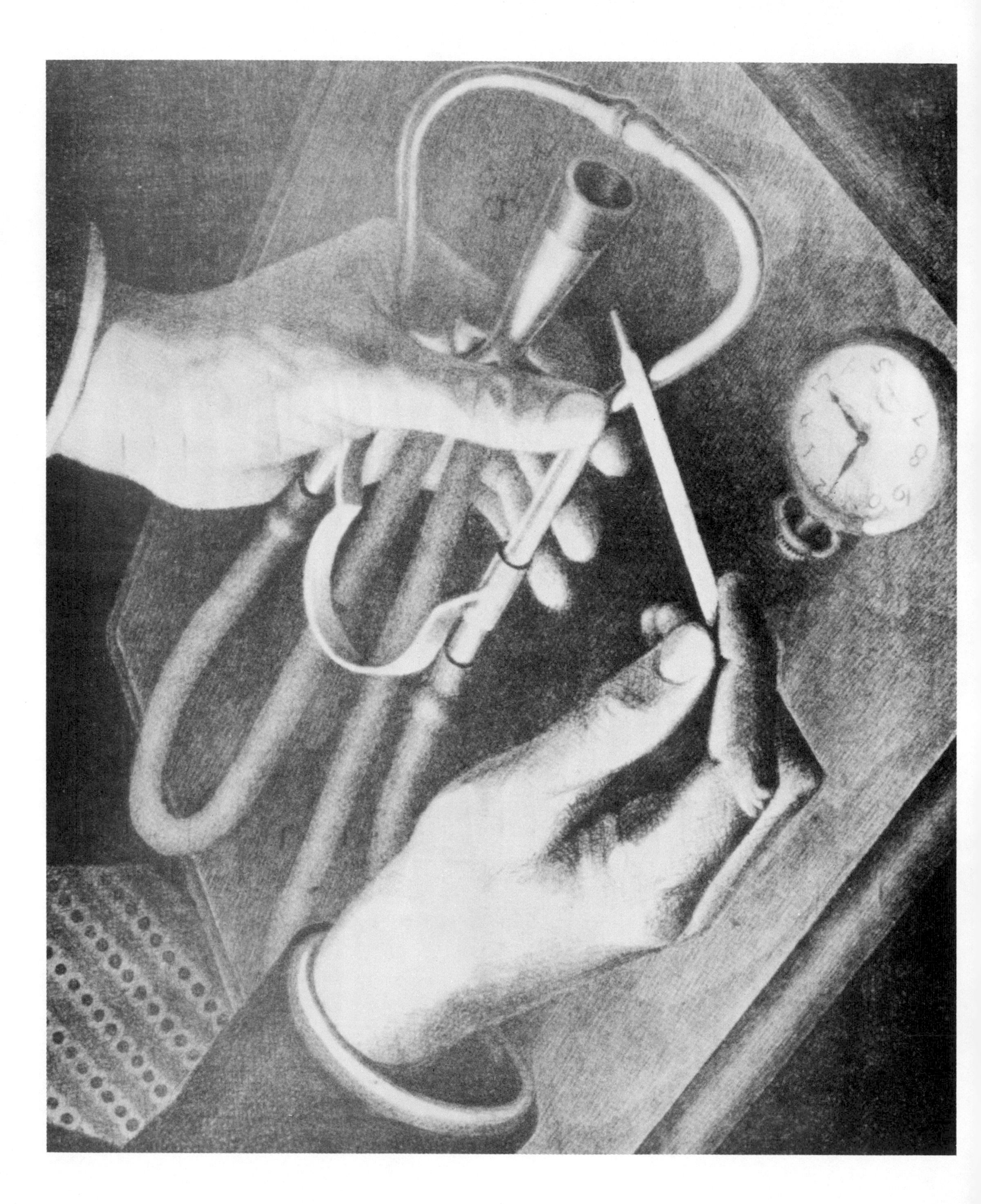

Irreproducible
information
from the
art (?) or
science (?)
of

MEDICINE

A Short Guide to Doctors

JOHN J. SECONDI, M.D.

Medicine, like every other field these days, is so overspecialized that even a card-carrying doctor like me has trouble telling who's who. The layman, I imagine, is almost helpless to distinguish the forest from the tree surgeons. I have noticed, however, that my colleagues have a tendency to run to type. So, in an effort to clear up the confusion, I have compiled a little list so simple that the most naive patient can spot at a glance which doctor is which.

The General Practitioner: These gentlemen used to be the ones you saw most often, when you lived back in Nebraska and watched the cars go by from your front porch for entertainment. Now they are nearly extinct, like the buffalo and the stork, although there are a few left in a preserve in Iowa. Most of them looked like a cross between Charlie Ruggles and Colonel Sanders; they were warm, wonderful, and always had time for you, even if they slept only three hours a night. They knew you inside and out from the moment they delivered you until their ink dried on your death certificate, and they were always there to help you push your car out of the mud. If anybody knows where there's one of these left, please drop me a note. I could use a good doctor myself.

The Internist: This is a general practitioner with more diplomas on the walls and without house calls. (He also has money in the bank.) By the age of thirty at the latest he becomes obese, sallow, and emphysematous. Usually bald, he is always found sitting and smoking a pipe. (The pipe is a deliberate attempt to evoke the Delphic Oracle, which also simmered and steamed with ideas. The internist is nothing if not oracular.) As opposed to the surgeon, who carries no equipment at all except the keys to his Rolls-Royce, the internist can be seen with a stethoscope protruding from one of thousands of pockets in his clothing. Really big stethoscopes are worn to give the impression of expertise in heart disease.

In his desk the internist stocks lifetime supplies of sample drugs; when you are in his office he may pick one or two at random and give them to you with alarming liberality. But don't worry; he won't let you know what they are. The internist is really happy only when deciding how to cope with some chronic incurable disease, preferably in a case some colleague has botched. The longer the name of the disease, the happier he is; and if it's in Latin he's ecstatic. An internist is required by law to have his phone ring twice an hour at least while he is at home, and he can never vacation. No internist's children ever become doctors.

The General Surgeon: These are the prima donnas of the trade. Today's surgeon is descended from the barbers of the Middle Ages, but washes more often. He may be fat or slim, but he is always loud, noticed, and in a hurry. He dashes dramatically in and out of rooms (whether patient, operating, or bath) and never lets you finish a sentence. To probing questions he nods wisely, smiles enigmatically, and runs off. (Never say "Now cut that out" to a surgeon.) He generally visits patients cloaked in green from head to toe to give the impression of being fresh from an operation, when probably (unless he is over fifty) he has been idling in his office all day waiting for someone — anyone — to call. After he does a surgical scrub, he raises his arms, which drip from the elbows. This posture serves both as a gesture to God for the usual assistance, and as a method to keep the bacteria flowing away from the surgeon — towards the patient.

Surgeons are taught early to rip off bandages as quickly as possible, pulling as much hair as they can get out with one clean tear. They have a distinctive jargon: For instance, they speak of healing by "primary intention" (which is to make lots of money), or by "secondary intention" (which means the wound got infected). If anything goes wrong during operations, surgeons are unanimous in blaming it on the anesthesiologists. Surgeons are the only doctors left who haven't cut out smoking, because they are confident they can cut out the cancer. If this description still doesn't make a surgeon flash in your mind, recall James Coburn in *Candy*. Absolutely accurate.

The Gastroenterologist: Gastrointestinal doctors, or "GI men," have had oral fixations since childhood. This means they are always talking a mile a minute, and at mealtimes they ingest like Electroluxes. They're usually roly-poly, literary, and very pleasant to gossip with, as a consequence. The unsavoriness of their work is grossly exaggerated; nevertheless, they do receive a lot of cologne for Christmas. As kids they were the ones whose parents always had to bang on the bathroom door to get them out. If Alexander Woollcott had become a doctor, he would surely have been a gastroenterologist.

The Obstetrician-Gynecologist: The real wise guy in medicine. Sitting on their high stools day after day with their patients in that absurd saddle, these comedians see the funny side of life. They have to have a good sense of humor because otherwise they would be so nauseated by some of the things that come along they would swear off sex forever. Always ready with a wisecrack or a foul story (depending on whether you are a patient or another doctor), they are universally popular, except with pediatricians. Child doctors blame every childhood disease from thumb-sucking to Mongoloid idiocy on the anesthesia the obstetrician used. It is not true that all obstetrics is done at three o'clock in the morning. I personally recall one case in 1968 that was done at six in the morning, and others may have had similar experiences. Many Ob-Gyn men are now getting crash courses in abortion, which was never part of the medical curriculum before. These are the ones with the sterile coathangers.

The Urologist: Urologists do for men what gynecologists do for women — more or less. They are drawn to their specialty irresistibly by its identification with the masses of tight curly hair they all have. Many urologists are now growing beards and muttonchops so you can spot them more easily than ever. Walking around weighted down by their waterproof rubber aprons and all those whiskers they resemble Noah before the flood. At home they putter around a lot with the kitchen sink to keep in practice. Like other plumbers they work good hours and make a good living.

The Anesthesiologist: Anesthesiology is the Tower of Babel of Medicine. There are a total of four English-speaking anesthesiologists in America: two on the East Coast, one in Chicago, and the other in L.A. All the others communicate with frenzied nasal accents or sign language. They are short, shy, retiring types who hide behind the sterile barrier during surgery and squeeze contentedly on their little black respiration bags. You will recall the bags from 1930's movies because when Lionel Barrymore came too late they quit moving and you knew the patient had died.

The night before your next surgery an anesthesiologist may mince into your room, unheralded and uninvited. He will never show up after five o'clock, however. Without bothering to identify himself or even to ask for an interpreter, he will quiz you on all the allergies you may have. This is done to choose exactly the right toxin to put you to sleep with the next day; for God's sake, don't forget anything that might be relevant. Then he will bow graciously and back out of the room, and you'll never see him again. (Whether you'll see anyone else again is another question.) If you wake up from surgery, you may have a sore throat, even after an abdominal operation. This complication occurs because the anesthesiologist routinely puts the rubber airway down your esophagus six times before he finds the right hole. Anesthesiologists fear surgeons the way helpless children fear angry fathers. They are very sensitive and feel left-out enough, so be kind to them.

The Pediatrician: All pedi-pods, as they are called on the wards, act and look like Peter Pan. They wear saddle-shoes and bow ties, have cherubic faces, and wear crew cuts or pageboys, depending on whether they are over thirty. They are shorter than most other doctors, although they can be told from anesthesiologists because pediatricians are slightly taller, speak English, and have horizontally placed eyes. They never use words longer than two syllables or sentences of more than four words. Generally they sound as if they are doing Jonathan Winters imitations. Many have a lilting gait, and a few skip during clinic. About the age of forty they lose patience with all those frantic mothers and either commit suicide, go into research, or start child labor camps.

The Orthopedist: All bone doctors without exception are former college jocks or team managers; i.e., they are big brutes or mousy types who wish they were. They *all* wear white athletic socks. (This is the one infallible rule of medicine and makes it a snap to recognize an orthopedist.) They usually have plaster of paris splattered on their clumsy-looking shoes.

The Ophthalmologist: (This is the real "eye doctor" and is not to be confused with optometrists or opticians, who aren't M.D.'s at all.) Ophthalmologists, despite being constantly misspelled and mispronounced, are the happiest men in medicine. They work laughably few hours, make extravagant fees, and are adored by their patients, who understandably value their sight above all else. Thus ophthalmologists are always well-tanned and talk knowingly of Tahiti and the Riviera. They also play a great deal of golf. Curiously, they are uniformly tall, slim, and vaguely ethereal. A good example would be Pope Pius XII with a suntan. They use a jargon so technical and so infinitesimally detailed they cannot even make small-talk over drinks with other doctors.

The Otolaryngologist: When you've been to an ear, nose, and throat man you'll never forget it. These are the true sadists in medicine. Children despise them; they're the only doctors who make you feel worse than when you came in. What with the nausea produced by cocaine sprayed in the nose, and all those tiny little probes poking God-knows-where back in your sinuses, and all that blood swallowed after a tonsillectomy, going to an ENT man is like being used for an experiment by Edgar Allan Poe. The only common physical characteristic by which these gentlemen can be spotted is that they all still have their own tonsils.

The Plastic Surgeon: Immediately identifiable. They all wear handtailored clothing of vast expense and have sculptured features worthy of a Phidias or a Michelangelo. Gorgeous from every angle, they look years younger than they are, and boy do they know it. They were the big face men of college fraternities, the ones who were put strategically at the front door during rush week and in the first row in the yearbook picture. They are very rich because there are a lot of jealous women who will pay *anything* to look as good as the plastic surgeon. They have a tendency to get very snotty when you ask them what kind of plastic they're going to put in, and lecture you on the origin of the word "plastikos" in the traditions of Greek sculpture. Nevertheless, they have terrific guilt complexes about spending all their time on frivolous surgery, so they occasionally take on a burn victim to soothe their consciences. The patron saint of plastic surgery is Narcissus.

The Psychiatrist: Spotting a psychiatrist on the street is easy enough, but as he wanders on the wards of a state hospital he may need a nametag. Psychiatrists either avert their eyes from you or stare right through you, whichever makes you more uncomfortable. If they sense you're going to ask a question, they slip one in first. They never use complete sentences, only clauses and long words. I know a psychiatrist who begins every sentence with the word "that" and ends it with an exact quotation of Plato. The main object a shrink has in mind when he sees a patient is not to rescue the patient's sanity but to prove his. After all, how many surgeons do you know who have five years of operations on themselves before they can practice?

Today's psychiatric resident may have elbow-length hair, wear rings in one ear, and go to work in purple satin capes. This kind of psychiatrist has not hit Park Avenue yet, but it's only a matter of time. Incidentally, the reason there are so many Jewish psychiatrists is that they are basically yentas with M.D.'s.

The Radiologist: Radiologists hide in dark places and never come out, like other rodents. They make creepy-crawly gestures and rub their noses frequently. Their world is one of shadows and they detest the light of day. They hole up in leadened tunnels and, though they tell you that X-rays are harmless, they generally have their offspring as early as possible. (By the time they reach fifty irradiation has given their skins the texture of refried beans). Like Whistler, they view everyone as a study in black, white, and gray. The thought of being responsible for a live human being is abhorrent to them. Yet they give off smiles of delicious perverse pleasure as they make patients swallow thick white slime and poke around in their bowels so their guts will show on film. Peter Lorre would have played a perfect radiologist.

I would like to complete the list, but I just got a frantic phone call from A.M.A. headquarters, and I have to run. Something about an emergency protest march against socialized medicine. ■

ADJUSTMENT OF MONKEYS TO FIVE CONTINUOUS DAYS OF WORK

Science 1962, *138*, 43.

****

EFFECT OF AUDITORY STIMULATION ON REPRODUCTION .IV. EXPERIMENTS ON DEAF RATS.

Proc. Soc. exp. Biol. & Med., 1964, *116*, 636

..."The experiments show that auditory stimuli during the copulation period do not cause a decrease in fertility in deafened rats in contrast to normal animals."

****

SUGGESTIONS ON ADULT CIRCUMCISION

A.K. Swersie
New York State J. Med., 50, 1108, 1950

"...the patient is admonished to confine the function of the operated appendage to micturation only for 2 weeks after operation. After that period, if he is so disposed, he may without pain, expand its field of usefulness."

****

SURVEY OF MEDICAL CARE IN A WAR INDUSTRY AREA

M.H. Merrill and M. Mills
J.A.M.A. 1944 *126,* 892

"During June a 24-hour service was initiated, seven days a week with 2 physicians and two nurses, one physician sleeping in the center."

****

THE RAT: AN IMPORTANT SUBJECT.

J. exp. Analysis Behavior, 1964, *7*, 355

"In addition to living a relative short time rats tend to become ill and sometimes interrupt experiments by dying."

CECIL TEXTBOOK OF MEDICINE

5th Ed. page 1031

"in a series of patients studied post-mortem more than 90 percent complained of shortness of breath."

****

ESTIMATION OF VENTILLATORY FUNCTION BY BLOWING OUT A MATCH

A.D. Cavilli and J.R. Henderson
Amer. rev. Resp. Dis., 1964, *89;* 680

"(2) This paper represents the personal viewpoints of the authors and should not be construed as a statement of official Air Force policy.

Acknowledgement: Staff Sergeant Salustro and A/1C M. Penland rendered excellent technical assistance."

(lighting the matches? ...Editors)

****

THE CHEMOTHERAPY OF BURNS AND SHOCK. VI. STANDARDIZED HEMORRHAGE IN THE MOUSE. VII. THERAPY OF EXPERIMENTAL HEMORRHAGE.

H. Tabor, H. Kabat and S.M. Rosenthal
Public Health Repts. *59*, 639, 1944

"The majority of mice died at the completion of, or shortly after a fatal hemorrhage."

CASE REPORT OF UAA—A NEW CLINICAL SYNDROME

Steven Seifert
University of Arizona
Tucson, Arizona

THE SENATE, by William Gropper, The Museum of Modern Art, NY

Case Presentation

A 27 year old male presented himself to the emergency department with a loud ringing emanating from his axillae. Both he and the emergency department staff were unable to stop the sound although several methods were tried. Initially, a roll-on deodorant was employed, followed by a Betadine wash, but without success. Excision was considered but the patient was admitted to the ICU for more conservative therapy. The noise bothered the other patients and he was placed in auditory isolation. Masking techniques, such as those used to treat tinnitus, were used but were also unsuccessful. Finally, the distraught patient committed suicide by jumping off the hospital roof. The ringing noise, however, continued for some time.

DISCUSSION

Recently, a new clinical syndrome has been described—the "Underarm Alarm"(UAA).[1] You may have seen the commercial dramatizing this condition in which a man raises his arm at a party to lean up against a wall and is surprised and embarrassed to have a very loud alarm go off. Despite his lowering the arm, however, he is unable to terminate this unfortunate reaction and herein lies the problem for medical personnel. I suspect that soon we may be confronted with patients presenting themselves to the emergency department requesting assistance in shutting off their underarm alarm. And, yet, how many physicians are familiar with the pathophysiology of the UAA and indeed know how to shut it off? Thus, the basis for this brief report.

The basic mechanism of the UAA is believed to be a hypersecretion of normal body products in an otherwise normally clean individual, or a normal level of secretion in a fastidious one. It is postulated that the excited sweat glands vibrate so vigorously that they bang against one another and create the observed ringing sound.[2]

In some respects, UAA may have some features in common with toxic sock syndrome (TSS),[3] although the auditory component is missing and a generally different patient population is involved. Physicians have long since discovered the best method for dealing withTSS—send the nurse in to perform a cleansing ritual. If no nurses are available, of course, a medical student will do. In TSS, Betadine washes have been found to be generally effective although reapplication of shoes, amputation, and the burning of incense have all been reported to be effective.[4] In UAA, these techniques may also prove effective, but it should be stressed that excision of axilla is to be considered a last resort.

These patients can wreak havoc in a busy emergency department as they frantically run around trying to shut off the alarm. Most disconcerting to the staff is the impression that an ambulance is arriving, the phone ringing, the fire alarm going off, and the sensation that it is time to get up, all occurring simultaneously. The clinician will be disappointed to learn that no simple "off-button" has yet been located.

A Variant Syndrome: *Poly-neuropathica Underarmus*

Some patients present with their alarms ringing loudly, yet seem entirely unaware of it and are, in fact, presenting with an unrelated complaint. I have observed a number of such cases and feel that this represents a distinct variation of UAA, namely *poly-neuropathica underarmus* (PU). It is mainly distinguished from uncomplicated UAA by the existence of three, separate, cranial nerve defects.

First, in PU, the patient has a defect in the eighth cranial nerve in that they cannot hear their own underarm alarm ringing, although everyone else certainly can.

Second, there must be a defect in the first cranial nerve because if they could smell what everyone else was smelling they would try to get as far away from the source as possible.

This may be related to the "flatus and progeny syndrome"[5] —they only like their own.

Third, there must be a defect in the second cranial nerve, because these people also fail to see the "alarm" in the faces of those around them.

Although an anatomic model has not yet been developed, we are hoping for an NIH grant to further study UAA and PU. Certainly, such distressing syndromes deserve at least as much funding support as, say, AIDS. Private sources of research money are not being overlooked, however. As a result of our experience here, particularly the persistence of the alarm even after an eight story fall, Timex has expressed considerable interest. ■

[1]Old Spice deodorant commercial.
[2]Odorus, Mal, Chitty, Chitty, Bang, Bang: Mechanisms of audible sweat; J. Smell, 67(2):353-357, 1980.
[3]Rubin, M, personal communication.
[4]Hearsay, innuendo, rumor and personal opinion.
[5]Airwick, JJ, Anyone who hates kids and flatus...; Stench, 806(12)5:1463-1469, 1885.

PRINCIPLES OF RESEARCH ADMINISTRATION

Michael B. Shimkin

The Clean Canary Principle. It is axiomatic that research scientists are a valuable commodity. The Director, therefore, must protect them carefully, including from themselves, humor their tempers, and coddle them in order to maintain the highest possible production rate. Like canaries, research scientists must be kept clean and ignorant so that they will best sing their little songs. Naturally, no one tells canaries how to sing, especially not those who have never been canaries but are experts on bird seed. It follows that there must be Freedom of Science. This freedom is a delicate thing and can easily be perverted to license.

****

IMMUNOLOGICAL CONTROL OF FERTILITY IN FEMALE MICE

Nature, 1964, *201*, 582.

"Unless the auto-antigenic factor from the sperm tail could be isolated in a effective form, immunization against other antigens present on spermatozoa, such as blood group factors might cause trouble...it would be desirable to avoid the formation of anaphylactically sensitizing antibodies, since one would not want to become allergic to the husbands as a result of a procedure designed to limit fertility."

A BIOLOGICAL VIEW OF TOPLESSNESS

New Scientist, 1964, *23*, 558

"The breasts do form an important part of woman's biological equipment for courtship and it is a question not so much of morals as of tactics to consider at what stage in the proceedings they are to be deployed to the best advantage.

****

GASTRIC HEMORRHAGE CURED BY TOTAL GASTRECTOMY

N.Y. State J. Med., 1964. Absolutely cured.

****

APPARENT SIZE OF HOLES FELT WITH THE TONGUE

Nature, 1964, *203*, 792

"...a size illusion is found when the tongue feels 1/8 - 1/4 inch holes but not when it feels 3/8 - 7/16 inch holes." (This paper comes from Cambridge, England).

Medical Practice in America, a Transcultural Study

FESOJ GOTRAH*
Department of Transcultural Scientific Junkets
University of Liechtenstein
Liechtenstein

This is the first in a series of studies conducted by Full Bright scholars loaned to the United States by the Government of Liechtenstein as part of the program of the Ministry of Foreign Affairs to help less fortunate countries, especially the emerging nations of the Orient. Because the United States is representative of the Orient and its culture (and east of Liechtenstein, too), the author (an Oriental behavioral scientist) eagerly took advantage of this exchange program to make his 3-month in-depth study. The people of the United States manifest the typical Oriental characteristics: a peaceful temperament, inscrutability, disinterest in worldly goods, xenophobia, a skill for making toys cheaply (and dangerously), a general disregard for human life, crowded living conditions, a dependence on drugs, a love for nature—especially flowers, a shrewd business sense, excessive pride in their national and racial identities, a love of dogs and children, a concern about face-saving (hence the Vietnam quagmire), and are often described with some apprehension in Liechtenstein as the aggressive Golden Horde.

Although my purpose in the United States was to bring modern Liechtensteinian medicine to the Americans, and the need and justification for this cannot be doubted, I did not want to be considered an Ugly Liechtensteinian. Therefore I agreed to teach at the American University, where fortunately I had the time and staff to complete a few publications, review some reviews, and do new research as well (for the sake of my back-home sponsors and academic status). Occasionally I was able to do some field work, talking with the natives (i.e., other members of the University faculty) and living among them briefly when the weather permitted. Most of my data is based upon the random sampling of students and faculty (50% of whom were other Full Brights). This scientific endeavor produced sufficient data for 10 research papers. Some of the forthcoming papers are: "The Intelligence Quotient of Americans", "The Kinship System of American University Students and Faculty", "Can Americans Compete in the Modern Liechtensteinian World?", "General Washington's Crossing the Delaware as a Prototype of Military and Political Tradition in the U.S.", "American Responses on the Liechtenstein Water Blot Test", "The Attitude of Americans toward Mental Health", and "Quaint American Social Habits as Observed from a Hovering Helicopter en route between Palm Springs, California, and Seattle, Washington."

The purpose of this paper is to describe in depth the most significant aspects of American healing ceremonies. My description centers on the practices of two types of widely used healers — the Surgeon and the Psychiatrist. It appears that the Surgeon deals with bodily problems and the Psychiatrist with mental or spiritual problems.

Surgical Procedure begins when the patient seeks help for some physical pain. The Surgeon typically examines the patient by touching various parts of the body and taking X-ray pictures. After a fee is agreed upon, the patient is taken to his accountant and then to the operating room. There the Surgeon performs a hand-washing ceremony and then dons a special costume (called a gown) with cap and face mask. The costume is always a single color (white, green, or blue)

---

* Dr. Gotrah was Visiting Professor of Spaced Age Behavior at the American University in San Francisco at the time this study was made.

This work was supported by the poor, the aged, and the disabled of Liechtenstein, who generously relinquished their financial assistance, school improvements, and medicines so that the frontiers of science might be advanced further out into space.

and bears no designs, pictures, or inscriptions. No typical amulets were noted. However, rubber gloves and floppy cloth overshoes were required (an interesting reversal of the no-shoes rule of some cultures). Several male and female assistants were dressed similarly. In the next step, the Surgeon or his assistant shaved the patient and annointed him with a colored liquid applied in a clockwise circular fashion without incantation, though I often heard references to giants, dodgers, pirates, and braves; possibly local spirits. On one occasion an assistant deviated from the proper annointing practice by moving counterclockwise and was scolded by the Surgeon. In fact, scolding by the Surgeon continued throughout the ceremony. This custom may be related to exorcism rites seen in other exotic cultures. The Surgeon cut the patient open, removed a part of the body, and sewed him up again. As in augury, the Surgeon examined the "specimen" and sometimes sent it to a specialist for soothsaying. Predictions of the future were made by means of this procedure. At the end of this ceremony the patient either lived or died. Generally the patient's relatives showed great respect for the Surgeon regardless of the outcome.

This extensive study of the Surgeon was based on observation of three procedures.

The Psychiatrist, who usually works alone, does not use tools and dresses only in the national costume. His office accouterments and hanging decorations, however, compensate for his conventional costume. He often refuses to talk to relatives of the patient and sometimes even to the patient. The example of the psychiatrist is based upon observation of three and several others I heard about.

The patient consults the Psychiatrist for psychic pain, then makes a confession. This is followed by prolonged talking "therapy" over a period of months. The longest I saw was 1 hour during which the patient demonstrated a wide range of emotions. The actual content of the talk I observed was unknown to me since I did not speak the native language at the time of the study. But this should not be construed to devalue my in-depth observations, since I am a keen Liechtensteinian observer with serveral advanced Liechtensteinian diplomas.

Another healer, for my benefit demonstrated intimate intrapsychic mechanisms by treating one patient in a lecture hall with 200 observers and another patient in his office, to which I was denied entrance. Another Psychiatrist, who appeared much more confident (and I think intimidated the patient a little) demonstrated his "electric treatment". Again, this healer wore the national costume though it was covered with a white coat. He was very willing to talk with students, visitors, and relatives of the patient, even while "demonstrating". He administered intermittent, convulsive electric shocks to the patient, who was restrained and silent. The healer seemed more pleased than the patient.

Evaluation of the results of these therapies must await a return visit to America.

CONCLUSION

This experience clearly demonstrates the value of in-depth overseas behavioral research. Also it shows that physical and mental illness exists in America, as does a long-established healing system that incorporates confession, punishment, exorcism, ritual, suggestion, and faith. Results of this study clearly support the views that primitive medicine has universal characteristics and that the people of America are rather superstitious and backward, though emerging.

* * *

M. R. Abeyaraten, W. A. Aherne and J. E. S. Scott
THE VANISHING TESTIS
Lancet 1969, Oct. 18, 822–824.

"Cabral (1964)[1] is said to have been the first to describe bilateral absence of the testis, ironically in the cadaver of a man who had been hanged for rape."

* * *

How Medical History is Made

(To the tune of "Little Boxes")

You observe some
groups of patients
and send data to a statistician
then you wait for his analysis
hoping that it will explain

some phenome-
non of nature
which will make your reputation
but it all seems to be due to chance
and the groups are all the same.

Then you sit there
and you think that
some nonparametric test
will reveal a hidden difference
which will someday bear your name

And you test your
observations
using binomial probabilities
but it all turns to ticky tacky
and it all looks the same.

Both your tabling
and your graphing
can simply be exaggerated
but your conscience (and the editor)
say it all looks the same.

So at last in
desperation
you add 50 observations
and you swear that
now you've got it
and you're back in the game.

But—
the statistician
states succinctly
that the damned null hypothesis
is again not to be rejected
so it all looks the same.

So you fire your
statistician,
publish in Reader's Digest,
and the Ladies Home Journal
puts you in the Hall of Fame.

(By William F. Taylor)
Div. of Medical Statistics
Epidemiology & Population Genetics

A Case History from the Hospital for Acrobatic Pathology

DR. TREMOR TAKE-OUT and DR. VERTIGO PUT-IN*
Megalomaniopolis, In-Continent

Whereas organ transplantation, kidney:, heart-, liver-, pancreas-, lung-, skin and hairy scalp-, has become nowadays the daily bread and wine of self-conscious medical centers, brain transplantation in a rural general practice-directed hospital with a modest staff like ours seems to be sufficiently rare to justify publication, so as to serve both warning and stimulus, the more so since thus far the human brain has been considered too intelligent to easily submit to the procedure, and the medical staff involved is of untouchable ethical motivation. In the present communication we therefore truthfully describe the erroneously performed brain-cross transplantation between a professor and a student.

Case Report

At 19 p.m. on ...., ..., 19.... (exact indication omitted in order to avoid recognition of donor, recipient and transplanters), a graying, male, sad-looking professor of undefined age and a fresh-looking student, both of the political science department of the nearby parochial university, were admitted following a car accident to the emergency room, in a state of questionable death. Since optimal ressuscitation procedures applied for an optimal period were of no avail, and since two live car accident-induced decerebrate adult candidates were abiding brain reception (today's common jargon), and transplantation team members were convocated by computer alarm reaching into their homes, all movie theaters, concert halls, discoteques, bridge clubs and medical libraries. The complete team convened in the transplantation board room within the prescribed minimum time and was ready for action, i.e. scrubbed and tranquilized, within the prescribed additional minimum time. There was a slight delay due to the wavering of the hospital ethical committee members whose opinions were divided as to the possible reaction of the press, but this was soon overcome by the firm attitude of three of the members, a medical student, the elevator official and the cafetaria vice-cook on service.

At .... p.m. (publication of exact time withheld for political reasons) both professor and student were proclaimed officially dead by three appointed staff members, dermatologist, psychiatrist and gynecologist, on the basis of the criteria determined by the Superior Council of the Superior Board of the Superior National Medical Organization, recognized by all political parties and religious currents. One second later, the permanently ready action flow sheet was started to be read over the hospital loud speakers and the drama became irrevocable.

On .... p.m. four brancards, supporting the two prospective donors and the prospective recepients, were wheeled into the mega-theatre and the transplantation team started its activities coordinated by the transplantation-programmed computer, to complete the assignment of two brain takeouts and two brain inputs with remarkable precision within the record time of three minutes (time data released by the hospital indirector). While listening to the regular tick-tack of the two recipient-electro-encephalographic registrations, derobing, and raising at the same time the now traditional post-transplantation dry Martini, the team members' later movement was suddenly arrested by the announcement of the chief anesthetist: "Gentlemen, congratulations-transplantation succeeded, however forgive me for slightly damping your elation by informing you that, due to a slight administrative error, the brains of the two prospective donors were cross-transplanted, instead of to the two intended recipients. Whereas the condition of the latter remains unchanged, both professor and student appear vital and show signs of intellectual revival.

The two steps taken next are noteworthy for their rapidity: shock-treatment to the transplantation team, and clamping down by the hospital indirector on publicity, both actions successful from the medical and hush-hush point of view, respectively.

After a few days, when the hospital staff had regained its normal metabolic steady state, the objective bystander could observe two main lines of consequence — political religious reaction to the transplantation and university riots.

Since the erroneous cross-transplantation was converted by the intensive publication care unit with shrewd diplomacy into a blilliant feat of intended professor-student brain exchange aiming at improvement of teacher-student relationship, no ethical criticism was expressed, neither in the daily commercial nor in the weekly political journals. On the other hand, the party of extremist Guru's disavowed the transplantation because of neglect by the team of sufficiently premeditated yoga exercise, whereas the party of extremist existentialists disapproved of the transplantation in view of the diversion of public interest from non-urgent affairs of state such as international clownery and space exhibitionism. The party of wavering liberals made pro-transplantation statements which however were drowned in the noise created by the broadcasted parliamentary brawl which followed the nasty remark of the vice-minister of cookery to the effect that he preferred his own brain above that of his colleague of the ministry of fashion clothing. The excitement was finally quietened down by a lengthy article from the deputee minister of paganism in the local journal of atheism, who concluded with the proposal to leave the issue open to further private initiative, in as far as guaranted by scientific progress on both sides of the Iron curtain.

More consequential was the development on the university campus. First there arose the matter of definition: — are a student with a professor's brain and a professor with a student's brain professor or student? Following a lengthy discussion in the senate, moderated by the dean of anthropology, the issue remained undecided since it was felt that only the future behaviour of these creatures would determine their species. Until then, as proposed by the department of applied mathematics, the original professor would be designated P_{sb} (professor with student brain) and the original student S_{pb} (student with professor brain). The importance of these decisions was stressed in the senate protocol in view of the possibility of future mass cross-transplantations, foreseen to be demanded by the student organization in case of potentially ensuing abolishment of examinations. All these deliberations were held in secrecy in the hermetically sealed laboratory of hibernative psychology.

The atmosphere, however, became tense when P_{sb} and S_{pb} were confronted on .... at 9 a.m. in the classroom at the regularly scheduled lecture on political science. The class was surprised to see the professor (P_{sb}) to sit down amongst them and their fellow students (S_{pb}) to climb the podium. All aroused and eagerly awaiting the announcement of some student festivity or strike, the class was surprised to here S_{pb} lecture in the professor's uninteresting dragging drawl, repeating exactly last decade's text. Matters became hot when, P_{sb} attacking S_{pb} on his stupidity and boredom, S_{pb} reacted with throwing the slide projector at P_{sb}. The class, now ready to continue their late at night interrupted discoteque experience, became soon heavily involved and from then on the fun spread as a fire over the campus in Berkeley-Columbia-Nanterre-Heidelberg-Tokio style.

Further developments may be read in Time, Life, Express, der Spiegel and some Near Eastern Journals. Suffice it to state that P-S brain cross transplantation was soon accepted by the International Student Organization and also, somewhat reluctantly, by the International Organization of University Professors, as a legitimate mass procedure. No doubt, our hospital has done much to promote academic proclivity, nationally and internationally. ■

* Our thanks are due to the hard- and soft ware nurses, the hospital indirection and the intensive public relation care unit.

A Montague

NATURAL SELECTION AND THE FORM OF BREAST IN HUMAN FEMALE

J.A.M.A., 1962, *180*, 826–827.

"A factor that may have correlatedly assisted in the development of the female breast is the fact that in most nonliterate societies, and especially in areas in which the nights are cold, fat women would tend to be preferred to thin ones. Love in a cold climate is considerably assisted by central heating. Eskimos, for example, always prefer fat women to thin ones. The forerunners of central heating in prehistoric times were fat ladies. Fat ladies would have large breasts, and the premium placed on fatness itself would in this way have further contributed to the development of the breast."

Umbilectomy: An Experimental Surgical Panacea

CARL JELENKO, III, M.D.

With the exception of two individual cases reported in earlier literature,[1] a common affliction of mankind has been the so-called umbilicus. This structure, situated, on the average, 7° 18′ N., and 80° 13′ W. of the pudenda, has received much scientific attention. Cullen[2] embraced it, and even published his embrasure. The structure has been variously contemplated, covered, adorned, and exposed. In 1864 it became known as the "Belly-Button," being so named by Sir Mount Anyodds[3], who perforce devised a technique for hanging his trousers from this structure after an unfortunate incident in which he lost his remaining underanatomy in a craps game.

It must be noted, however, that Sir Mount had a construction later characterized by Pupick[4] as "An Outer." Anatomic configuration of the type Pupick II: "an Inner," however, would have made that earlier work by Anyodds impossible.

A recent survey by an independent New York accounting firm has revealed that 24% of umbilici are "Inners," and, furthermore, worthy of economic consideration. These structures provide a rich harvest of lint with a refractive index of 1.495. This is identical to that of Noselite (Ger. *Naselint*) from which, by corruption, the term "navel lint," also called "belly-button lint," is derived. This material is useful for stuffing pillows and for packing chromatographic columns. Swine fed a steady diet of the material, however, develop gallstones, trichinosis, kwashiorkor, and tsutsugamushi fever. The matter is listed in the Index Medicus under "Page 7491."

It should be noted that lint gleaned from "Inners" will not sink in water. For this reason, the material was used to construct sea-going vehicles. Such vessels, in due course, became named for the material of their origin, and, in groups, are called "Navy."

Since 1960, emphasis has been placed upon economic augmentation; and Federal grants-in-aid have enabled the cultivation of the productivity inherent in "Inners." It is, therefore, no surprise that the 76% of the population that are "Outers" are walking poverty pockets, and psychologically unstable.

It was proposed that, for the foregoing reason, in the interest of a populace that would be uniform and psychologically sound, all umbilici be surgically removed. The procedure returned the general status to a state of quo. It was found to be an economically sound venture, since the surgical fees charged were equal to the monentary gain from the local lint harvest. In addition, lint was able to be obtained from a variety of oranges; and it was therefore possible to continue to float a Navy.

The project has, however, been discontinued due to unfortunate and unexpected legal action by the D.A.R. who obtained a Federal injunction against the operation on the grounds that removal of the umbilicus is the ultimate rejection of mother. ■

REFERENCES

1. Bible, The: Genesis 2:7, 22.
2. Cullen, W., *Diseases of the Umbilicus*. The Johns Hopkins Press, Baltimore, 1906; Vol. I & II.
3. Anyodds, Sir Mount, "On the Levitation of the Pantaloons in lieu of Re-buttal, & c." Brit. Jour. Plast. Surg., 4; 119, 1846.
4. Pupick, Scherner, "A Look at the Umbilicus as a Whole," Gut; 9: 714, 1958.

From the British Medical Journal

Treatment of Spinal Subdural Haematoma With Ferret's Ear

F. B. RUMBOLD
JAMES TELLEMENT-FOU
A. HOWARD GARKE

BMJ, 1972, 3, 257

Spinal subdural haematoma is one of the rarer causes of acute spinal compression and is usually associated with injury or a bleeding diathesis (Stewart and Watkins, 1969).

A case is presented here in which the patient's serious plight was alleviated by treatment with the left ear of a dead ferret, specifically prepared.

Case Report

A man aged 59 presented with violently severe pain in the limbs and abdomen, accompanied by weakness of the lower legs. A lumbar puncture yielded bloodstained fluid with xanthochromic supernatant containing 800 mg of protein and 20 mg of sugăr per ml. CSF pressure was 45 mm of CSF, with no rise on jugular compression. There was severe flaccid paraparesis.

We waited until the night of the full moon, and slaughtered a male ferret in rut by striking it with the femur of a defrocked sexton. We then removed its ears. the right ear was nailed to the door of Ely Cathedral as a precaution (see *Treatment of Calcinosis Circumscripta With Mole Soup, BMJ*, 1972, 2, 499), and the left ear was brought to the Orthopaedic Research Unit of the Royal Camden Hospital in a teapot. It was then swung round three times.

We drew a chalk circle on the floor of the operating theatre and took our trousers off.

The patient was then premedicated by the anaesthetist, who shook a beaker of dried peas over him, murmuring "Tu ne quaesieris, scire nefas", twice, and brought into theatre. He was painted blue, and the left ear of the ferret was then pushed up his right nostril at the exact stroke of midnight.

By morning, the patient was his old self again, i.e. suffering from spinal subdural haematoma. In the post-clinical discussion no clear explanation emerged, but the fact that the teapot had once stood on a shelf next to a garlic plant (a fact nor previously known to the surgical team) was held to be of prime significance.

REFERENCES

Stedman, D.H.F., and Merrill, P. (1968) *Journal of Neurosurgery*
Wrigley, King Norman, (1966) *Ferret Remedies And Songs*
Siggs, F. (1957) *Whistle Away Your Fracture*

RARE DISEASES

OLE DIDRIK LAERUM, M.D.
Institute for General and Experimental Pathology,
University of Oslo, Rikshospitalet, Oslo 1.

Rare diseases are not commonly encountered in medical practice. When a patient, who is suffering from a rare disease is discovered, it is therefore customary to write a paper and publish it as a case report.

Such case reports often have a peculiar feature: One writes about one or two observed cases of a disease because it is so rare. Then the author(s) conclude that this disease nevertheless may be of importance and is probably not so infrequent as it may appear. One must even be aware of this condition in general practice. Quotation: "During the last 30- years period we have had two such cases in this country."

There is therefore a need for a survey of the different categories of rare syndroms and disorders. An analysis of pecularities and common features of such maladies will be given, followed by a discussion of their prevalence in relation to our concept of the term, "rare diseases".

TYPES OF INFREQUENT DISORDERS

1. Rare diseases which in reality are not so infrequent because they are often not diagnosed.

People are not made aware of the fact that they may suffer from such disorders and walk about totally ignorant of their presence. It can be said that these diseases have not been subject to a proper marketing. In this connection radio, television and newspapers can do much. And they do, indeed.

Every time there has been an information week with campaign for a certain disease, people afterwards run to their doctor, believing that they suffer from it. If we in addition could make an information week for rare diseases, one could probably catch a lot of infrequent disorders which are normally hidden.

A case report can illustrate this:

NN, male, born 1940, at that time a medical student, read in the newspapers during the spring 1963 that a contagious horse cough (equine influenza) had appeared in the stables of a hippodrome in Oslo.

This passed without complications. During the winter 1964 he read in the local newspapers that another epidemic of horse cough had come, this time in the stables of another hippodrome in Oslo.

After the last newspaper message he felt unwell, and the following days he developed increasing malaise with rhinitis, fever and a strong, continuous cough. The coughing seizures started pianissimo in tempo di andante, followed by a strong crescendo up to forzato (NN was a passionated amateur musician) and then terminating in ritardando with utterly *neighing* coughing fits.

NN sought a general practitioner, who had never heard about horse cough in human beings. NN therefore had to leave the office without any approbation, but later spontaneously recovered. NN is convinced that the disease was precipitated and thus directly caused by that newspaper article.

Another disorder which is never diagnosed is volvulus of the pineal body. Today this disease is without known existence because our diagnostic methods are so limited. Still many persons may suffer from it and have problems because they are not able to convince their doctor.

2. Diseases which are rare because they depend on certain periods of history.

The term "an asthmatic locomotive" is well-known in the literature. However, only few people know why asthma in locomotives was not observed in the 17th century. The reason is that Stephenson first had to invent the steam-locomotive. Nowadays, the condition is again becoming rare. In any case it is not seen in our hospital wards. This is obviously because the steam engine has been gradually replaced by diesel- and electric locomotives.

3. Conditions which are seldom seen because they are rare.

University teachers commonly present such diseases in a double lecture. The lecture is often terminated by the following statement: "The condition is not infrequent at all and is therefore important to be aware of. Any person who spends his whole life in a department of internal medicine will have ample chance of observing one case. Personally I have seen one patient with the disease. That was before the war, — the first world war".

On this background we can formulate the following problem as a basis for the present study: *Does there exist diseases which are so rare that they have not yet occurred?* If it does, and that we have reason to assume, there are another two categories of rare diseases:

4. Diseases which have yet not occurred, but will be discovered some time in the future, sooner or later.

A typical example of such a disorder is moon-dust poisoning, which may be a very serious condition. Another important, hitherto not described disease in this group is podocytoma of the kidney. The podocytes are known as epithelial cells which adhere to the glomerular membrane with numerous foot processes. They particpate in the urinary filtration (Fig. 1).

These cells have hitherto never developed any tumour. But some time in the future they will. That makes a podocytoma, a highly differentiated, benign tumour, which was predicted and first described in this article ("Lærum's tumor"). The characteristic electronmicroscopical picture of this condition is that the foot processes march out of step (Fig. 2).

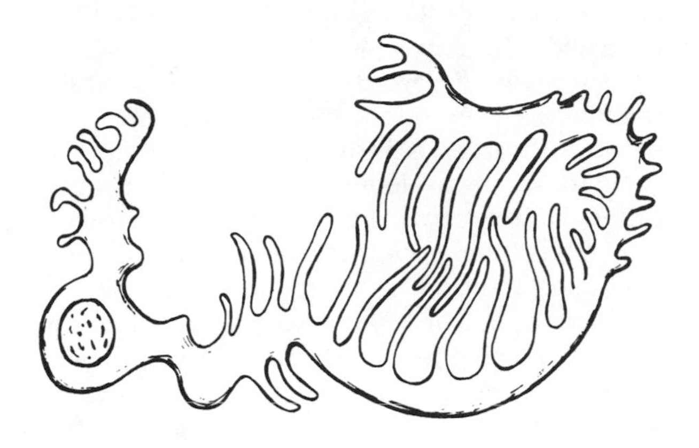

Fig. 1
Normal podocyte.

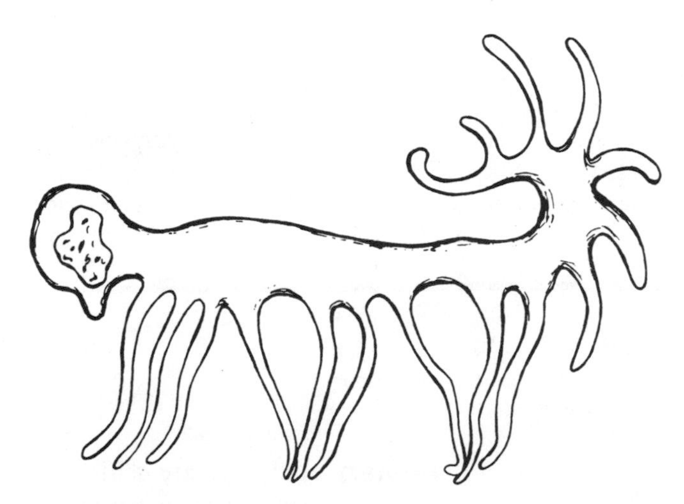

Fig. 2
Detail from podocytoma. Foot no. 4 and 7 are out of step.

5. Diseases which have not yet occurred and will never occur in the future either.

These are extremely rare and have very little practical importance from a clinical point of view. It is not necessary for a general practitioner to know anything about them.

DISCUSSION

Is our concept of the term "rare diseases" becoming outdated? In any case the term is ambiguous. As earlier shown the word "rare" in the medical world is mainly used in publications which are written because a disorder is infrequent. But at the same time the authors try to demonstrate that in reality the condition is frequent. The article is therefore accepted for publication *both* because the disease is so rare, and also because it is so frequent.

Our concept of "rare diseases" also lacks precision and is changing as a function of time. New rare diseases are coming. Other infrequent diseases disappear and are not seen any more. It would therefore be an idea to take care of certain rare diseases before they disappear for ever.

It is also important to realize that the medical use of the word "rare" has a tendency to become so common that it is losing its practical meaning. We therefore must not use the word so often that it becomes too common.

It is also misleading that the word "rare" is often used when quite banal disorders occur in an unusual manner, such as caries in toothless persons.

If we take all of the presently known diseases and group them according to their prevalence, the group "common diseases" will include only a relatively small number of unities. The group "uncommon" diseases is far larger. But if we take the group "rare diseases", then only the list of the names comprises a voluminous book.[1] This there exist a lot of rare and obscure diseases and syndroms.

Although each of them is very seldom seen, they are together so many. As a group they therefore make a large entity, which altogether perhaps is the most important we have to deal with in medicine.

Conclusion: Rare diseases are no longer uncommon. What are they then?

SUMMARY

A survey of our concept of rare diseases is given. The author demonstrates that there exist diseases which hitherto never occurred. Some of these will never occur in the future either. A new tumour, *podocytoma* ("Lærum's tumor"), a benign lesion of the kidney, is described to illustrate the pathology of a clinical entity without known existence.

It is concluded that because there are so many different types of them, rare diseases should no longer be considered as uncommon. ■

1 Leiber, B. and Olbrich, G.: Wörterbuch der klinischen Syndrome. Urban und Schwarzenberg, München — Berlin 1959, 730 pp.

LETTERS TO THE EDITOR

83 Rodney St.
Glen Rock, N.J.

Sir:

Letters to the editor apparently are seldom if ever published in your vital publication and therefore infrequently appear in its pages. Nevertheless, it would seem to me that any journal of scientific inconsequence owes it to posterity to allow space for diffusionary views.

Take the article in the June, 1972, issue, which I received in October, which refers to the infrequent occurrence of rare diseases. Balderdash, I say unabashedly to reflect my ferment. It is common knowledge among those who consistently deal with rare diseases[1] that their occurrence is quite frequent among those afflicted with them. In fact, the rate of occurrence among those who have such diseases is 100 per cent.[2] Conversely, such diseases are rare *only* among those to whom they seldom if ever manifest themselves,[3] i.e., those to whom they are rare.

It seems incumbent upon you, therefore, in the interest of scientific divisiveness, to present a forum less infrequently than not in which some of the more obfuscated views of your learned contributors could be unsolidified, i.e., be made more gaseous rather than be presented for consumption whole. And vice versa, of course.

In the interests of creative knowledge, I remain,

Sincerely yours,
John H. Lavin

1. Ibid
2. Ibid
3. Ibid

PROGRESS REPORT NO. 9

Fund for the Replacement of Animals in Medical Experiments, (May, 1974)

MYRNA currently projects four personalities: 1. MACMAN a digital computer which models the heart and circulatory system, 2. MACPUFF which provides information about the lungs and respiratory system, 3. MACPEE which simulates the functions of the kidneys and 4. MACDOPE which deals with the interaction of drugs.

* * *

Lancet, ii:1425 (1973)

The concept of the bladder as an inert container of urine no longer holds water.

* * *

Reich, W. J., Nechtow, M. J.

CANINE GENITAL MONILIASIS AS A SOURCE OF REINFECTION IN THE HUMAN FEMALE

J.A.M.A. 141:991 (1949)

An adult white woman had been treated for moniliasis for over a year without abatement of her subjective symptoms . . . On detailed history taking, it was disclosed that the patient possessed dogs which on occasion would sleep with her under the covers. Also, when she left her home the dogs were locked up in the bathroom and would void in the same tub in which the patient took a daily bath. Culture of the material found in one dog's inflamed vagina and around the other dog's inflamed penis showed dense growth of *Candida*.

* * *

T. S. Sasz

THE SECOND SIN. THE ETHICS OF PSYCHOANALYSIS

Routledge & Kegan, Paul (1973)

Psychiatric diagnoses are stigmatising labels phrased to resemble medical diagnoses and applied to persons whose behavior annoys or offends others.

* * *

SYMPOSIUM ON THE MECHANICS OF CONTACT BETWEEN DEFORMABLE BODIES

Enschede, Netherlands, Inst. Technology, 20-23 (August 1974)

1. Theoretical and fundamental experimental studies of contact
2. Mechanics of friction
3. Pure sliding
4. Steady and unsteady rolling

(Attendance 100 [by invitation only])

* * *

E. M. Hildebrand

SWEETPOTATO INTERNAL CORK VIROSIS INDEXED ON SCARLETT O'HARA MORNING GLORY

Science, *123*: 506-7 (1956)

* * *

LEDERLE SURVEYS ATTITUDES TOWARD "THE PILL"

News from Lederle

Pearl River, N.Y. (May 1, 1974)

Do women resent the responsibility involved in taking the pill? Not as much as many of us have thought, say women taking part in research on today's attitudes toward oral contraception, currently being conducted by Lederle Laboratories, a division of American Cyanamid Company.

In fact, some women indicated they would be hesitant to relinquish their responsibility for birth control. If there were a similar pill for men, they agreed nearly unanimously that they wouldn't trust men to take it.

At last! Blessed relief for the pain and discomfort of Percentorrhea

MICHAEL A. LaCOMBE, M.D.
Assistant Resident In Medicine
Strong Memorial Hospital
Rochester, New York

House officers bored by endless percentages and professors prone to recite them will benefit alike by committing to memory the indispensable rule presented in this article.

LaCombe's Rule of Percentages (my name is LaCombe) will, when properly applied by house officers, force rounders and attendings to teach rather than perpetuate what Orr calls "shifting dullness."[1] The rule is: *The incidence of anything worthwhile is either 15-25 per cent or 80-90 per cent.*

With this rule in mind, you need only know whether something is common or uncommon in order to affect profound medical knowledge when replying to percentage questions. Thus, you'll force those who want to teach you to deal in pathophysiology rather than percentages, and you may learn something.

Let me illustrate with a clinical example.

ROUNDER: What percentage of patients with Hodgkin's Disease experience pain at the site of active foci following the ingestion of alcohol?

HOUSE OFFICER (thinking to himself): I remember hearing about this once as I was falling asleep in a hematology lecture . . . I've never seen it, so it can't be too common . . . He's smiling at me as if I should know, so it can't be too rare. (Aloud, loudly): About 15 per cent to 25 per cent, depending on the study.[2]

ROUNDER: Excellent! Actually, it's 17 per cent. Would you like to become an Associate Professor of Medicine like me?

HOUSE OFFICER: I'm afraid I don't know enough percentages yet, sir.

Now, don't get the idea that the house officer was being dishonest, specious, or glib. He was using LaCombe's Rule as a weapon, fully aware of the tremendous power of the phrase "depending on the study," which can mean anything upward from "I know as many references as you do."

There exist, naturally, exceptions to my rule, and two corollaries must be remembered. The first called the Scrinch Amendment, refers to that instance when the rounder screws up his face (scrinches) and searches from student to intern to resident before asking for a really bizarre percentage. The corollary: A *scrinch is less than 1 per cent.*

The second, called Dudenhoefer's Corollary after an intern who used it twice, is: *An answer of 50 per cent will suffice for the 40-60 range.* It is to be applied rarely, and only by a master at reading nonchalance in facial expression, verbal nuance, and voice inflection. Why? It takes rare cool for a rounder to ask a percentage question when the answer desired is in the 40- 60 realm. Lucky and blessed is the man who once in his lifetime successfully executes Dudenhoefer's Corollary.

With practice, LaCombe's Rule will prove indispensable. Like "The Barnes Manual of Medical Therapeutics," it is an acquaintance that once made will never be shunned. However, for those in medical subspecialty fellowships who aim for dead accuracy, I suggest study of the accompanying table. I would hope that anyone who commits the figures to memory would be right 80-90 per cent of the time, and that pedantry will be reduced by them to a scrinch. ■

Withering and the Quarterly Journal of Negative Results*

RICHARD J. BING, M.D.

The current trend by some of us to stress the scientific side of medicine to the exclusion of everything else has prompted the following editorial. It is written by a man who envies both the true scientist and the good clinician, but who would have their personalities and ambitions wedded into one unit. The wondrous physician who emerges from this union would have his research sponsored by the granting agencies, and would be beloved by his patients, a fortunate combination indeed.

On a recent trip to Shropshire, England, I discovered some documents which are worth discussing. Since they concern Dr. Withering, the discoverer of the therapeutic effects of digitalis, I thought them of sufficient interest to publish them. Although they were written in the 18th century, they appear so modern that they could have been written today.

Following are the discovered documents.

Dear Dr. Withering:

We regret that your article on "The Account of the Foxglove and Some of its Medical Uses," which you submitted to us for publication in the *Quarterly Journal of Negative Results,* is unacceptable for publication. The reviewers, whose comments are enclosed, feel that although it is an interesting clinical observation, it lacks the quantitative approach with which clinical studies should be supported. In addition, in the hands of one of the reviewers who had the opportunity to try Foxglove on a patient of his, the drug proved fatal. May we also draw your attention to the fact that the Food and Drug Administration does not sanction the use of any drug without having full information on the toxicity of the drug as obtained in animal experiments.

Thank you for submitting this manuscript to us.

Sincerely yours,
Editors.

Comments of the first reviewer:

Dr. Withering's data are entirely empirical. There are no studies on the chemical composition of the drug; it is particularly difficult to understand why a paper has been submitted to this journal on a drug which purports to influence renal function without the author's including more results on renal clearances, blood potassium, sodium, and blood rhubarb levels. The use of radioisotope scans of the kidney would also have been useful. The absence of statistical evaluation is regrettable. What are the *p* and *t* values of these observations? The clinical material is poorly selected.

*Reprinted with permission of the author from Archives of Internal Medicine 1963, *111*, 143.

Was the patient population homogenous; what type of heart disease were the patients suffering from?

One of the most serious criticisms is directed toward a statement of the author, "I cannot pretend to charge my memory with particular cases, or particular effects, and I had no leisure to make notes." This lack of scientific documentation is regrettable. How does the author expect his paper to be of value without necessary documentations? The whole work demonstrates a degree of unfamiliarity with scientific methods which may well be responsible for the amateurish approach of the author to this complicated subject.

A second reviewer's remarks are as follows:

> The author has attempted to convince us of the medical value of Foxglove. Although he attempts in his case histories to furnish proof of his contention, his scientific evidence for the value of Foxglove is lacking. In the first place, the author writes, "My opinion was asked concerning a family receipt for the cure of the dropsy. I was told that it had long been kept a secret by an old woman in Shropshire who had sometimes made cures after the more regular practitioners had failed." The attitude of the author to commence scientific studies because of an old wives' tale is to be regretted. In addition, this journal does not accept personal communication without written approval by the scientist whose work is quoted. The author writes, "I could not take notes of all cases which occurred." If the author had studied the papers in the *Quarterly Journal of Negative Results,* he would have discovered that no paper is complete without detailed tables, charts, and statistics. Why did the author make no attempt to isolate the active principle of Foxglove? Why were such obvious methods as column chromatography, electrophoresis, and ultracentrifuge not carried out? It is quite clear that Foxglove can have no therapeutic value unless its active principle is scientifically defined. Despite the recommendation of a Dr. Darwin, which the author includes, the paper appears unacceptable for publication.

So much for the comments of the learned journal.

When the writer of this editorial discovered these documents, he felt that they were provocative and deserving of comments. It occurred to this author that the correspondence between Withering and the *Quarterly Journal of Negative Results* illustrates the growing schism between the two modern types of medicine. One is exemplified by Withering's article, as rejected by the *Quarterly Journal of Negative Results.* Withering's article and the work underlying it are the outgrowth of pure clinical observation, not watered down or enriched (according to the point of view) by quantitative analytical examination. This type of thinking has served medicine well, and no attempts should be made to consider it of less quality than any other type of medical research. The physician in academic medicine considers the busy but observing clinician often less than his equal; his attitude has introduced a dangerous schism in modern medicine. The academically inclined physician, usually working at a University Hospital, whose papers are acceptable to scientific journals, often lacks the "feel" for the patient as well as the natural gift for simple observation which so often characterizes the practicing physician.

This division which is opening in modern medicine can only be overcome by a sincere effort of the man in academic medicine to learn from his purely clinical colleague. He must look at the whole patient, observe him, and understand his personality. He can still pursue his researches. Love for the patient is not incompatible with love for science. On the contrary! Love for the patient inspires interest to help, and therefore interest in the science of medicine. If the academic teacher of medicine does not make these adjustments, his ivory tower will become his prison. ■

=========

FURTHER READING

A. G. Foraker: If Robert Koch had applied for a Research Grant, SCIENCE, 1963, *142*, 11

R. Past: The Editors Desk, Amer. J. Pathology, 1962, *40*, 13

A. G. Foraker: By the River of Babylon, Jour. Med. Education, 1960 *35*, 868

A. G. Foraker: Latinonian Exchange, J.A.M.A., 1960, *172*, 168

A. G. Foraker: Icarus, Kiwi and John, A Parable of Three Acolytes in Medical Research, New Engl. J. Med., 1960, *262*, 81.

The Incidence & Treatment of Hyperacrosomia in the United States

Journal of Facetious Pathology

EDMOND A. TOURE
M.D., M.B.A., B.E.E., F.O.H., R.S.V.P.
Faculty of Medicine, University of Nicarobia

It has been estimated that one in every three Americans suffers from hyperacrosomia.[1] The extremely high incidence of the disease is thought to be due to the effect of diet aggravating existing genetically-derived predisposition.

Some very famous Americans have indeed been afflicted with acute hyperacrosomia, among them Abraham Lincoln, George Washington, and Lyndon Johnson.[2] Their condition is readily apparent upon comparison with such normal individuals as Napoleon Bonaparte, Truman Capote, and Dick Cavett.[3] Populations with a high frequency of hyperacrosomic individuals can be quickly detected by examining the behaviour of the normal population, which can frequently be observed making ducking[4] movements, sidestepping, and jumping up and down in crowds.[5] The disease, which warps the lives of all its victims and those normal individuals who live in proximity, is seldom fatal except when complications set in. Several historical figures died of secondary causes as a result of their hyperacrosomic condition, among them Louis XVI of France and a certain Philistine warrior who was superficially wounded by a sling-accelerated projectile.[6] Indeed, the pathological nature of the condition has long been recognized in folklore, as evidenced by the epigram "the bigger they are, the harder the fall", an American cliché.

Treatment of hyperacrosomia can proceed along a number of lines depending on how advanced the disease has become. An affected child can be restored by simple adjustments in diet, in particular a therapeutic deficiency in one or more essential amino acid, or by prophylactic administration of nicotine, ethanol, caffeine, or sleep deprivation. The affected adult, however, must be helped by surgery, usually the application of the advanced femurectomy technique pioneered by Andreas Procrustes of the University of Peloponnesus.[7]

The most promising avenue of control, however, most certainly lies in genetic counseling. While diet and other environmental factors play a major role in the degree to which the condition is expressed, the individual's genetic potential for the disease is a multifactorial inherited trait with incomplete sex limitation. The disease is most strongly manifested among males. Unfortunately there exists a cultural bias in the United States favoring the symptoms of hyperacrosomia, and thus a comprehensive attitudinal change must take place through education and personal consideration. Since the male population does express the condition to a higher degree, it falls primarily to the female population to objectively consider the risks of involving themselves with hyperacrosomic males. One would hope that as the female population grew more enlightened they would implement the necessary shift in attitude and come to favor normal individuals consistently over those with manifestations of the pathology.[8,9] ■

REFERENCES AND NOTES

[1] R. Clocher, *Mensonges et fraudes de science* (Presse du Hôpital, Charenton, 1832)

[2] It is sobering to realize that both major candidates in the 1972 American Presidential election suffered from the affliction.

[3] As far as is known, neither Art Garfunkel or Sam Yorty is affected.

[4] "Ducking" is a verb describing the action of lowering the body to avoid an approaching object likely to encounter one's head, and does not imply any imitation of the movements of a bird of the same root name (family Anatidae).

[5] G. de Bâtard, "Les américains croissent comme herbes mauvaises", *Acta Med. et Vét. de Gr. Fenwick* 7 (1934) 62.

[6] E. Smith, "Twelve Ways You Can Improve Your S*x Life", *Readers Digest* 34 (1912) 23.

[7] A. Procrustes, "Selective Extension and Truncation of the Skeletal Frame", *Annals of Orthopaedic Mythology* 81:(523 B.C.) 71.

[8] L. Strauss, *Keep Your Jeans Clean* (Levi Strauss & Co., San Francisco, 1873).

[9] I thank N. Mailer, J. Susann, J. Carson, A. Wellington, S. Douglas, M. Berle, the Army of the Republic of Argentina, and the entire population of the United States of America from 1723 to the present, without whose help this work could never have been completed.

This study was supported by the Southern California Fair Young Lady Rating and Escort Society, Committee for Equal Opportunity by Height, grant #WYB002. I thank the staff of the Broadway & 59th Field Research Station of the University of Nicarobia Institute of Natural Ecology for the use of their facilities in the preparation of this manuscript.

12 October 1971; revised 16 September 1965.

AN EMPIRICAL VALIDATION OF AN ENGLISH PROVERB

Robert R. Hill,
Jane E. Budnek,
Linda K. Wise,

"An apple a day keeps the doctor away" has become a familiar, if not trite, aphorism. Has its continual propagation been merely the result of a highly successful campaign undertaken by members of the apple industry, or can this sententious precept be empirically validated? The authors have sought to verify empirically the relationship between apple consumption and the need for professional medical attention. The results of this study have impelled the authors to evaluate possible ramifications of these findings as they may affect an overtly health-conscious American society. For indeed, as the infallibility and conclusiveness of modern inferential statistics have indicated, an apple a day—at the appropriate significance level—does keep the doctor away.

To verify any statement, one should first evaluate its origin. This becomes somewhat difficult when dealing with the proverb in question. Books of quotations denote the saying as being both anonymous[1] and an "English proverb, not recorded before the XIX century."[2] An earlier version is recorded as "Ait a happle avore gwain to bed, and you'll make the doctor beg his bread."[3] Later anonymous references include these definitions: "Apple—a fruit you eat to keep the doctor away. The more you eat and the more doctors you keep away, the better."[4] or yet, "Apple—a fruit that keeps the doctor away unless you get the seeds in your appendix."[5] Further still, author Thomas Lamont recalls this stanza from his boyhood:

When I was young and full of life
I loved the local doctor's wife
And ate an apple every day
To keep the doctor far away.[6]

But perhaps the earliest testimony to the validity of this saying may be found amidst the remains of possibly the oldest apple orchard in the West. The orchard lies in Manzano, a small Mexican village, whose name is Spanish for apple tree. The orchard is over 300 years old, begun in the late Sixteenth Century by Franciscan padres, who brought with them apple seeds.[7] The authors make no claims that the orchard's fruit was responsible for the exceptional health of the padres; however, these apple-eating settlers could have proved to be a formidable power in Mexico had they not all been slain by Apache Indians in 1669.

Alleged validity such as this, however, can in no way supplant quantitative analysis in determining the Truth. "I eat apples, therefore I am healthy," simply does not lend itself to inferential statistics. Therefore, appropriate quantitative data have been gathered and a least-squares linear regression model painstakingly constructed. Five random samples of ten apples each produced a mean of 2.857 apples in one pound.[8] The per capita, in pounds, civilian

FRUIT, by James Peale, The Corcoran Gallery of Art

consumption of apples[9] for each year in the analysis was multiplied by 2.857 apples per pound to determine the mean number of apples a person consumed in any given year. Dividing this number by 365 results in the mean number of apples consumed per capita per day (x).

In 1946 the mean number of physician contacts, per physician, was 6000.[10] However, productivity has increased since the turn of the century and doctors may accommodate more patients now than in previous years. To account for this an output index was calculated using 1946 as a base year assuming that physician's productivity followed the same pattern as that of the overall U.S. economy. The output per man-hour labor productivity index[11] was converted to a 1946 base year was used as an estimate of the change in physician productivity over time.

This resulting index for each year was then multiplied by 6000—the number of contacts in base year 1946—to produce an accurate estimate of the number of physician contacts in each year. This figure was then multiplied by the number of physicians in that year[12] to calculate the total number of visits in each year. This figure was then divided by the population for that year,[13] giving the mean number of physician visits per capita per year (y).

The least-squares line for the data was computed with x the mean number of apples consumed per capita per day and y the mean number of physician visits per capita per year. The result is $y = 16.676 - 28.145x$, where $n = 37$and contains selected years from 1909-1969. Then the null hypothesis that B is greater than or equal to 0 was tested. The calculated value of t was -7.33. Subsequently, at an alpha level of .0005 (which should suffice for even the most stringent of statisticians) one may reject the null hypothesis and conclude that a significant negative linear relationship does exist.

The theory "An apple a day keeps the doctor away," is equivalent to a y-value of 0 (keeping the doctor away), at an x-value of 1 (an apple a day). Accordingly, a 99% confidence interval for y at an x-value of 1 is computed. The resulting confidence interval for y at an x-value of 1 is computed. The resulting confidence interval is $-23.57 \leq y \leq .585$. This interval does indeed include 0, ergo, an apple a day does keep the doctor away.

The implications and potential uses of these findings are manifold. If apple consumption can be substantially increased, the world situation will be vastly improved upon. Mortality rates in underdeveloped and developing nations could be greatly reduced if the United States sent apples rather than dollars in foreign aid. In America, national health care would no longer be an important matter of public concern, as few people would be demanding medical attention. Doctors may find themselves forced to advertise, competing among each other for patients' patronage while extracting lower fees. Clearly the consumer has only to reap benefits from increasing his apple consumption; not only will he be healthier and require fewer doctor visits, but also he will derive immeasurable pleasure in the knowledge that somewhere there exists a doctor whose meager subsistence the consumer has personally effectuated.

In order to increase apple consumption, two areas of basic economics must be attended to if substantial price increases of apples are to be prevented—an increased demand, and a greater than or equal to increase in supply. As is always the case, this can be brought about most efficiently through government intervention in the free-market system. Various options exist to induce increased apple production. By far the most effective means is the availability of low cost federally-insured loans to apple producers. In Great Britain, long-term loans were offered for "Planting new fruit areas; restocking old orchards; laying service pipes for water and spraying; and erection of gas stores for fruit,"[14] at the attractive interest rate of 3¼%. The authors believe that a program such as this, particularly at such an alluring interest rate, would be successful here in America, despite the fact that this British program was sponsored in 1936. Moreover, this should be well worth the added loss to the government. This program, coupled with ever-effective price supports and government subsidies, should provide sufficient incentives to increase apple production.

However, increased production alone will not guarantee increased supply. Particular attention must be paid to ensure that the apples remain free from damage and disease, and are therefore made available for human consumption. Many insects detrimentally affect the apple fruit, including the codling moth (Cydia pomonella), the plum curculio (Conotrachellus nenuphar), the green-fruit worm (Xylina), and of course, the dreaded apple maggot (Rhagoletis pomonella).[15] In addition, severe damage to apple orchards can result from wind, mice, rabbits, and deer.[16] Devastatingly dangerous diseases that threaten the apple include black rot, pink rot, bitter rot, spongy dry rot, sooty blotch, and fly-speck fungus.[17] Apple producers must have adequate funds, no doubt at government expense, to prevent effectively the occurrence of these damages.

The land grant universities, such as Texas A&M, have a long and successful history of performing research on such topics. If such universities are provided sufficient funding they will certainly step forward to aid in finding methods to control or eliminate such pests. Perhaps Centers for Research on Apple Pests (CRAP) can be established at the major land grant institutions across the nation.

As supply is increased, demand for apples must be augmented. Here is where the private sector can offer beneficial assistance. Life insurance companies may find it advantageous to institute an apple eaters' premium discount, much as is applied to non-smokers. This should provide incentive for value-conscious families to increase their consumption of apples. Furthermore, various apple producers' associations might sponsor television commercials recounting new and imaginative uses for apples, such as has been done with eggs and baking soda. Moreover, organizations with a public health concern, the Red Cross, for example, might institute various programs to educate the general public on the medicinal value of apple consumption.

The perceptive reader, no doubt, in considering the implications of these findings, will realize the potential threat that this empirical evidence poses for the medical profession. Doctors will undoubtedly, and in their own best interests, attempt to undermine the validity of this research. The American Medical Association can be expected to prove themselves formidable opposition against which apple proponents must be prepared to battle. Actual evidence exists whereby physicians have claimed that apples are effective treatment against both constipation[18] and diarrhea.[19] The authors do not wish to imply that physicians are deliberately attempting to damage the apple's fine reputation by causing them to be associated with not one, but two, very unpleasant intestinal disturbances. These conflicting attestations do, however, seem to be of questionable integrity.

Likewise, the Food and Drug Administration, motivated solely by their concern for human life, health, and the pursuit of an acceptable artificial sweetener, will certainly test this proverb themselves. Care must be taken to ensure that the FDA studies are performed without pressure from or threat of retaliation by the AMA. Should the FDA find heavy apple consumption to be dangerous to rats, a required disclaimer such as "Use of this product may be hazardous to your health. This product contains apple pectin which has been determined to cause cancer in laboratory animals." could adversely affect the consumption of apples. However, the authors refuse to consider even the possibility of such an event, and will therefore not refer to it again.

"An apple a day keeps the doctor away"—once an oft-heard English proverb, now an empirically validated truth. The statistical approach upon which this conclusion is based has been presented; so, too, have the implications of this finding. Whether or not one will live to a "ripe old age" depends on a multitude of factors; however, consumption of an apple every day can provide that added assurance that although all else may fail, one will still have one's health. ■

NOTES

[1]J. Bartlett, ed., *Familiar Quotations,* 1955, p. 1006.
[2]H.L. Mencken, ed., *A New Dictionary of Quotations,* 1966, p. 52.
[3]B. Stevenson, ed., *The Home Book of Quotations,* 1949, p. 91.
[4]H.W. Prochnow & Prochnow, H.W., Jr., eds., *A Dictionary of Wit & Satire,* 1962, p. 11.
[5]Ibid, p. 11.
[6]T.W. Lamont, *My Boyhood in a Parsonage,* 1946, p. 29.
[7]"The West's Oldest Apple Orchard?" *Sunset,* (November 1972), p. 34.
[8]L.K. Wise, Skaggs Albertson's Grocery, College Station, Texas, 1979.
[9]U.S. Department of Agriculture, *Agriculture Statistics* 1972, p. 305.
[10]M.Y. Pennel & Alternderfer, M.E., *Health Manpower Source Book I. Physicians,* 1952, Section 1.
[11]U.S. Department of Commerce, Bureau of the Census, *Historical Statistics of the U.S., Colonial Times to 1970,* 1975, p. 162.
[12]*Ibid.,* pp. 75-76.
[13]*Ibid.,* p. 8.
[14]Great Britain Ministry of Agriculture and Fisheries, *Bulletin No. 84,* 1936.
[15]A.E. Wilkinson, *The Apple,* 1915, pp. 196-208.
[16]*Ibid.,* pp. 260-265.
[17]*Ibid.,* pp. 221-229.
[18]F. Lape, *Apples & Man,* 1979, p. 75.
[19]AMA Council on Food, *Accepted Foods and their Nutritional Significance,* 1939, p. 208.

In a review of the book GENES, MIND AND CULTURE by C.J. Lumsden and E.O. Williams (Harvard Univ. Press, 1981) writes the reviewer, Professor of Biology at Harvard, R.C. Levontin:

"Despite its ambitious, if not pretentious title, Genes, Mind and Culture contains no information about genes, mind, or culture and certainly nothing about their interconnections. The second author is a past master of rhetorical inflation and that would seem to be his major contribution to the present book."

It is reported that, to monitor homosexual soliciting in men's toilets in New York subway stations, the police have installed one-way mirrors. These are of course peer review panels.

(R.A. Lewin)

PRESSURIZED HYDROGEN MAY RETARD CANCER

In an attempt to repeat the experiment with a different type of tumor, Texas A&M's Dr. Fife used a group of mice with Erlich's ascites carcinoma, an internal cancer. Although the results were somewhat equivocal (one of the three mice in the test chamber drowned in its water bowl) and the hydrogen therapy did not prevent the other mice from dying, it appeared to prolong their lives by about 30%.

BRAIN AND LANGUAGE
Vol. 22, No. 1 May 1984

Andrew W. Young, Andrew W. Ellis, and Pamela J. Bion.

Hemisphere Superiority for Pronounceable Nonwords, but Not for Unpronounceable Letter Strings.

D. Brahams
Informed consent does not demand full disclosure of risks
The Lancet 2 July, 1983 p. 58

"..." in a case in the 1940' ... where a boy was admitted...for tonsilectomy and due to administrative error was circumcised instead.

In this case "the doctor was as much victim of the error as the boy."